INDUCTION MOTORS - APPLICATIONS, CONTROL AND FAULT DIAGNOSTICS

Edited by **Raúl Igmar Gregor Recalde**

Induction Motors - Applications, Control and Fault Diagnostics
http://dx.doi.org/10.5772/59637
Edited by Raúl Igmar Gregor Recalde

Contributors

Rafael Rodriguez-Ponce, Fortino Mendoza-Mondragón, Moises Martinez-Hernandez, Marcelino Gutierrez-Villalobos, Nordin Bin Saad, Muawia A. Magzoub, Rosdiazli B. Ibrahim Ibrahim, Muhammad Irfan, Yu Zou, Raúl Igmar Gregor, Jorge Rodas, Derlis Gregor, Federico Barrero, Raul Gregor, Hugo Guzmán, Mario Durán, Ignacio González, Javier Riedemann, Rubén Peña, Ramón Blasco-Gimenez, Takeo Ishikawa, Oscar Barambones, Patxi Alkorta, Jose M. Gonzalez De Durana, Jose A. Cortajarena, Fevzi Kentli, Rudolf Riehl, Fernando de Souza Campos, Alceu Ferreira Alves, Ernesto Ruppert, Wilson Wang, Ali Saghafinia, Mohammad Ali Sandidzadeh, Amir Ebrahimi, Amir Heydari

CBS, Edition **2017**

Published by InTech

Janeza Trdine 9, 51000 Rijeka, Croatia

Publishing Process Manager Iva Lipovic

Technical Editor InTech DTP team

Cover InTech Design team

Additional hard copies can be obtained from orders@intechopen.com

Induction Motors - Applications, Control and Fault Diagnostics, Edited by Raúl Igmar Gregor Recalde

p. cm.

ISBN 978-953-51-2207-4

INTECH

open science | open minds

Contents

Preface

AC motors play a major role in modern industrial applications. Squirrel-cage induction motors (SCIMs) are probably the most frequently used when compared to other AC motors because of their low cost, ruggedness, and low maintenance. The material presented in this book is organized into four sections, covering the applications and structural properties of induction motors (IMs), fault detection and diagnostics, control strategies, and the more recently developed topology based on the multiphase (more than three phases) induction motors. This material should be of specific interest to engineers and researchers who are engaged in the modeling, design, and implementation of control algorithms applied to induction motors and, more generally, to readers broadly interested in nonlinear control, health condition monitoring, and fault diagnosis.

Section I gives an introduction of squirrel-cage induction generators (SCIGs) and doubly fed induction generators (DFIGs) with regard to modelling and control. Furthermore, a power control system applied to the induction generator is explained. This section concludes with anthIM sttural properties of IMs, emphasizing a methodology to determine experimentally the parasitic capacitances in (VSI)-fed IM drives based on the (PWM) techique.

Section II focuses on the health condition monitoring of induction motors and fault diagnosis of squirrel-cage induction motors with broken rotor bars and end rings. Condition monitoring of electric machines has received a strong impulse from the industry to ensure consistent and reliable operation of the modern industrial systems.

Section III reviews the fundamentals of the modeling, simulation, and control of IMs. It focuses mainly on the direct torque control (DTC), hybrid fuzzy–fuzzy controller (HFFC), adaptive fuzzy sliding-mode controller (AFSMC) for an indirect field-oriented control (IFOC), and robust control. In general, some of the requirements of the motor control system developed and evaluated in this section are accuracy, dynamic performance, and robustness.

Multiphase induction motors have become one of the main research topics in the field of IM during the past decade for various applications, especially in those where high reliability and fault tolerance are needed. Different topologies of multiphase IMs have been recently developed mainly for high-power applications, where an increase in the number of phases enables reduction in the power per phase, which leads to a reduction in the power per inverter leg. Section IV explains the fundamentals of the modeling and control of two different topologies of multiphase motors. On the one hand, regarding the five-phase IM, some simulation and experimental results are presented to show the behavior of the entire system in healthy and faulty conditions. On the other hand, with regard to the asymmetrical six-phase IM, the implementation of the model-based predictive control (MBPC) techniques as well as a comparative study of reduced-order observers used to estimate the rotor currents in an MBPC current control are presented.

Finally, I wish to express my gratitude to the authors of the chapters as well as to the Engineering Faculty of the National University of Asuncion (FIUNA).

Raúl Igmar Gregor Recalde
Engineering Faculty of the National University of Asuncion
Asuncion, Paraguay

Section 1

Applications and Structural Properties

Chapter 1

Induction Generator in Wind Power Systems

Yu Zou

Additional information is available at the end of the chapter

http://dx.doi.org/10.5772/60958

Abstract

Wind power is the fastest growing renewable energy and is promising as the number one source of clean energy in the near future. Among various generators used to convert wind energy, the induction generator has attracted more attention due to its lower cost, lower requirement of maintenance, variable speed, higher energy capture efficiency, and improved power quality [1-2]. Generally, there are two types of induction generators widely used in wind power systems – Squirrel-Cage Induction Generator (SCIG) and Doubly-Fed Induction Generator (DFIG). The straightforward power conversion technique using SCIG is widely accepted in fixed-speed applications with less emphasis on the high efficiency and control of power flow. However, such direct connection with grid would allow the speed to vary in a very narrow range and thus limit the wind turbine utilization and power output. Another major problem with SCIG wind system is the source of reactive power; that is, an external reactive power compensator is required to hold distribution line voltage and prevent whole system from overload. On the other hand, the DFIG with variable-speed ability has higher energy capture efficiency and improved power quality, and thus dominates the large-scale power conversion applications. With the advent of power electronics techniques, a back-to-back converter, which consists of two bidirectional converters and a dc-link, acts as an optimal operation tracking interface between DFIG and loads [3-5]. Field orientation control (FOC) is applied to both rotor- and stator-side converters to achieve desirable control on voltage and power [6,7].

In this chapter, a brief introduction of wind power system is presented first, which is followed by introduction of SCIG and DFIG from aspects of modeling and control. The basic FOC algorithm is derived based on DFIG model in *dq* reference frame. At last, the power generation efficiency is considered through different Maximum Power Point Tracking (MPPT) methods that have attracted a lot of attention in the variable-speed operation systems. A comparative analysis involving advantage and disadvantage of the methods is conducted.

Keywords: wind power systems, SCIG, DFIG, back-to-back converter, FOC, MPPT

1. Introduction

The core component of a modern induction generator wind power system is the turbine nacelle, which generally accommodates the mechanisms, generator, power electronics, and control cabinet. The mechanisms, including yaw systems, shaft, and gear box, etc., facilitate necessary mechanical support to various dynamic behavior of the turbine. The generator is dedicated to the conversion between mechanical energy, which is captured by turbine rotor, and electrical energy. The generated electrical energy then needs to be regulated and conditioned to be connected to the power grid for use. In this section, the wind power system layout and classification are introduced first, which is followed by the outlining of the feasible power electronic converter interface between generators and loads. Lastly, the control scheme is briefly addressed and discussed in detail in section 2.

1.1. Overview of wind power systems

Figure 1 shows the general layout of a wind turbine nacelle. The generator is either driven (in generation mode) or propelling (in motoring mode) the turbine blades through a shaft. The gearbox can be used to facilitate the speed difference between turbine and generator. The blade stall and pitch mechanisms are also involved to limit the power as well as the turbine plane yawing and tilting. By these means, the blade effective aerofoil cross section and thus the interface with wind pressure can be controlled. The performance coefficients responding to different yaw angle and pitch angle show significant variations [1-3]. In addition, as the most dynamically efficient choice, three blades connected through a hub with flanges is the commonly used topology in the front of the nacelle. The flanges are designed to enable the pitch angle adjustment. In most of the variable-speed wind systems, the high-efficiency operation always relies on the wind speed information. As a result, the anemometer can be used as one of the solutions. The basic function of the tower is to reach a higher position in order to obtain more airstream and wind speed. The tower can be constructed in either soft or stiff ways. A stiff tower has a natural frequency which lies above the blade passing frequency. Soft towers are lighter and cheaper but have to withstand more movement, and thus suffer from higher stress levels [2].

There are a number of classifications that group the wind power systems into different categories. According to the loads, grid integrated system and islanded system are employed to feed power grid and isolated load, respectively. According to the generators used, popular options are SCIG wind system, DFIG wind system, and Permanent Magnet Synchronous Generator (PMSG) wind system. Other alternative generator systems are also mentioned in

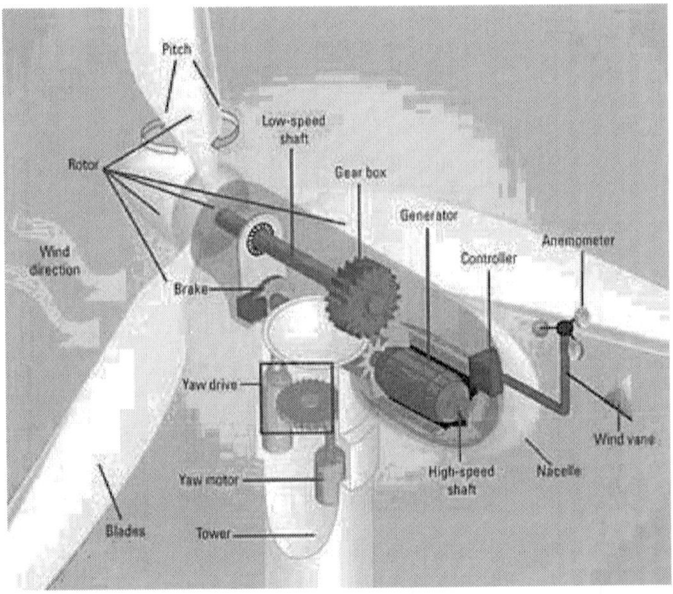

Figure 1. Wind power system nacelle [8]

the literature, such as brushless DFIGs (BDFIG) system [5,6], direct-drive synchronous generator (DDSG) system [7,9], switched reluctance generator (SRG) system [10], multiple-stage geared SCIG system [10], and radial/axial/transversal-flux PM generator systems [7,12-14]. These solutions generally require relatively complex operation principle and equipment assembly. According to the presence of the gear box, there are multistage gear box wind system, single-stage gear box wind system, and direct drive wind system (without gear box) in where the Synchronous Generator (SG) qualifies the system to have a simpler and more reliable drive train. However, the lower generator speed, and thus larger torque, requires more poles, larger diameter, and volume, and hence higher cost.

The most promising classifications in induction generator wind systems are fixed-speed, limited-variable-speed, and variable-speed wind systems, according to the operations of induction generator speed. Comparisons between these wind power systems have been intensively conducted, based on different speed variation levels [12,15-19]. A summary of their advantages and disadvantages is presented in Table 1. The fixed-speed concept has been successfully applied in SCIG wind systems. The drive train applies multiple-stage gearbox and a SCIG is directly connected to the grid via a transformer. To support the grid, external reactive power compensation and soft starter are necessary [5,6]. The limited variable-speed system is an improved version of the SCIG type but it uses a wound rotor induction generator instead, which allows the stator to be connected to the grid, and the rotor to have a variable resistance controlled by a power converter. Through the control of rotor resistance, the slip of the generator is varied. The variable-speed system is a concept commonly used in large power rating applications (>1.5 MW). Different combinations among DFIG, SCIG, partial or full converters would lead to variable-speed operation systems. The control system maintains the optimal generator speed, thus the optimal output power, through controlling the generator currents and voltages. Due to the high efficiency and capability of Faults Ride Through (FRT), this type of wind power system dominates the high-capacity power market nowadays.

	Advantages	Disadvantages
Fixed-speed system	a. Simple construction and robust b. Low cost and maintenance c. Easy control	a. Not optimal operation, thus low efficiency b. Easy power fluctuation caused by wind speed and tower pressure c. External reactive power compensation is needed d. Weak capability of FRT
Limited-speed system	a. Limited speed variation is implemented b. The slip ring may be replaced by optical coupling	a. Speed variation range depends on the size of the variable rotor resistance (<10%) b. The controlled rotor power must be dissipated by heat in the resistor c. Still need reactive power compensation and cannot support the grid alone
Variable-speed system	a. Large range of speed variation b. Appropriate control enables optimal operation for maximum power extraction c. No external power compensation is needed and is able to support the grid d. High FRT capability e. Suitable and commonly used for large-scale wind farms	a. Relatively complicated control system b. Higher converters and control costs c. May need a multistage gearbox and slip ring in DFIG system d. May need expensive PM material and large diameter design in direct drive

Table 1. Comparison among different wind power systems

1.2. Power electronics interface topologies in wind power systems

Power electronics is the key element enabling the regulation and conditioning of the power, voltage, and frequency with high efficiency and flexibility. In addition, more involvement of distributed power systems nowadays emphasizes the crucial role of power electronics interface among energy generation, storage, and transmission.

Due to the developments in semiconductor switches and microprocessors, many power electronics techniques have been developed during the past decades [20,21]. Besides the diode converters, line-commutated thyristor converters and self-commutated IGBT/MOSFET converters are found applicable to wind power systems. The line-commutated converters are generally used in high-power applications but they are incapable of controlling the reactive power. The self-commutated converters are able to transfer and control power bidirectionally because of the capability of controllable switch turning-off. Nowadays, wind power systems, especially the variable-speed wind power system, primarily rely on the converters that implement full power control. Different converter topologies and combinations have been successfully employed in this field, as shown in Figure 2.

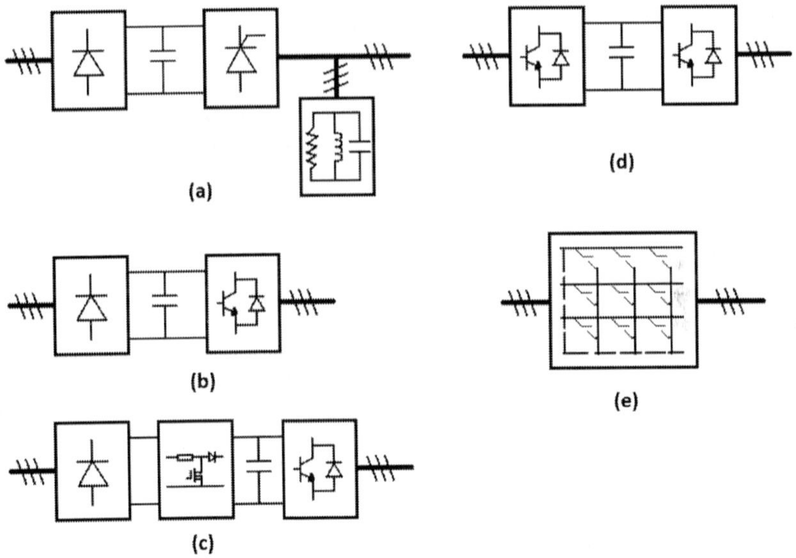

Figure 2. Commonly used power electronics converter topologies for wind power system ((a) diode and line-commutated converter, combined with reactive power compensation; (b) diode and PWM VSI converter; (c) diode and DC/DC chopper and PWM VSI converter; (d) back-to-back PWM VSI converter; (e) matrix converter)

Due to the employment of diode rectifier, the topology in Figure 2(a) is uncontrolled and a thyristor inverter is used to regulate the generator speed through dc-link voltage to obtain firing angle commands. Obviously, this scheme is simple for control and costs less than self-commutated converter. More importantly, it is suitable for high power rating applications. However, the weakness is that extra reactive power compensation is required, which contains a voltage source converter (VSC). The grid voltage may be regulated to obtain reference current for the compensator and the control signal comes from the regulation of the compensator current [22,23]. To remove the compensator, a self-commutated converter could be used to take the place of thyristor inverter, as shown in Figure 2(b). Again, the regulation of dc-link voltage can provide current reference, which is controlled to generate control signals for the PWM inverter [24]. Two self-commutated converters connected through a dc-link, as shown in Figure 2(d), enable bidirectional power flow, which is the key to ensuring high efficiency in motoring operation of generator. The FOC is applied on both sides of converters based on *dq* reference frame [23,25]. The grid-side converter keeps a constant dc-link voltage, while the generator-side converter is responsible for both active and reactive power control [23]. In the generator-side converter control, the *d*-axis current could be set at zero to maximize the torque, while the *q*-axis current is derived from power regulation [26,27]. An alternative topology of Figure 2(d) is shown in Figure 2(c), where the generator-side self-commutated converter is replaced by a diode rectifier connected to an intermediate chopper [28]. This configuration is impossible for bidirectional power flow caused by the diode rectifier. But it can achieve a similar wide range of speed variation as two self-commutated converters. The grid-side converter controls the dc-link voltage for *d*-axis current reference and controls the reactive power for *q*-axis current reference. The active power regulation and thus the speed control are carried out to generate reference dc-link current. The duty cycle of chopper switch can be obtained using current regulation. The converter configurations discussed are actually multistage implementation of AC conversion. An intermediate DC stage is needed to assist conversion and associated control. In recent years, such procedure has been investigated by a single-stage converter, the matrix converter, which performs the energy transformation without help from a bulky storage stage. The controllable switches are arranged in such a way that any input phase may be connected to any output phase at any time. The matrix converter may be applied to the DFIG system, like the topology in Figure 2(e) [29,30]. According to the stator flux FOC, the reactive and active power can be regulated by *d*- and *q*-axis current, respectively [30]. An alternative control strategy is by regulating the rotor winding voltage to control the power factor (PF) and applying the double space vector PWM technique [29]. It is worth noting that the SCIG system has high starting currents. One effective way to limit the starting current is by using the soft-starter that applies thyristors to limit the RMS starter current below rated current. The starter is shorted after the full load is reached. The torque peak can be decreased as well, which reduces the gearbox pressure [9,10].

The high-efficiency energy conversion and full control of power exclusively rely on the power electronic converter and the control scheme applied on. Consequently, the broadly accepted total wind power system topologies subject to above power electronics are reviewed as follows. Since the high-efficiency variable speed systems are the primary focus, Figure 3 summarizes the feasible variable-speed system topologies, for induction generator systems as well as synchronous generator systems for systematized purpose. Due to the low demand on the converter power rating of approximately 30% of the total power rating, the DFIG with partial converter, shown in Figure 3(a), is a widespread topology for wind power systems. Also, due to the presence of a rotor-side converter, the rotor power is fed back to the grid without dissipation in the resistor. Instead of a partial converter, PMSG or SCIG can be connected to a full rating converter, as shown Figure 3(b). This topology has better grid FRT ability because the generator-side is totally independent of the grid-side. However, the converter rating and loss are high. Figure 3(c) shows the direct drive system, which is aimed at removing the gearbox and associated loss [16]. The generator rotor is connected to the turbine shaft directly and runs at the same very slow speed. Therefore, a high torque and a large machine radius are required to transfer the same amount of power. Fewer components enable less loss and thus more reliable performance in this type of system. To compromise between machine size and spinning speed, the single-stage gearbox, shown in Figure 3(d), is applied [15]. Figure 3(e) shows the electrically excited synchronous generator (EESG) system, which has a rotor-side converter to provide DC excitation while the stator is connected to a full converter like the case in Figure 3(c). Although there is an increase in cost due to the extra winding for excitation and it also requires more maintenance, the EESG could minimize the loss through controlling the flux via rotor converter [7,11,12].

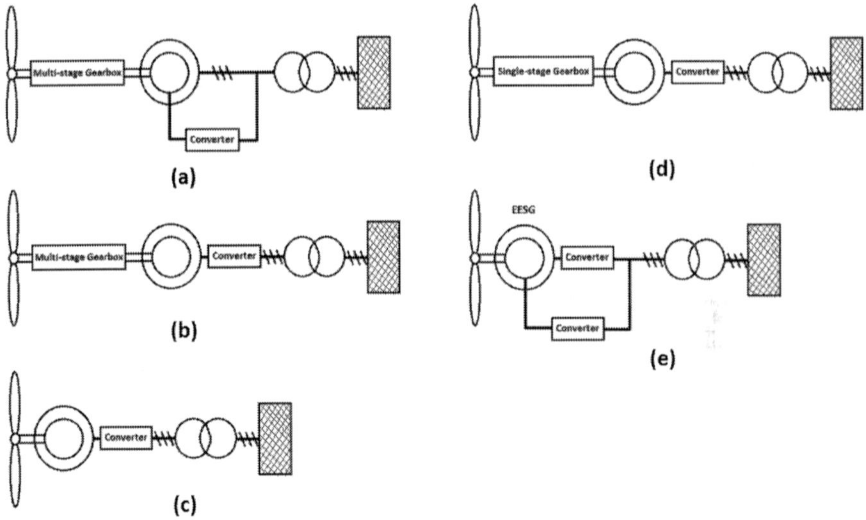

Figure 3. Commonly used wind power system topologies ((a) DFIG with partial/matrix converter; (b) PMSG/SCIG with full converter; (c) direct drive; (d) PMSG with full converter and less stage gearbox; (e) EESG direct drive)

1.3. Generator control schemes in wind power systems

SCIG and DFIG are used almost exclusively in the energy conversion stage of the induction generator wind power system. The most commonly used system topologies are SCIG directly connected into the power grid and DFIG fed by back-to-back converter (Figure 3(a) and Figure 3(d)). The first topology implies a constant frequency and voltage of the SCIG that establishes a fixed-speed operation. In such system, the SCIG relies on the grid (or capacitor bank) to provide reactive power which is necessary to build electromagnetic excitation for rotary field. The generating mode of SCIG is triggered by driven torque which acts opposite to the generator speed within the super-synchronous speed operation region. Due to the absence of the power electronics interface, such system can only serve the grid support applications, wherein just limited control (pitch angle control) can be applied.

The DFIG system, on the other hand, enables the flexible and efficient operations with FOC applied on the rotor-winding-side power electronics interface. The FOC is an instantaneous control that effectively manipulates the position-dependent variables, such as torque and power, in induction generator wind power systems. By aligning a particular space variable with d-axis, stator currents could be decoupled into flux component and torque component in dq rotating frame. The currents can be thus controlled separately like in DC motor drive. To implement the control in hardware, PWM technique is generally employed based on Space Vector Modulation (SVM). The SVM is based on space reference voltage vector and associated switching logics. Any space vector can be comprised of vector sum of two adjacent voltage vectors, and the duty cycles of three-phase voltages are calculated based on the dwelling time of two voltage vectors. This is the method widely used in standard industry applications.

The following section examines the detailed modeling and control strategies of both systems.

2. Model and control of induction generator in wind power systems

2.1. Model of wind power and wind turbine

As a typical kinetic energy, wind energy is extracted through wind turbine blades and then transferred by the gearbox and rotor hub to mechanical energy in shaft. The shaft drives the generator to convert the mechanical energy to electrical energy. According to Newton's law, the kinetic energy for the wind with particular wind speed V_w is described as:

$$E_k = \frac{1}{2} m V_w^{\ 2} \tag{1}$$

where m represents the mass of the wind, and its power can be written as:

$$P_w = \frac{\partial E_k}{\partial t} = \frac{1}{2}\frac{\partial m}{\partial t} V_w^{\ 2} = \frac{1}{2}\left(\rho A V_w\right) V_w^{\ 2} = \frac{1}{2}\rho A V_w^{\ 3} \tag{2}$$

where ρ and A are the air density and turbine rotor swipe area, respectively. The extracted mechanical power can thus be expressed as:

$$P_m = C_p\left(\lambda, \beta\right) P_w = C_p\left(\lambda, \beta\right)\frac{1}{2}\rho A V_w^{\ 3} \tag{3}$$

where P_m is the mechanical output power in watt, which depends on performance coefficient $C_p(\lambda, \beta)$, C_p depends on tip speed ratio λ and blade pitch angle β, and determines how much of the wind kinetic energy can be captured by the wind turbine system. A nonlinear model describes $C_p(\lambda, \beta)$ as [3]:

$$C_p\left(\lambda, \beta\right) = c_1\left(c_2 - c_3\beta - c_4\beta^2 - c_5\right)e^{-c_6} \tag{4}$$

http://dx.doi.org/10.5772/60958

where, c_1=0.5, c_2=116/Λ_i, c_3=0.4, c_4=0, c_5=5, c_6=21/Λ_i and

$$\lambda_i = \frac{1}{\lambda + 0.08\beta} - \frac{0.035}{\beta^3 + 1} \tag{5}$$

With the dependence on the λ and β, maximum value of C_p could be reached and maintained through controlling the pitch angle and generator speed at particular wind speed. A group of typical $C_p - \lambda$ curves for different β is shown in Figure 4 and there is always a maximum value for C_p at one particular wind speed. Correspondingly, the output power is determined by different C_p and also the generator speed at different wind speed, as shown in Figure 5, where there is always one maximum power value for each wind speed, which is the goal of the MPPT control.

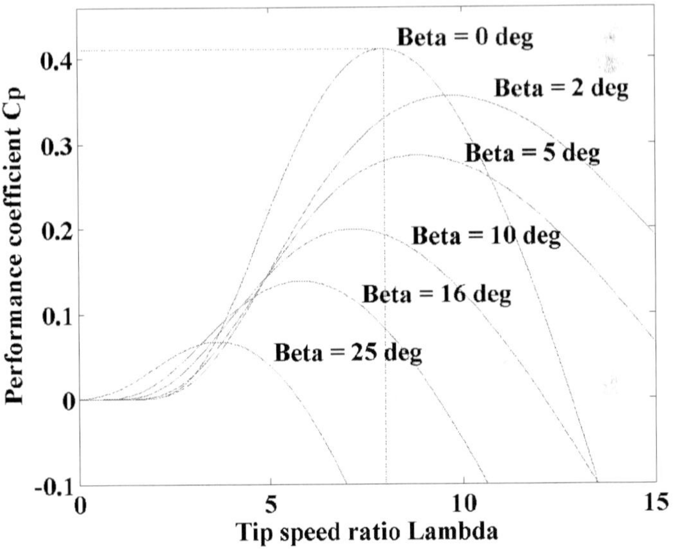

Figure 4. C_p versus λ curve for a wind turbine (β is the pitch angle) [23]

2.2. Model and control of SCIG

As a fixed-speed wind power system, SCIG is directly connected to the grid through transformer and thus operates at almost constant speed without controlling from power electronics interface. It was commonly used in Denmark during 1980s and 1990s and thus is also called "Danish Concept" system. The robust and simple configuration qualifies such system for many applications where the cost is a higher priority concern than efficiency. Figure 6 shows the schematics of entire SCIG wind system including the wind turbine, pitch control, and reactive

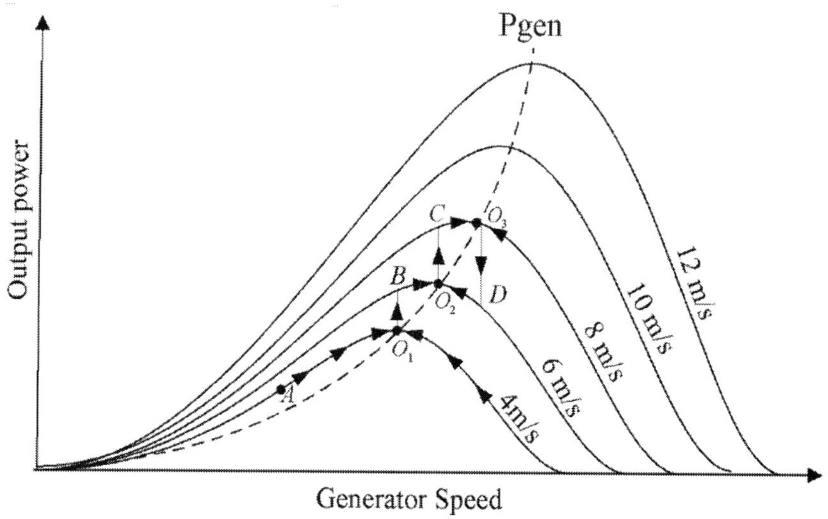

Figure 5. Power versus generator speed curve for wind turbine [31]

power compensator. The entire system includes three stages for delivering the energy from wind turbine to the power grid. The first one is wind farm stage which handles with low-voltage V_{wt}; the second is distribution stage which has medium-voltage V_{dis}; the third is grid transmission stage which has high-voltage V_{grid}. The three-phase transformers take care of the interface between two stages [10]. The nominal power is considered as active power reference to regulate the pitch angle, while the distribution line-to-line voltage and phase current are monitored to favor the reactive power compensation for distribution line. This fairly straight-forward technique was firstly used since it is simple, with rugged construction, has reliable operation and is low cost. However, the fixed-speed nature and potential voltage instability problem severely limit the operations of SCIG wind system [1,3].

It is clear from Figure 5 that at a particular wind speed, the output active power is also a fixed value in the case of fixed generator speed. Thus, the output power is exclusively wind speed dependent until the nominal power is reached. The wind speed at nominal power is called nominal wind speed. Beyond this wind speed, the pitch angle system will prevent the output power from exceeding the nominal value. The pitch angle is determined by an open-loop control of regulated output active power and, as shown in Figure 7. Due to the huge size of blade and thus the huge inertia, pitch angle has to change at a slow rate and within a reasonable range. It is also worth noting that without reactive power source, the SCIG system tends to a voltage droop in distribution line which will cause overload problem.

Simulation in [23] illustrates the operation of a 0.855MW SCIG system. From Figure 8, the initial generator speed is set at slip s = -0.01 p.u. with respect to synchronous speed and then response to the wind speed input disturbance. Since the power is lower than the nominal value (0.855

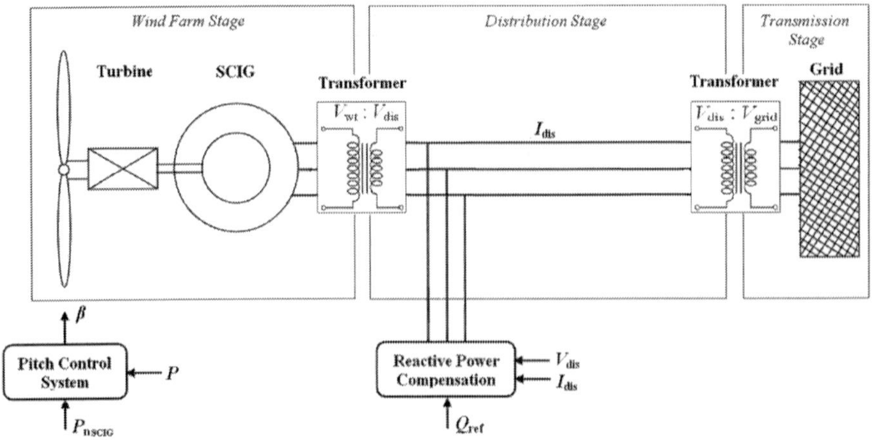

Figure 6. SCIG wind power system configuration

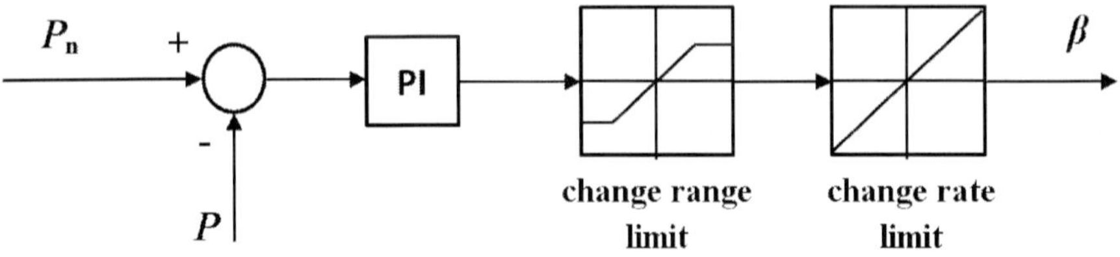

Figure 7. Pitch angle control

MW) before t = 10 s, pitch angle control is not online. Since that moment, the wind speed increases and so do the generator speed and power until the wind speed exceeds the nominal value (11 m/s) at where the pitch control is triggered to block the further increase of output power. In this way, the output power persists at nominal value thereafter.

It is noted that the generator speed can only vary in very small range around 1 p.u. and thus it is impossible to attain the optimal output power. Also, without independent control ability, SCIG system consumes reactive power of 0.41 Mvar at the steady state, which will lead to line voltage droop. To provide necessary reactive power, a Static Synchronous Compensator (STATCOM) is applied in distribution line. As in Figure 9, distribution line voltage can drop by approximately 0.055 p.u. in SCIG system without STATCOM, which will be a potential induction of overload in system. In contrast, SCIG system with STATCOM can hold distribution voltage at 0.99 p.u., which is favorable to grid system stability. The compensated reactive power from STATCOM is shown in Figure 10 and is equal to 0.3 Mvar at the steady state. Although STACOM provides impressive help to a constant distribution line voltage, DFIG wind system presents more attractive attributes.

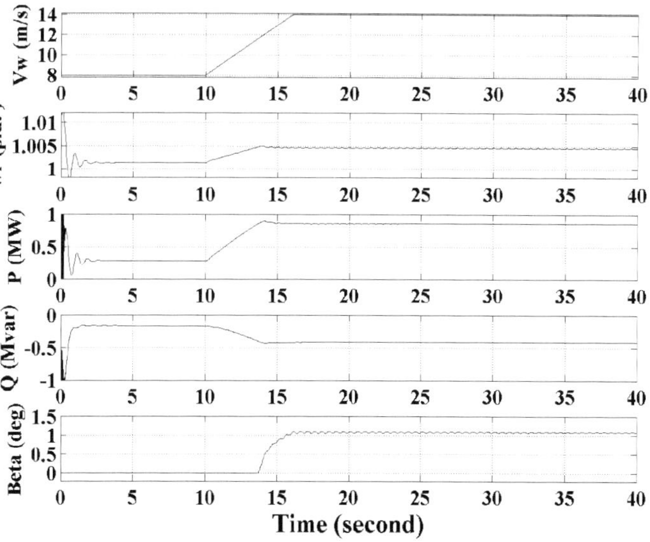

Figure 8. Pitch angle control for SCIG system [23]

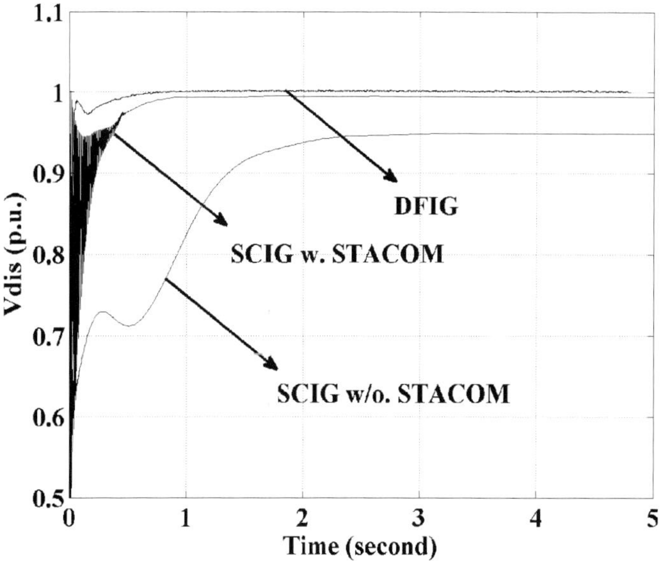

Figure 9. Grid voltages comparison between SCIG w/o. STACOM, SCIG w. STACOM and DFIG [23]

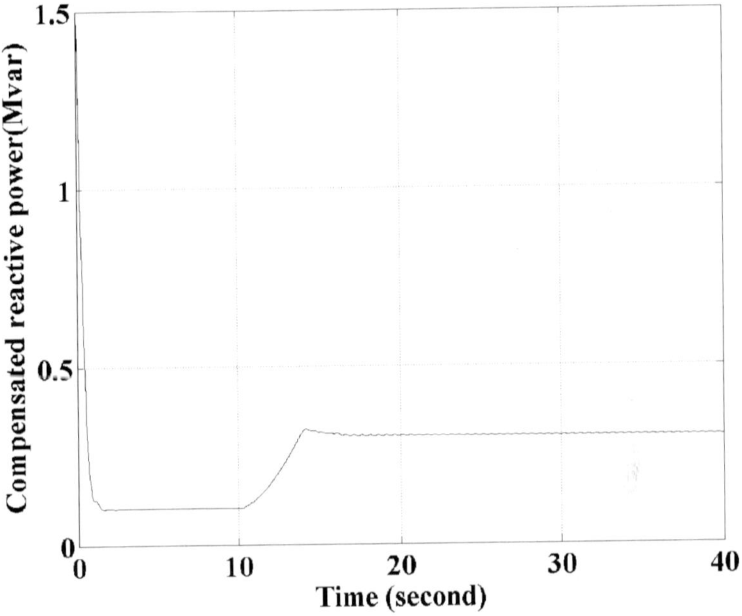

Figure 10. Compensated reactive power from STATCOM [23]

2.3. Model and control of DFIG

Traditionally, the dynamic slip control is employed to fulfill the variable-speed operation in induction generator wind system, in which the rotor windings are connected with variable resistor and control the slip by varied resistance [3,11]. This type of system attains limited variations of generator speed but external reactive power source is still necessary. In order to completely remove the reactive power compensation and control both active and reactive power independently, DFIG wind power system is one of most popular methods in wind energy applications [1,3,7]. The DFIG wind power system with associated back-to-back converter is a typical variable speed system as shown in Figure 11, which complies with the topologies in Figures 3(a) and 2(d). The generator stator windings are connected directly to grid (with fixed voltage and frequency of grid) while the rotor windings are fed by an AC/DC/AC IGBT-based PWM converter (back-to-back converter with capacitor dc-link), at variable frequency through slip rings and brushes. Although such system needs the gearbox and slip rings to function, many advantages enable DFIG system to dominate most wind market nowadays. It facilitates variation of a wide speed range (±30% around synchronous speed), the lower rating requirement on power converters (30% of generator power), and thus lower cost. Also, it has high efficiency induced by bidirectional power flow, and the ability to perform reactive power compensation and smooth grid integration. In this configuration, the back-to-back converter consists of two parts: the stator/grid-side converter and the rotor-side converter. Both are voltage source converters while a capacitor bank between two converters acts as a dc voltage interface.

In this section, the modeling of DFIG is introduced first and followed by the consequent FOC algorithm which is divided into two parts: stator-side converter control and rotor-side converter control. The SVM method and islanded operation control are also addressed.

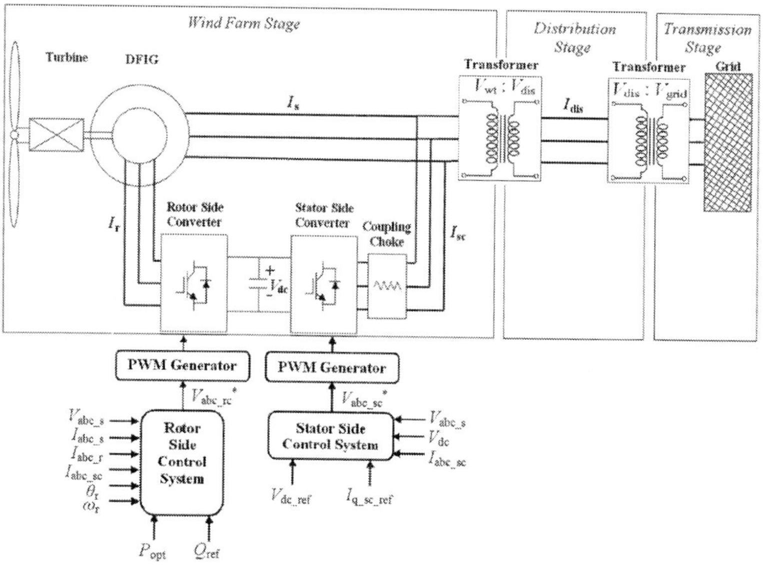

Figure 11. DFIG wind power system configuration

2.3.1. *dq model of DFIG*

The modeling is conducted under the *dq* reference frame. The equivalent circuits of DFIG in the *dq* reference frame are depicted in Figure 12(a, b) and the relationships between voltage V, current *I*, flux Ψ, and torque T_e can be derived by writing KVL equations. For stator-side, the *d*- and *q*-axis voltage components are given as:

$$V_{ds} = R_s I_{ds} - \omega_s \Psi_{qs} + \left(L_{ls} + L_m\right)\frac{dI_{ds}}{dt} + L_m\frac{dI_{dr}}{dt} \quad a$$
$$V_{qs} = R_s I_{qs} + \omega_s \Psi_{ds} + \left(L_{ls} + L_m\right)\frac{dI_{qs}}{dt} + L_m\frac{dI_{qr}}{dt} \quad b \tag{6}$$

And similarly, the *d*- and *q*-axis voltage components in rotor-side are given as:

$$
\begin{aligned}
V_{dr} &= R_r I_{dr} - s\omega_s \Psi_{qr} + \left(L_{lr} + L_m\right)\frac{dI_{dr}}{dt} + L_m \frac{dI_{ds}}{dt} \quad a \\
V_{qr} &= R_r I_{qr} + s\omega_s \Psi_{dr} + \left(L_{lr} + L_m\right)\frac{dI_{qr}}{dt} + L_m \frac{dI_{qs}}{dt} \quad b
\end{aligned}
\tag{7}
$$

Because the flux linkage along d- and q-axis follow:

$$
\begin{aligned}
\Psi_{ds} &= L_s I_{ds} + L_m I_{dr} \quad a \\
\Psi_{qs} &= L_s I_{qs} + L_m I_{qr} \quad b
\end{aligned}
\tag{8}
$$

$$
\begin{aligned}
\Psi_{dr} &= L_r I_{dr} + L_m I_{ds} \quad a \\
\Psi_{qr} &= L_r I_{qr} + L_m I_{qs} \quad b
\end{aligned}
\tag{9}
$$

The reorganized DFIG stator voltages in d- and q-axis, respectively, are presented as:

$$
\begin{aligned}
V_{ds} &= R_s I_{ds} - \omega_s \Psi_{qs} + \frac{d\Psi_{ds}}{dt} \quad a \\
V_{qs} &= R_s I_{qs} + \omega_s \Psi_{ds} + \frac{d\Psi_{qs}}{dt} \quad b
\end{aligned}
\tag{10}
$$

And the DFIG rotor voltages in d- and q-axis, respectively, are presented as:

$$
\begin{aligned}
V_{dr} &= R_r I_{dr} - s\omega_s \Psi_{qr} + \frac{d\Psi_{dr}}{dt} \quad a \\
V_{qr} &= R_r I_{qr} + s\omega_s \Psi_{dr} + \frac{d\Psi_{qr}}{dt} \quad b
\end{aligned}
\tag{11}
$$

The generator electromagnetic torque is correspondingly given as:

$$
T_e = \frac{3}{2} n_p \left(\Psi_{ds} I_{qs} - \Psi_{qs} I_{ds}\right) \tag{12}
$$

where $L_s = L_{ls} + L_m$; $L_r = L_{lr} + L_m$; and $s\omega_s = \omega_s - \omega_r$ represents the difference between synchronous speed and generator speed; subscripts r, s, m, d, q denote the rotor, stator, magnitizing, d-axis and q-axis components, respectively; T_e is electromagnetic torque; L_m and n_p are generator mutual inductance and the number of pole pairs, respectively.

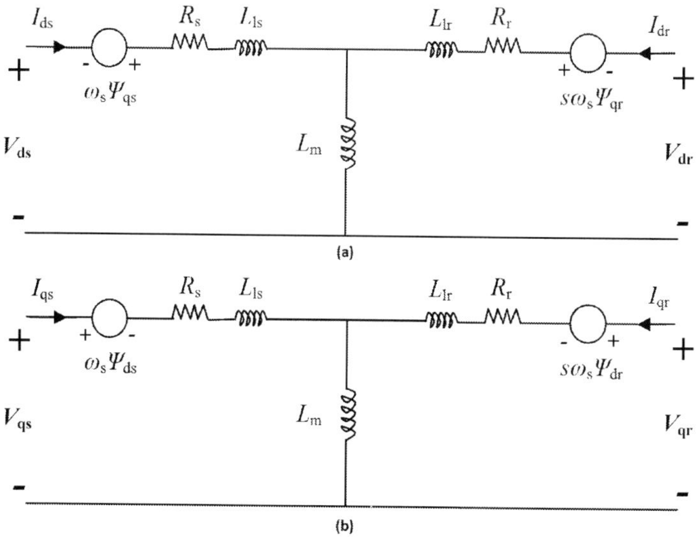

Figure 12. Equivalent circuit of DFIG ((a) *d*-axis; (b) *q*-axis)

2.3.2. Control of rotor-side converter

The control of DFIG modeled above is applied on back-to-back converter and is therefore also divided into rotor-side control and stator-side control.

First, the rotor-side converter is studied. To *d*-axis, the rotor flux linkage Ψ_{qr} in Equation (7a) is substituted by Equation (9b), resulting in:

$$V_{dr} = R_r I_{dr} - s\omega_s \left(L_r I_{qr} + L_m I_{qs} \right) + \frac{d\left(L_r I_{dr} + L_m I_{ds} \right)}{dt} \tag{13}$$

By substituting the I_{ds} by Ψ_{ds} in Equation (8a), the Equation (13) can be expressed as:

$$V_{dr} = R_r I_{dr} - s\omega_s L_r I_{qr} - s\omega_s L_m I_{qs} + L_r \frac{dI_{dr}}{dt} + \frac{L_m}{L_s} \frac{d\left(\Psi_{ds} - L_m I_{dr} \right)}{dt} \tag{14}$$

Because it is directly connected to the grid, the stator voltage shares constant magnitude and frequency of grid voltage. One could make the *d*-axis align with stator voltage vector, and it is true that V_s=V_{ds} and V_{qs}=0, thus Ψ_s=Ψ_{qs} and Ψ_{ds}=0, which are stator voltage-oriented vector control scheme, as depicted in Figure 13. Therefore, Equation (14) can be organized as:

http://dx.doi.org/10.5772/60958

$$V_{dr} = [R_r + \left(L_r - \frac{L_m^{\ 2}}{L_s}\right)\frac{d}{dt}]I_{dr} - s\omega_s\left[L_r I_{qr} + L_m I_{qs}\right] \tag{15}$$

Equation (15) implies that the *d*-axis rotor voltage consists of two voltage components V_{dr}^{1} and V_{dr}^{2}:

$$\begin{aligned} V_{dr}^{\ 1} &= [R_r + \left(L_r - \frac{L_m^{\ 2}}{L_s}\right)\frac{d}{dt}]I_{dr} \quad a \\ V_{dr}^{\ 2} &= -s\omega_s\left[L_r I_{qr} + L_m I_{qs}\right] \quad b \end{aligned} \tag{16}$$

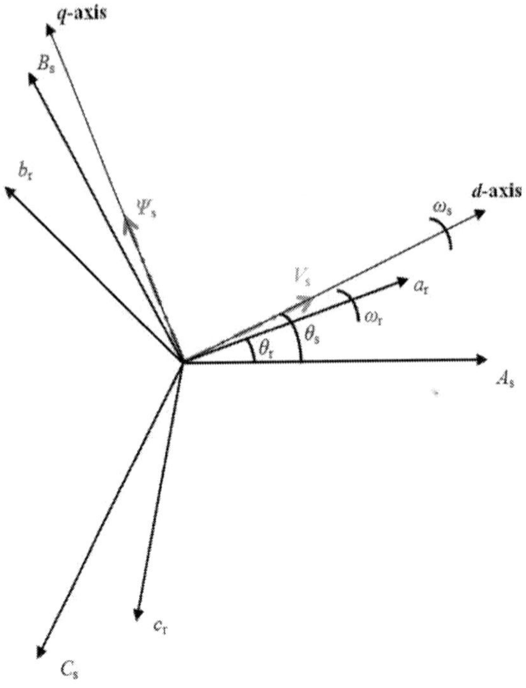

Figure 13. Stator voltage FOC reference frame

The V_{dr}^{1} is called current regulation part and depicted by Figure 14, where $\sigma = L_r - L_m^2/L_s$. Due to the linear relationship between V_{dr}^{1} and I_{dr}, the PI controller is employed. Besides, V_{dr}^{2} is the cross-coupling part and requires feedforward compensation for a complete control. Eventually, the rotor-side converter voltage in *d*-axis is derived as:

$$V_{drc} = V_{dr} = V_{dr}^{\ 1} + V_{dr}^{\ 2} \tag{17}$$

where subscript rc denotes the rotor-side converter. After the conversion of *dq-abc*, the rotor-side converter voltage V_{abc_rc} can be obtained, which is used to generate PWM control signals for rotor-side converter.

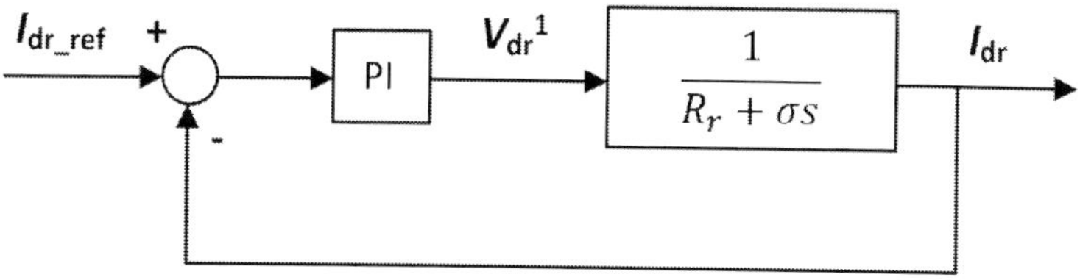

Figure 14. Current regulation part of *d*-axis rotor-side converter voltage

If only steady-state is considered, the derivative parts in Equation (10) are neglected and one can obtain stator flux as:

$$\begin{aligned} \Psi_{ds} &= \frac{V_{qs} - R_s I_{qs}}{\omega_s} && a \\ \Psi_{qs} &= (V_{ds} - R_s I_{ds}) / (-\omega_s) && b \\ \Psi_s &= \sqrt{\Psi_{ds}^2 + \Psi_{qs}^2} && c \end{aligned} \tag{18}$$

According to Equations (8), (10), and (12), the rotor-side converter reference current is derived as:

$$I_{dr\_ref} = -\frac{2L_s T_e}{3n_p L_m \Psi_s} \tag{19}$$

where

$$\begin{aligned} P_{e\_ref} &= P_{opt} - P_{loss} = T_e \omega_r && a \\ P_{loss} &= R_s I_s^2 + R_r I_r^2 + R_c I_{sc}^2 + F\omega_r^2 && b \end{aligned} \tag{20}$$

where I_{sc}, R_c, and F are stator-side converter current, choke resistance, and friction factor, respectively. P_{opt}, P_{e_ref}, and P_{loss} are desired optimal output active power, reference active power, and system power loss, respectively. Combining Equations (8), (10), and (11), the active power is used as command inputs to determine current references I_{dr_ref}.

Similarly, the q-axis rotor-side converter voltage consists of current regulation and cross-coupling parts too:

$$V_{qr}^{1} = [R_r + \left(L_r - \frac{L_m^2}{L_s}\right)\frac{d}{dt}]I_{qr} \quad a$$
$$V_{qr}^{2} = s\omega_s\left[L_r I_{dr} + L_m I_{ds} + \frac{L_m}{L_s}\frac{d\Psi_s}{dt}\right] b \tag{21}$$

$$V_{qrc} = V_{qr} = V_{qr}^{1} + V_{qr}^{2} \tag{22}$$

where the derivative of stator flux in Equation (21b) is considered as zero at steady-state. Also, the current regulation part is illustrated in Figure 15. If the stator-side converter's reactive power is controlled to be zero, the output reactive power is stator reactive output power. Then, one has:

$$Q_o = Q_s + Q_{sc} = Q_s = = -V_{ds}I_{qs} = -V_{ds}\frac{1}{L_s}\left(\Psi_s - L_m I_{qr}\right) \tag{23}$$

Thus, the regulation of reactive power can lead to I_{qr_ref}.

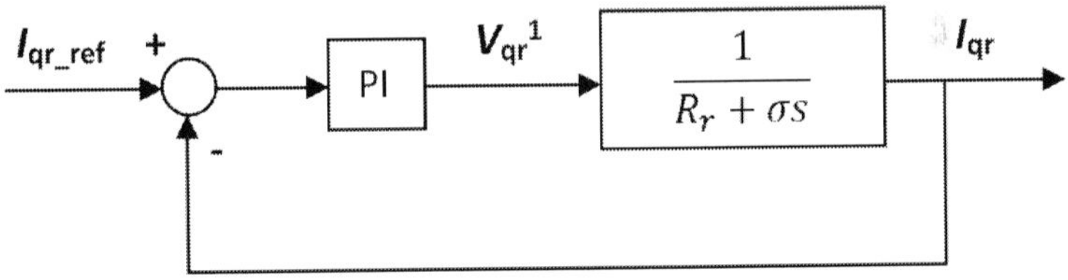

Figure 15. Current regulation part of q-axis rotor-side converter voltage

Involving the deviations of rotor voltage and reference currents in both d- and q-axis, Figure 16 exhibits the total control scheme for rotor-side converter, where the P_{opt} is obtained from MPPT.

2.3.3. *Control of stator-side converter*

The stator-side converter is controlled based on relationship between voltage, flux, and current of stator and choke, which is modeled by a cross-coupling model, as described in Figure 17. It

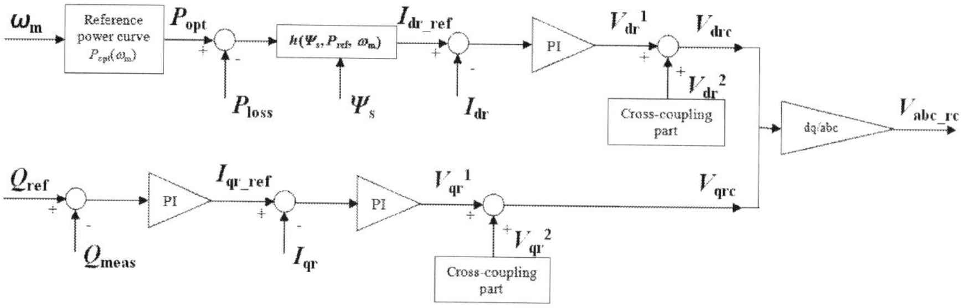

Figure 16. Total rotor-side converter control scheme

is seen that the grid (stator) voltage is equal to the sum of stator-side converter voltage and choke occupied voltage. By KVL:

$$
\begin{aligned}
V_{ds} &= R_c I_{dsc} - \omega_s \Psi_{qsc} + L_c \frac{dI_{dsc}}{dt} + V_{dsc} \quad a \\
V_{qs} &= R_c I_{qsc} + \omega_s \Psi_{dsc} + L_c \frac{dI_{qsc}}{dt} + V_{qsc} \quad b
\end{aligned} \tag{24}
$$

The flux linkage follows:

$$
\begin{aligned}
\Psi_{dsc} &= L_c I_{dsc} \quad a \\
\Psi_{qsc} &= L_c I_{qsc} \quad b
\end{aligned} \tag{25}
$$

Thus, the reorganized stator-side converter voltage in *d*- and *q*-axis, respectively, are presented as:

$$
\begin{aligned}
V_{dsc} &= V_{ds} - R_c I_{dsc} + \omega_s L_c I_{qsc} - L_c \frac{dI_{dsc}}{dt} \quad a \\
V_{qsc} &= V_{qs} - R_c I_{qsc} - \omega_s L_c I_{dsc} - L_c \frac{dI_{qsc}}{dt} \quad b
\end{aligned} \tag{26}
$$

where the subscripts sc and ch denote the variables of stator-side converter and choke, respectively. L_c and R_c are the inductance and resistance of the choke.

Based on the model in Equation (26a, b), the current regulation part of choke voltage in *d*- and *q*-axis are described as (27a, b) and Figure 18(a, b).

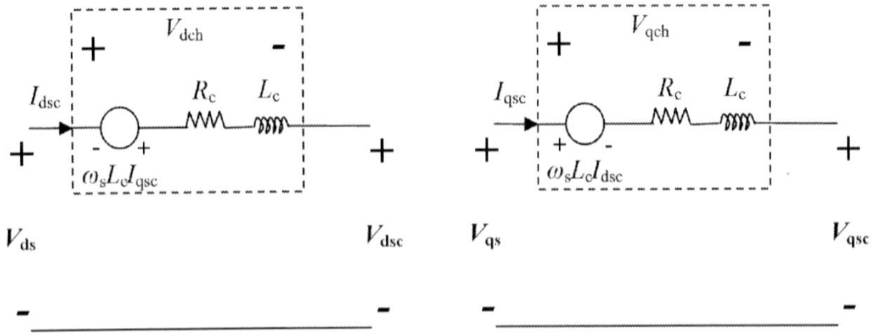

Figure 17. Equivalent circuit of stator-side converter choke [23]

$$V_{dch}^{1} = \left(R_c + L_c \frac{d}{dt} \right) I_{dsc} \quad a$$
$$V_{qch}^{1} = \left(R_c + L_c \frac{d}{dt} \right) I_{qsc} \quad b \tag{27}$$

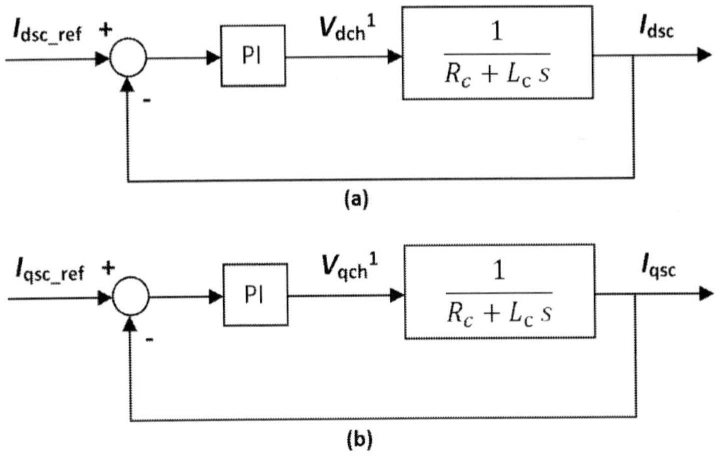

Figure 18. Current regulation part of choke voltage ((a) *d*-axis; (b) *q*-axis)

The cross-coupling part of choke voltage V_{dch2} and V_{qch2} are expressed as (28a, b) and the total stator-side converter voltage is derived as (29a, b).

$$\begin{aligned} V_{dch}^{2} &= -\omega_s L_c I_{qsc} \quad a \\ V_{qch}^{2} &= \omega_s L_c I_{dsc} \quad b \end{aligned} \tag{28}$$

$$\begin{aligned} V_{dsc} &= V_{ds} - V_{dch}^{1} - V_{dch}^{2} \quad a \\ V_{qsc} &= V_{qs} - V_{qch}^{1} - V_{qch}^{2} \quad b \end{aligned} \tag{29}$$

The current reference I_{qsc_ref} is generally set at zero for zero reactive power output from stator-side converter while I_{dsc_ref} is determined by the regulation of dc-link voltage V_{dc}. The stator-side converter voltage control is depicted in Figure 19.

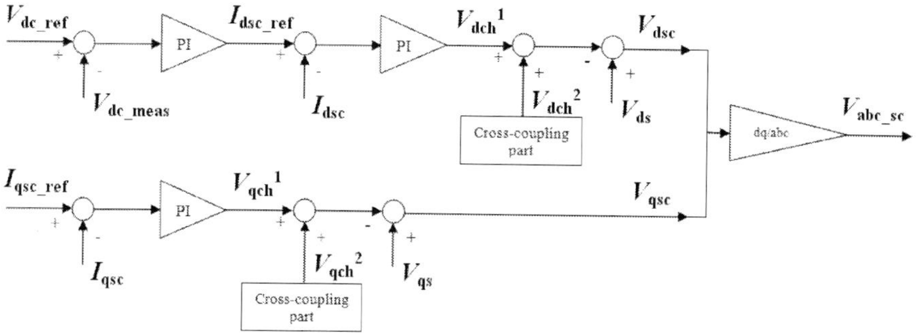

Figure 19. Total stator-side converter control scheme [23]

With both rotor- and stator-side converter controls, the simulation results [23] in Figure 20 present a stable and controllable dynamic response to a gusty wind speed. Also, an FRT capability is verified by a voltage droop happening within a constant wind speed. Figure 21 shows twice oscillations at two dynamic moments and the control system effectively recovers the system-regulated outputs in short amount of time.

2.3.4. *Space Vector Modulation (SVM)*

The purpose of both rotor- and stator-side converter controls is to obtain the reference voltages which are expected to be produced by the converter. The next step is obviously to generate the corresponding PWM gate signals for the converter. To a 2-level three-phase voltage source inverter, there are six switches of three legs in inverter controlling the phase voltage and thus the current of induction generator. By defining the "ON" and "OFF" states of upper switch by "1" and "0," respectively, for one leg, there exist up to eight different states for inverter outputs. They are summarized in Table 2 as well as the resulted phase voltage in *abc* and *αβ* frames. Eight inverter output voltages can be considered as eight voltage vectors [0, 0, 0] through [1, 1, 1] that are illustrated in Figure 22.

http://dx.doi.org/10.5772/60958

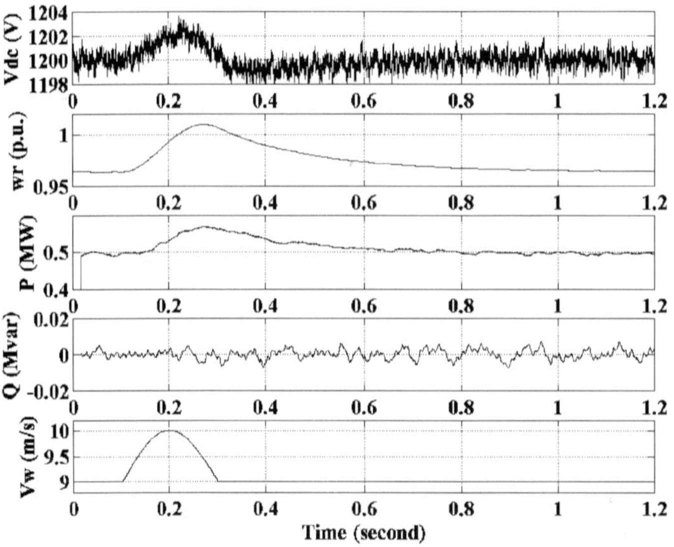

Figure 20. Gusty wind responses ((a) DC-link voltage V_{dc}; (b) generator speed ω_r; (c) active power P; (d) reactive power Q; (e) wind speed V_w)

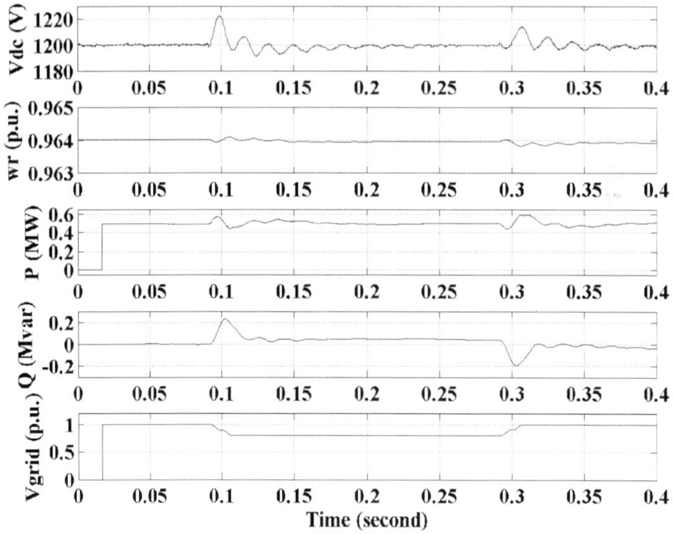

Figure 21. Dynamic responses to grid voltage droop ((a) DC-link voltage V_{dc}; (b) generator speed ω_r; (c) active power P; (d) reactive power Q; (e) grid voltage V_{grid})

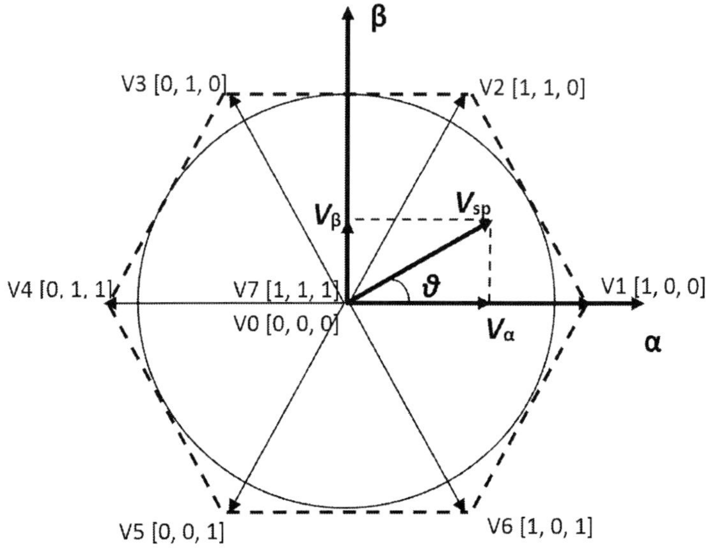

Figure 22. Eight inverter voltage space vectors

L_1	L_2	L_3	V_{an}	V_{bn}	V_{cn}	V_α	V_β
1	1	1	0	0	0	0	0
1	0	0	$2V_{dc}/3$	$-V_{dc}/3$	$-V_{dc}/3$	$\sqrt{(2/3)}V_{dc}$	0
1	1	0	$V_{dc}/3$	$V_{dc}/3$	$-2V_{dc}/3$	$\sqrt{(1/6)}V_{dc}$	$\sqrt{(1/2)}V_{dc}$
0	1	0	$-V_{dc}/3$	$2V_{dc}/3$	$-V_{dc}/3$	$-\sqrt{(1/6)}V_{dc}$	$\sqrt{(1/2)}V_{dc}$
0	1	1	$-2V_{dc}/3$	$V_{dc}/3$	$V_{dc}/3$	$-\sqrt{(2/3)}V_{dc}$	0
0	0	1	$-V_{dc}/3$	$-V_{dc}/3$	$2V_{dc}/3$	$-\sqrt{(1/6)}V_{dc}$	$-\sqrt{(1/2)}V_{dc}$
1	0	1	$V_{dc}/3$	$-2V_{dc}/3$	$V_{dc}/3$	$\sqrt{(1/6)}V_{dc}$	$-\sqrt{(1/2)}V_{dc}$
0	0	0	0	0	0	0	0

Table 2. Space vector states (L_1–L_3 represent inverter leg1–leg3)

Once the reference space vector voltage in $\alpha\beta$ frame is achieved by current regulation, the magnitude and angle of the voltage are used to implement the SVM. With constant PWM frequency, a space vector is always realized by a vector sum of two adjacent vectors in Table 2. Taking the space vector voltage (0 to 60 degree section) in Figure 23 as an example, it is equal to the vector sum of V_1 and V_2 with magnitudes of d_x and d_y, respectively, which are the duty cycles of two vectors [32]:

$$d_x = \frac{|V_{sp}|}{\sqrt{\frac{2}{3}}V_{dc}} \frac{\sin(60^\circ - \gamma)}{\sin 60^\circ} \quad a$$

$$d_y = \frac{|V_{sp}|}{\sqrt{\frac{2}{3}}V_{dc}} \frac{\sin\gamma}{\sin 60^\circ} \quad b \qquad (30)$$

$$d_z = 1 - d_x - d_y \quad c$$

where d_z denotes the duty cycle of zero vector. Generally, the zero vectors [0, 0, 0] or [1, 1, 1] contribute the remaining PWM period after d_x and d_y. The space vector voltages located in other sections can follow the same procedure to obtain the duty cycles of d_x, d_y, and d_z. Then, the Minimum-Loss Space Vector PWM (MLSVPWM) technique is applied to determine the sequence of vectors [32]. The PWM signals are eventually obtained based on computed duty cycles and sequence of vectors. Figures 24 and 25 show the simulation and experimental three-phase duty ratios for inverter phase A, B, and C, where no switching action happens if 0 or 1 duty cycle is the case. It is seen that there is always one phase being absent of switching at any moment, which minimizes the switching loss of the semiconductor switches. Also, the experimental results reveal the sinusoidal nature of the line voltage duty ratio that is expected for sinusoidal fundamental line voltage output of inverter. With this PWM SVM technique, the rotor- and stator-side converters are controlled by previously derived FOC.

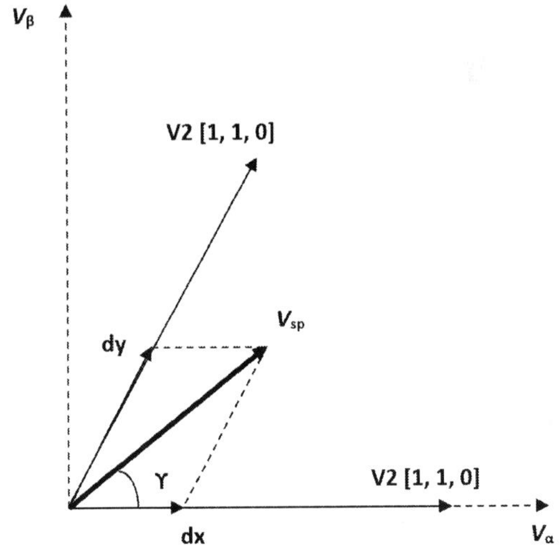

Figure 23. Duty cycles of vectors for reference space vector voltage

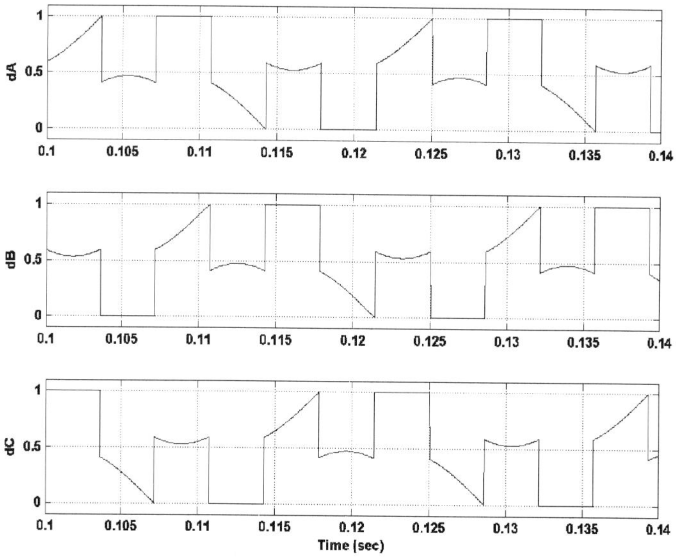

Figure 24. Three-phase duty cycles using MLSVPWM (simulation results [33])

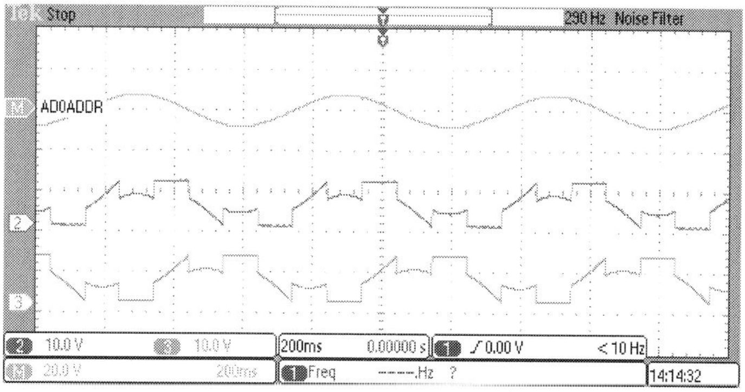

Figure 25. Phase duty cycles and phase-to-phase duty cycle using MLSVPWM (experiment results [33])

2.3.5. Islanded operation

Compared to the grid integration DFIG wind system, the isolated DFIG wind system operating at regulated voltage (magnitude and frequency) is also found applicable and valuable to some independent power subgrid or distributed power systems. One of the application examples – DFIG-Synchronous machine system configuration – is shown in

Figure 26. Modified FOC for power generation in Figure 27 is used and the line voltage magnitude and frequency are stabilized by extra variable load and synchronous machine. The line frequency is held by compensating the resistive load, while the line voltage is held by feeding controlled field voltage of synchronous machine. The proposed controller scheme in [34] is employed for synchronous machine field voltage controller. As shown in Figure 28(a, b), the line frequency is regulated at 60Hz with limited error while the line voltage is regulated at 1 p.u. The constant frequency and magnitude in transmission line voltage is the basic requirement for a controllable power delivery. Based on the regulations of frequency and line voltage, the active power and reactive power are also under controlled respectively [25]. The dc-link voltage is kept at nominal value, while the generator speed is controlled at optimal 0.95 p.u., as shown in Figure 29.

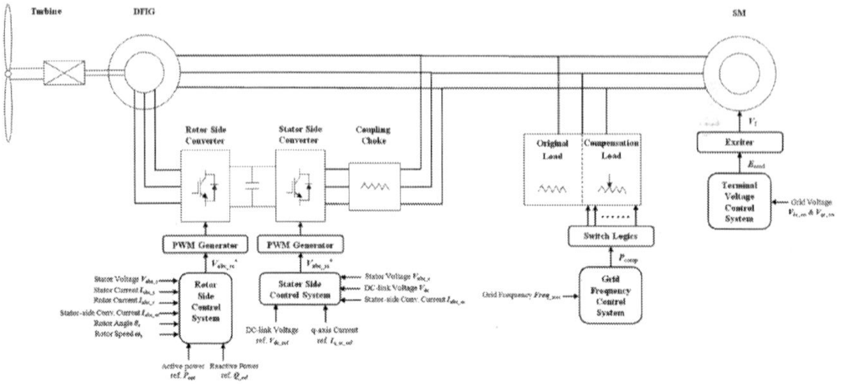

Figure 26. Islanded DFIG-Synchronous machine wind power system

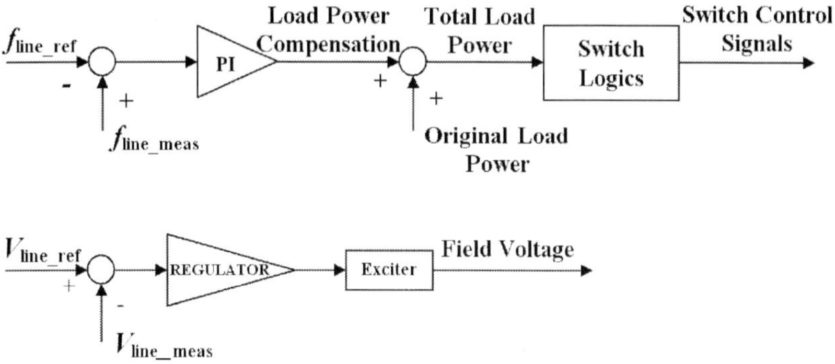

Figure 27. DFIG-Synchronous machine system control scheme

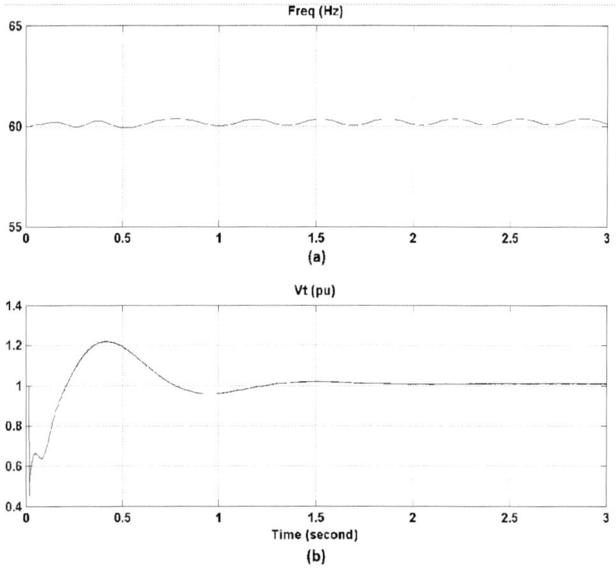

Figure 28. Line voltage and frequency in islanded DFIG system ((a) line frequency; (b) line voltage)

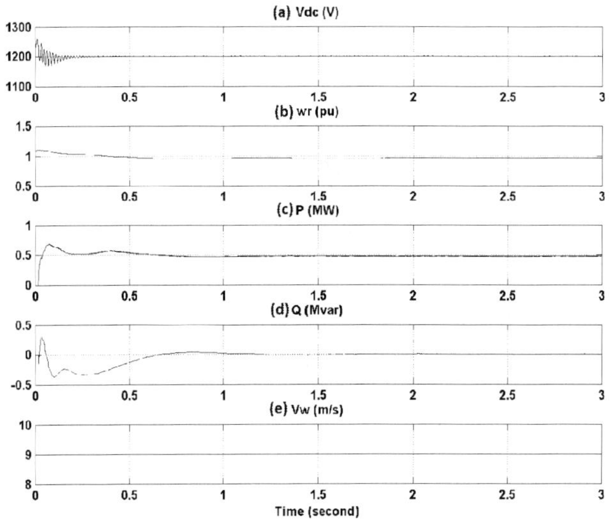

Figure 29. Simulation results for islanded operation ((a) dc-link voltage V_{dc}; (b) generator speed ω_r; (c) active power P; (d) reactive power Q; (e) wind speed V_w)

2.4. Grid synchronization

The key to performing FOC is to follow the position angle of the d-axis component so that the output can be synchronized with dq frame, especially in the grid integration operation mode. In order to operate in this mode, the induction generator voltage must be synchronized with the grid voltage by applying the Phase Lock Loop (PLL) technique. The technique takes the grid signal as input and keeps track of the position angle of the grid voltage for FOC as well as reproducing the grid voltage frequency as output, in a real-time manner. To introduce the algorithm, assume an estimated grid voltage angle (accurate grid voltage angle of θ); the resulting grid voltages in dq frame are written as:

$$\begin{aligned} V_d &= V_\alpha \cos\hat{\theta} + V_\beta \sin\hat{\theta}\, a \\ V_q &= -V_\alpha \sin\hat{\theta} + V_\beta \cos\hat{\theta} b \end{aligned} \tag{31}$$

where

$$\begin{aligned} V_\alpha &= V_m \cos\theta\, a \\ V_\beta &= V_m \sin\theta\, b \end{aligned} \tag{32}$$

and V_m denotes the magnitude of the voltage space vector. By substituting V_α and V_β in (31a, b) by (32a, b), the V_d and V_q can be organized as:

$$\begin{aligned} V_d &= V_m \cos\left(\theta - \hat{\theta}\right) a \\ V_q &= V_m \sin(\theta - \hat{\theta}) b \end{aligned} \tag{33}$$

It is seen that if is equal to θ, V_d is equal to V_m, and V_q is equal to 0. Therefore, the accurate grid voltage angle θ can be obtained by regulating the grid voltage V_q to zero. Assuming there is an error δ between and θ that $\delta = \theta$ - , due to the small value of δ, it is true that $V_q \cong V_m\delta$ and the PLL system in s-domain can be described as Figure 30, where (s) denotes the estimated grid voltage angular frequency and $K_c(s)$ is a PI controller. After removing the unknown accurate θ, the PLL scheme is essentially a regulation of V_q in Figure 31, where the measured V_q goes through a 1st-order low-pass filter whose cutoff frequency is ω_c. In this way, the noise is effectively eliminated.

The introduced PLL algorithm is simulated and shown in Figure 32, as well as zoom-in image in Figure 33 [33], where 0 radian grid voltage position coincides with the zero-crossing of phase voltage V_{an} and the frequency can be detected to be 60Hz after short transient (the initial grid frequency is assumed as 55Hz). These results indicate a successful "locking" of grid frequency and position angle, with which the FOC (Figures 16 and 19) are conducted on back-to-back converter in a real-time manner and can thus continuously "match" the generated voltage with the grid voltage.

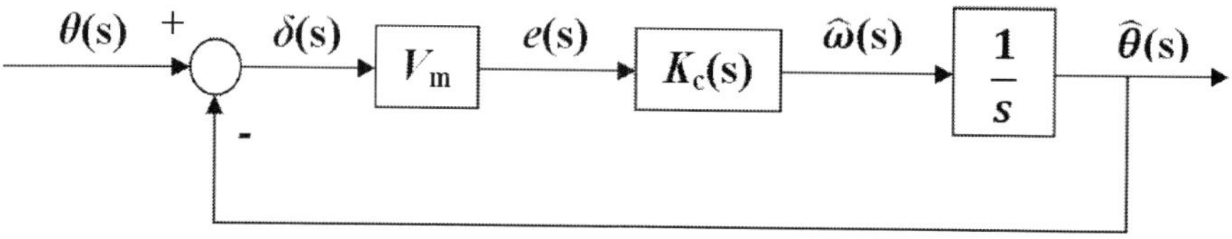

Figure 30. PLL scheme for grid voltage angle estimation

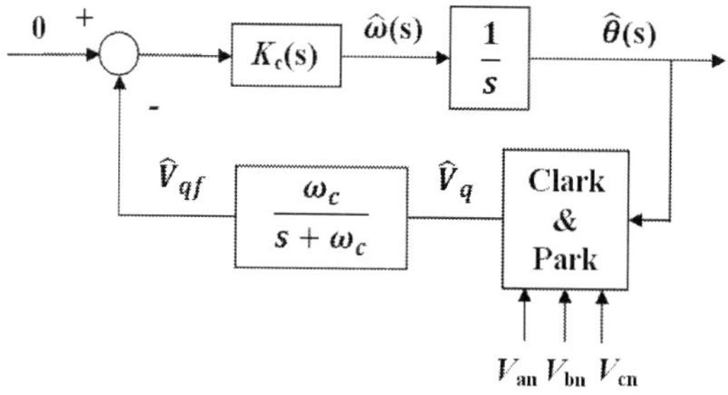

Figure 31. PLL scheme with regulation of V_q

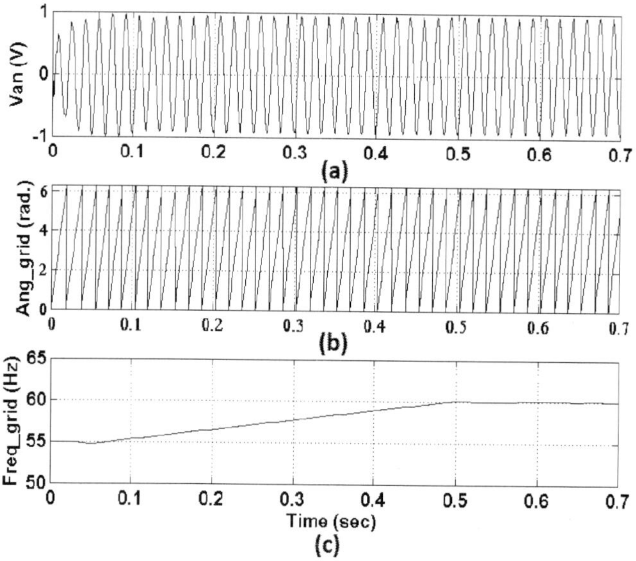

Figure 32. PLL results of grid voltage angle and frequency for grid integration operation ((a) phase A voltage V_{an}; (b) grid voltage angular position; (c) grid voltage frequency)

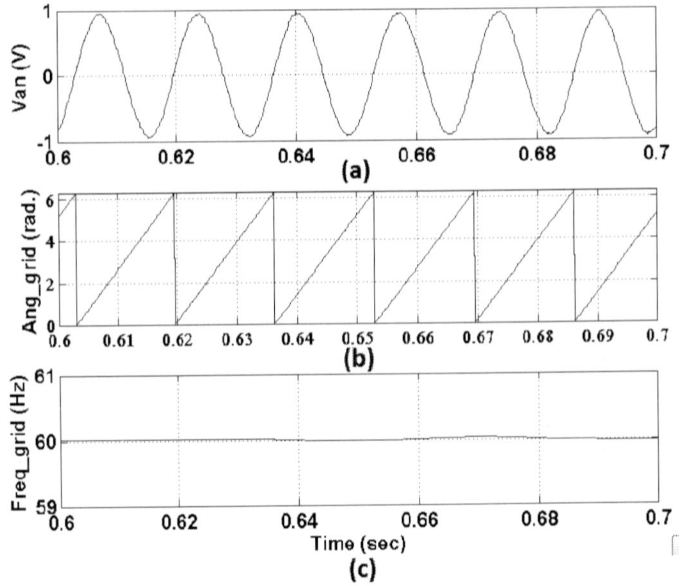

Figure 33. Zoom-in image of Figure 32

3. Maximum Power Point Tracking (MPPT)

Efficiency always plays an important role in induction generator wind systems. While the SCIG system loses precise control of power due to the fixed-speed operation, to achieve high efficiency in wind power conversion systems, the MPPT in variable-speed DFIG system has been intensively investigated. Basically, the studied techniques in MPPT include three strategies: (1) the methods relying on wind speed, (2) the methods relying on output power measurement and calculation, and (3) the methods relying on reference power curve.

3.1. Pitch angle control of induction generator wind systems

An overall picture of induction generator wind system operation versus wind speed is depicted in Figure 34, where the output power must be "truncated" after reaching certain level. Pitch angle control, as investigated in section 2.2, is used not only in SCIG system but also in DFIG system for this purpose. It is seen that the capability of pitch angle control in response to the increase of wind speed, on limiting the power output, is primarily dependent on turbine blade physical structure. Therefore, the system needs to be shut down by brake system in the case of wind speed cutoff. Figure 34 also emphasizes the augmented power output of MPPT operation over fixed-speed operation and this inspires the investigation of advanced variable-speed wind systems, where the induction generator speed can always be controlled in a large range to capture desired output power by combining the previously discussed FOC with MPPT strategies.

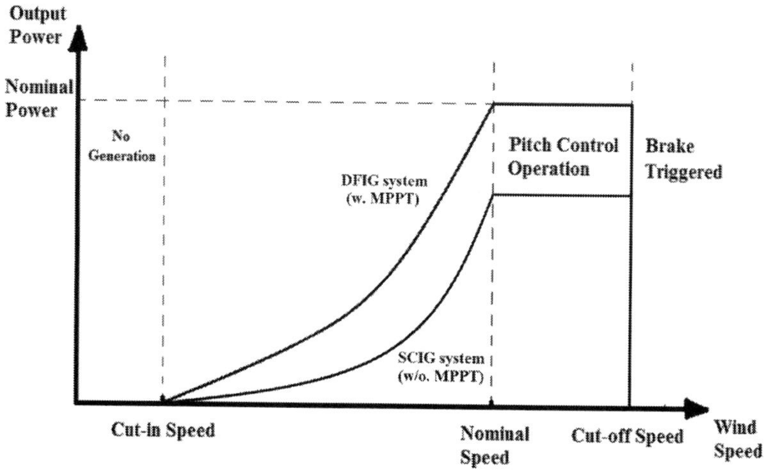

Figure 34. Wind power system operations

3.2. MPPT methods for DFIG wind systems

3.2.1. Wind speed based method

Most DFIG wind power systems are dependent of wind speed measurement [2,4]. In these systems, anemometers are applied to measure the wind speed and thus the systems suffer from additional cost of sensors and complexity. In order to solve this problem, wind speed estimation methods have been reported [25, 35-36]. Relying on the complex algorithms, the accurate wind speed can be captured for controlling the optimal tip speed ratio so that the MPPT can be performed accordingly, as shown in Figure 35. However, the wind speed information and associated efforts on software/hardware are still necessary and significant. To eliminate the dependence on wind speed, some sensorless control strategies have been developed [31,37-38]. These methods are in test for small-scale stand-alone systems and the complicated estimation algorithms remain, which will result in weakening of accuracy and control speed in real operating environment where the wind speed changes rapidly.

3.2.2. Power variation rate based method

Tracking the maximum power can also be accomplished through measuring the output power directly [39-42]. The idea of this method is through checking the variation rate of the output power with respect to that of generator speed ($dP/d\omega$), the power operation point location can

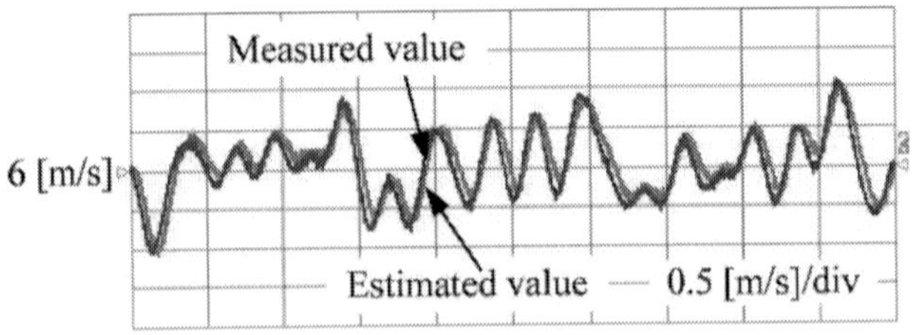

Figure 35. Estimated wind speed and real measured wind speed [35]

be determined and be accordingly controlled thereafter. Theoretically, the maximum power operating point can be reached when $dP/d\omega = 0$, as shown in Figure 36. A flowchart of the algorithm is shown in Figure 37, where the operation point (ω_m (k), P(k)) is measured and compared with (ω_m(k-1), P(k-1)) under wind speed V_w(i), where i is the index of wind speed; k is given as the test step index under particular wind speed V_w(i). Among all the tested points, only one point holds the truth that $dP/d\omega = 0$ and it is the optimal operation point (ω_{mopt}(i), P_{opt}(i)) that will be returned and saved. This procedure is required in a means of real-time for different wind speed (i=1,2,...,). According to the information from the optimal points, either the generator speed or the duty cycle of converter can be tuned.

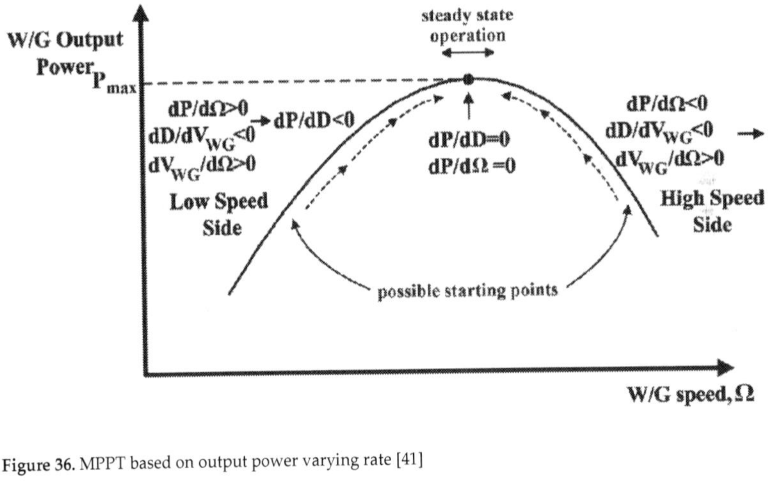

Figure 36. MPPT based on output power varying rate [41]

3.2.3. Reference power curve based method

Besides the above strategies, MPPT can be carried out by means of tracking the reference (optimal) power curve, which is the fitting curve going through all the maximum power points of all wind speeds [43-45]. A generalized reference power curve is given as:

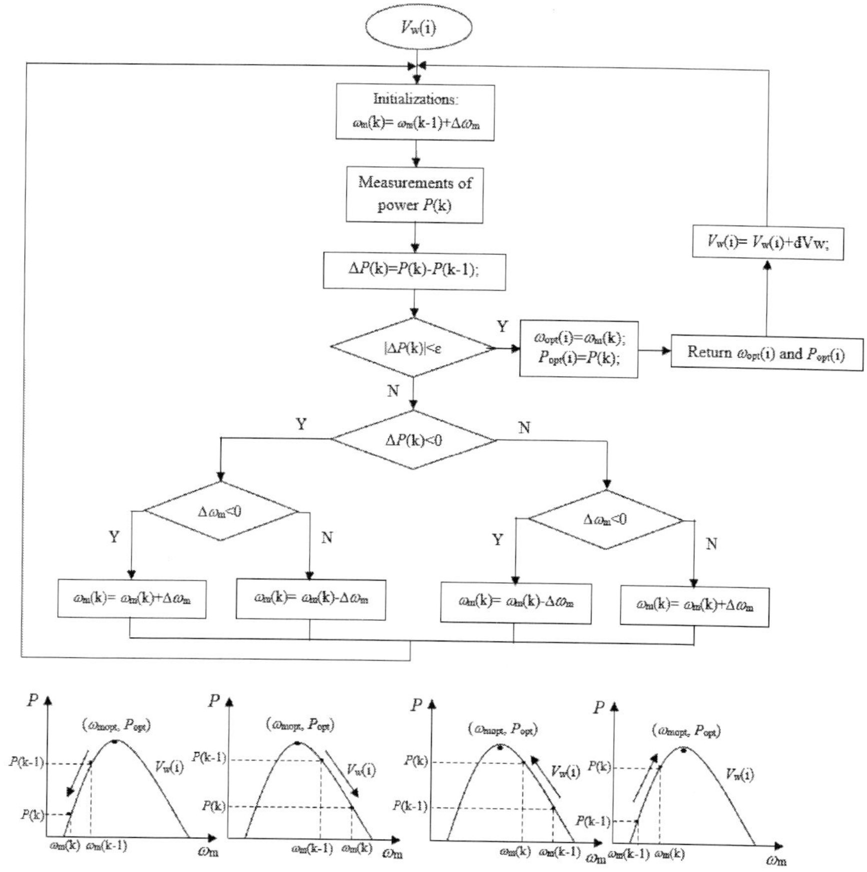

Figure 37. Flow chart of power variation rate testing algorithm

$$P_e = b_k \omega_m^k + b_{k-1} \omega_m^{k-1} + \ldots + b_1 \omega_m + b_0 \tag{34}$$

To determine the optimal degree of the polynomial, comparison is conducted for a 2.678 MW DFIG wind system [33]. Under a particular wind speed, four reference curves lead to four different operation points and the 3rd-order polynomial in Equation (35) leads to the most accurate reference curve along optimal operation points, as shown in Figure 38.

$$P_e = b_3 \omega_m^3 + b_2 \omega_m^2 + b_1 \omega_m + b_0 \tag{35}$$

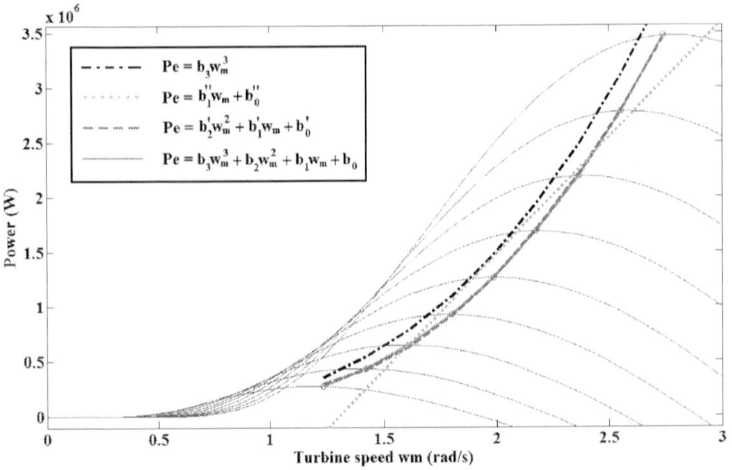

Figure 38. Comparison of reference power curve fittings

This method has been widely used due to its simple concept and absence of extra wind measurement costs. The optimal reference power curve is constructed according to the experimental tests and programmed in a microcontroller memory, to be used as a lookup table. The algorithm diagram is illustrated in Figure 39. Either the generator speed is measured to obtain power reference for power regulation, or the wind speed is measured to obtain generator speed reference for generator speed regulation. The former method produces more accurate output power, while the latter has faster control speed [25]. Some research works simply apply a cube function of generator speed as reference power or a square function of generator speed as reference torque. Despite these feasible solutions, the accurate maximum power and corresponding optimal generator speed are undervalued. Such approximation will obviously lead to harmed power generation efficiency. More importantly, analysis is necessary to verify the stability of the method in terms of varying wind speed and output power.

An evolved solution was proposed in [47] to effectively minimize the drawback of the above method. The real-time tuning of reference power curve coefficients is conducted and followed by updating the reference power curve. First, instead of disturbing output power directly, the most significant coefficient is incrementally disturbed by constant. This change of reference power curve induces the variation of the output power, which is measured and compared with previous step power. When the difference in output power between two consecutive steps approaches a small enough value, the disturbed coefficient is returned to update the reference power curve. The resulting reference power curve is the accurate optimal reference power curve. Due to the existence of reference power curve, such tuning

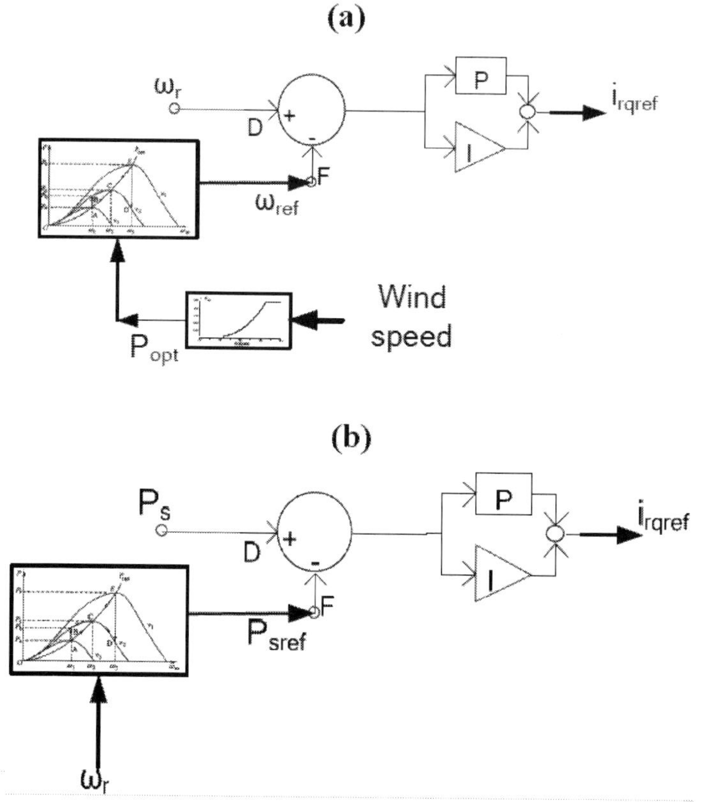

Figure 39. MPPT based on reference power curve [46]

calculation does not need to be conducted continuously with high frequency. In addition, without disturbing output power directly, this method can conduct updating and perturbation faster. Moreover, any deviation of system model will not give rise to deviation of optimal power generation because of the real-time tuning. Thus, the method is robust. As depicted in Figure 40, power variation is checked to capture the optimal coefficient and the reference power curve is updated accordingly to lead the system running in MPPT mode. The whole procedure is described in the simulation results in Figures 41 and 42, where the perturbation of coefficient b_3, the generator speed, and generated power halt after reaching the optimal values. No more perturbation and updating are needed, thus saving the calculation cost. Despite the oscillations at each b_3 perturbation step, the dc-link voltage and the reactive power remain at desired values while the output power and generator speed are updated, step by step, toward the optimal values. The generator speed and output power are generally measured with much higher frequency than that of perturbation. It is also worth noting that the bandwidth of b_3 updating must ensure that both generator speed and output power are able to reach their steady-states.

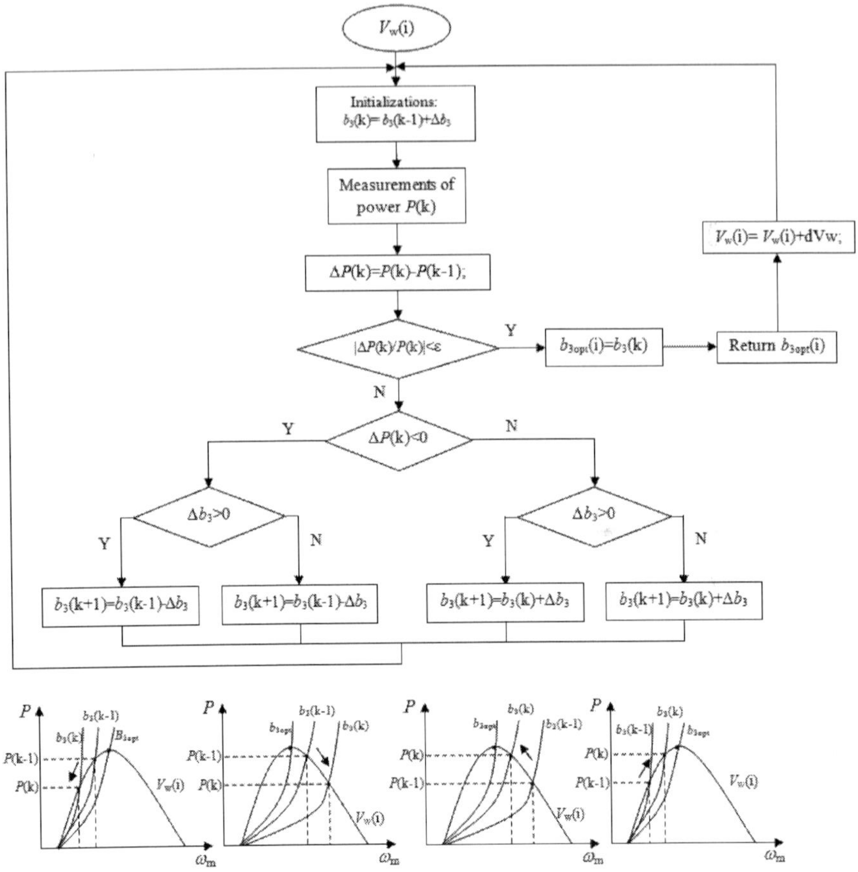

Figure 40. Novel MPPT algorithm proposed in [47]

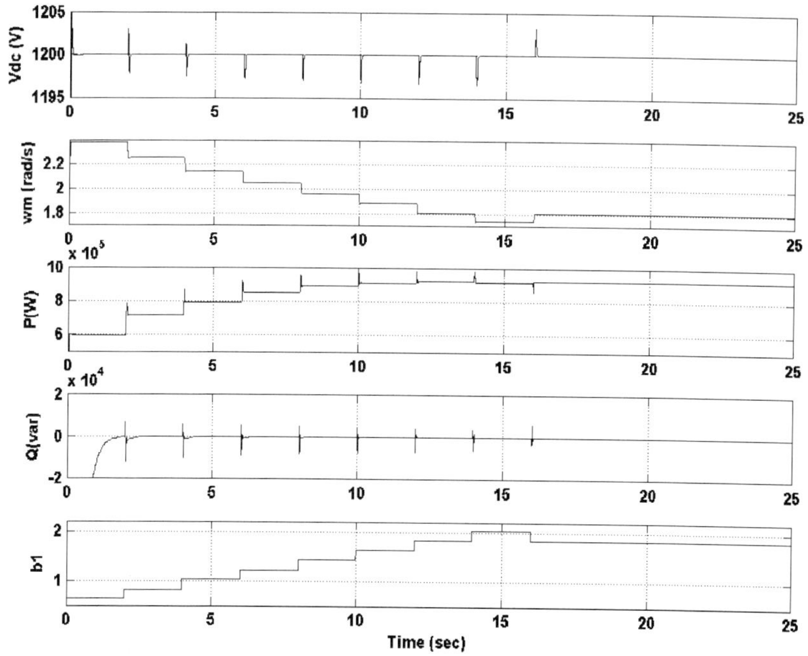

Figure 41. DFIG system operation applying the novel MPPT (simulation results)

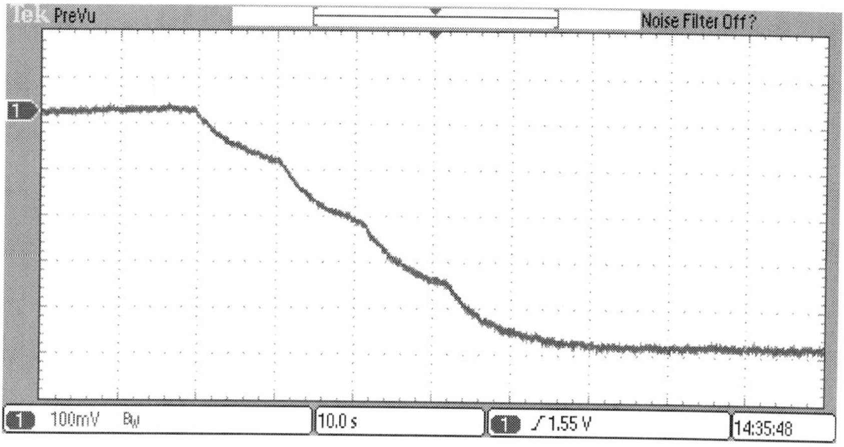

Figure 42. DFIG speed variation induced by the novel MPPT (experiment result)

4. Conclusion

Wind power systems have been widely studied and applied for years. By virtue of many advantages, induction generators are found to be suitable in this area. This chapter introduced and studied two popular types of induction generators – SCIG and DFIG. An overview of the generators, power electronics, and control strategies was presented first, followed by detailed modeling of entire wind system. Most importantly, the control algorithms were illustrated, ranging from FOC, SVM, PLL, to MPPT. Especially, different MPPT strategies were investigated and compared.

Acknowledgements

The contents of this chapter are the result of work at the Power Electronics Research Lab at the University of Akron, where my research was funded by Dr. Yilmaz Sozer; and work at the Renewable Energy Lab at Saginaw Valley State University, where my research was funded by the Faculty Research Grant.

I highly appreciate Dr. Yilmaz Sozer and Dr. Malik Elbuluk at the University of Akron for supervising my research work as well as guiding my progress in a peaceful and productive direction. I highly appreciate the support from Saginaw Valley State University that granted my start-up lab platform for long-term research commitment.

I am also very grateful to my wife, parents, and mother-in-law for their support. They established the foundation on which rests every success of my career.

Author details

Yu Zou*

Address all correspondence to: yzou123@svsu.edu

Saginaw Valley State University, Michigan, USA

References

[1] Stiebler M. *Wind Energy Systems for Electric Power Generation*. Rijeka: Springer; 2008, 14-26, 30-43.

[2] Ahlström A. *Simulating Dynamical Behaviour of Wind Power Structures*. Licentiate Thesis. Royal Institute of Technology; 2002, 7-20.

[3] Heier S. *Grid Integration of Wind Energy Conversion Systems*. Rijeka: John Wiley & Sons; 2006, 31-368.

[4] Walker J. Jenkins N. *Wind Energy Technology*. Rijeka: John Wiley & Sons; 1997.

[5] Boldea I. *The Electric Generators Handbook – Variable Speed Generators*. Rijeka: Taylor & Francis; 2006, I18- I28.

[6] Rüncos F., Carlson R., Sadowski N., Kuo-Peng P. Performance analysis of a doubly fed twin stator cage induction generator. In: Wiak S., Dems M., Komeza K. (eds.), *Recent Developments of Electrical Drives*. Rijeka: Springer; 2006, 361-373.

[7] Dubois M.R. *Optimized Permanent Magnet Generator Topologies for Direct-Drive Wind Turbines*. PhD Dissertation. Delft University Technology; 2004, 9-27.

[8] Rodrigo. Dissertation on Renewable Energy Sources. http://writepass.com/journal/2012/12/working-on-res-renewable-energy-sources/ (December 2012)

[9] Grauers A. *Design of Direct-Driven permanent-magnet generators for wind turbines. PhD Dissertation*. Chalmers University of Technology; 1996, 11-19, 37-55.

[10] Torrey D.A. Switched reluctance generators and their control. *IEEE Transact Indust Electron* 2002; 49(1) 3-14.

[11] Li H., Chen Z. Overview of different wind generator systems and their comparisons. *IET Renew Power Gen* 2008; 2(2) 123-138.

[12] Dubois M.R., Polinder H., Ferreira J.A. Comparison of generator topologies for direct-drive wind turbines. Proceedings of Nordic Countries Power and Industrial Electronics, NORPIE 2000 June 2000, Aalborg, Denmark, 22-26.

[13] Patil N.S., Bhosle Y.N. *A Review on Wind Turbine Generator Topologies.* International Conference on Power, Energy and Control, ICPEC 2013, 6-8 February 2013, Sri Rangalatchum Dindigul, 625-629.

[14] Lampola P. *Directly Driven, Low-Speed Permanent-Magnet Generators for Wind Power Applications*. PhD Dissertation. Helsinki University of Technology; 2000, 11-19.

[15] Siegfriedsen S., Bohmeke G. Multibrid technology – a significant step to multi-megawatt wind turbines. *Wind Energy* 1998; 1(2) 89-100.

[16] Polinder H., Van Der Pijl F.F.A., De Vilder G.J., Tavner P.J. Comparison of direct-drive and geared generator concepts for wind turbines. *IEEE Transact Energy Conver* 2006; 21(3) 725-733.

[17] Hansen L.H., Helle L., Blaabjerg F., Ritchie E., Munk-Nielsen S., Bindner H., Sørensen P., Bak-Jensen B. Conceptual survey of generators and power electronics for wind turbines. Riso National Laboratory Technical Report; 2001, 21-36, 46-50.

[18] Li H., Chen Z. Design optimization and evaluation of different wind generator systems. International Conference on Electrical Machines and Systems, ICEMS 2008, 17-20 October 2008, Wuhan, China, 2396-2401.

[19] Kanellos F.D., Papathanassiou S.A., Hatziargyriou N.D. Dynamic analysis of a variable speed wind turbine equipped with a voltage source AC/DC/A converter interface and a reactive current control loop. 10th Mediterranean Electrotechnical Conference, MELECON 2000, 29-31 May 2000, vol.3, 986-989.

[20] Chen Z., Guerrero J.M., Blaabjerg F. A review of the state of the art of power electronics for wind turbines. *IEEE Transact Power Electron* 2009; 24(8) 1859-1875.

[21] Baroudi J.A., Dinavahi V., Knight A.M. A review of power converter topologies for wind generators. IEEE International Conference on Electric Machines and Drives 2005, 15-15 May 2005, San Antonio, TX, USA, 458-465.

[22] Chen Z., Spooner E. Current source thyristor inverter and its active compensation system. *IEE Proc Gen, Trans Distrib* 2003; 150(4) 447-454.

[23] Zou Y., Elbuluk M., Sozer Y. A complete modeling and simulation of induction generator wind power systems. IEEE Industry Applications Society Annual Meeting, IAS 2010, 3-7 October 2010, Huston, TX, USA, 1-8.

[24] Tan K., Islam S. Optimum control strategies in energy conversion of PMSG wind turbine system without mechanical sensors. *IEEE Transact Energy Conver* 2004; 19(2) 392-399.

[25] Sun L., Mi Z., Yu Y., Wu T. Active power and reactive power regulation capacity study of DFIG wind turbine. International Conference on Sustainable Power Generation and Supply 2009, SUPERGEN 2009, 6-7 April 2009. Nanjing, China, 1-6.

[26] Schiemenz I., Stiebler M. Control of permanent magnet synchronous generator used in a variable speed wind energy system. IEEE International Electric Machines and Drives Conference, IEMDC 2001, 17-20 June 2001. Cambridge, MA, USA, 872-877.

[27] Hansen A.D., Michalke G. Multi-pole permanent magnet synchronous generator wind turbines' grid support capability in uninterrupted operation during grid faults. *IET Renew Power Gen* 2009; 3(3) 333-348.

[28] Ohyama K., Arinaga S., Yamashita Y. Modelling and simulation of variable speed wind generator system using boost converter of permanent magnet synchronous generator. European Conference on Power Electronics and Applications 2007, 2-5 September 2007, Aalborg, Denmark, 1-9.

[29] Keyuan H., Yikang H. Investigation of a matrix converter-excited brushless doubly-fed machine wind-power generation system. The 5th International Conference on Power Electronics and Drive Systems 2003, PEDS 2003, 17-20 November 2003, vol.1, 743-748.

[30] Zhang L., Watthanasarn C., Shepherd W. Application of a matrix converter for the power control of a variable-speed wind-turbine driving a doubly-fed induction generator. 23rd International Conference on Industrial Electronics, Control and Instrumentation 1997, IECON 1997, 9-14 November 1997, New Orleans, LA, USA, vol.2, 906-911.

[31] Singh M., Chandra A., Singh B. Sensorless power maximization of PMSG based isolated wind-battery hybrid system using adaptive neuro-fuzzy controller. IEEE Industry Applications Society Annual Meeting, IAS 2010, 3-7 October 2010, Huston, TX, USA, 1-6.

[32] Sozer Y. *Direct Adaptive Control Of Permanent Magnet Motors*. Ph.D Dissertation. Rensselaer Polytechnic Institute; 1999, 54-73.

[33] Zou Y. Modeling, *Control and Maximum Power Point Tracking (MPPT) of Doubly-Fed Induction Generator (DFIG) Wind Power System*. PhD Dissertation. University of Akron; 2012, 44-90.

[34] Bansal R. C. Automatic reactive-power control of isolated wind-diesel hybrid power systems. *IEEE Transact Indus Electron* 2006; 53(4) 1116-1126.

[35] Abo-Khalil A.G., Lee D.C. MPPT control of wind generation systems based on estimated wind speed using SVR. *IEEE Transact Indus Electron* 2008; 55(3) 1489-1490.

[36] Guo P. Research of a new MPPT strategy based on gray wind speed prediction. 2nd International Symposium on Knowledge Acquisition and Modelling 2009, KAM 2009, 30 November-1 December 2009. Wuhan, China, 120-123.

[37] Liu D., Wu Z., Wang H., Wang T. MPPT control strategy for off-grid wind power system. IEEE 2nd International Symposium on Power Electronics for Distributed Generation Systems 2010, PEDG 2010, 16-18 June 2010, Hefei, China, 759-764.

[38] Pan C., Juan Y. A novel sensorless MPPT controller for a high-efficiency microscale wind power generation system. *IEEE Transact Energy Conver* 2010; 25(1) 207-216.

[39] Jia Y., Yang Z., Cao B. A new maximum power point tracking control scheme for wind generation. International Conference on Power System Technology 2002, 13-17 October 2002. 144-148.

[40] Neammanee B., Krajangpan K., Sirisumrannukul S. Maximum peak power tracking-based control algorithms with stall regulation for optimal wind energy capture. Power Conversion Conference 2007, PCC 2007, 2-5 April 2007, Nagoya, 1424-1430.

[41] Koutroulis E., Kalaitzakis K. Design of a maximum power tracking system for wind-energy-conversion applications. *IEEE Transact Indus Electron* 2006; 53(2) 486-494.

[42] Wang P., Liu H., Guo C., Tao C. MPPT control algorithms for wind power generation based on voltage disturbance. 7th World Congress on Intelligent Control and Automation 2008, WCICA 2008, 25-27 June 2008, Chongqing, China, 7398-7402.

[43] Barote L., Marinescu C. Storage analysis for stand-alone wind energy applications. 12th International Conference on Optimization of Electrical and Electronic Equipment 2010, OPTIM 2010, 20-22 May 2010, Basov, 1180-1185.

[44] Fernandez L.M., Garcia C.A., Jurado F., Saebz H,R. Control system of doubly fed induction generators based wind turbines with production limits. IEEE International Conference on Electric Machines and Drives 2005, 15-15 May 2005, San Antonio, TX, USA, 1936-1941.

[45] Yuan X., Wang F., Boroyevich D., Li Y. DC-link voltage control of a full power converter for wind generator operating in weak-grid systems. *IEEE Transact Power Electron* 2009; 24(9) 2178-2192.

[46] Yang L., Yang G.Y., Xu Z., Dong Z.Y. Optimal controller design of a doubly-fed induction generator wind turbine system for small signal stability enhancement. *IET Gen, Trans Distrib* 2010; 4(5) 579-597.

[47] Zou Y., Elbuluk M., Sozer Y. *A Novel Maximum Power Points Tracking (MPPT) Operation of Doubly-Fed Induction Generator (DFIG) Wind Power System*. IEEE Industry Applications Society Annual Meeting, IAS 2012, 7-11 October 2012, Las Vegas, NV, USA, 1-6.

Chapter 2

Active and Reactive Power Control of Wound Rotor Induction Generators by Using the Computer and Driver

Fevzi Kentli

Additional information is available at the end of the chapter

http://dx.doi.org/10.5772/61130

Abstract

In this chapter, a power control system for a wound rotor induction generator has been explained. This power control system has realized a control method using a rotating reference frame fixed on the air-gap flux of the generator. Application of such a system allows control of the active and reactive power of generators independently and stably. So, a two-step process is presented here. The first step is to acquire the complex power expression (and thus the active and reactive power expressions) for an induction machine in space vector notation and in two-axes system. Then, a computer aided circuit is given to realize the power and current control by analyzing them. Also, the results of an experiment given in literature are shown to be able to compare the results.

Keywords: Doubly-fed wound rotor, Induction generator, Active and reactive power control

1. Introduction

Energy is defined as the capacity of a body to do mechanical work. Energy is preserved in earth in many different forms, such as solid (coal), liquid (petroleum), and gas (natural gas). Also, several resources are available such as solar, wind, wave, geothermal, and nuclear energy. On the other hand, energy cannot be found in nature in electrical form. However, electric energy has many advantages: easy to transmit at long distances and complying with customer's needs through adequate control.

More than 30% of energy is converted into electrical energy before usage by the help of electric generators that convert mechanical energy into electric energy [1]. There are two kinds of electric generators: synchronous generators and induction generators. Generally, synchronous generators are generally used in big power plants to produce electricity. On the other hand, induction generators are used in utilizing nonstable small energy resources such as uncontrollable and automatically load-regulated small running water and wind turbines (plants) [2, 3]. Wind turbines have become much more popular due to the increasing demand for clean energy. Due to expensive production and maintenance costs, multiwatt turbines/wind farms are preferred. Configurations of generators and their controllers differ. Squirrel cage generators, wound rotor generators, permanent magnet generators, DC generators, and variable reluctance generators are operated in these systems. But nowadays a kind of wound rotor, doubly-fed induction generator, has begun to be used more [4].

Both, the synchronous generator with rotating DC magnetic field and the induction generator, have similar fixed stator winding arrangement, which, when energized by a rotating magnetic field, produces a three-phase (or single phase) voltage output. However, the rotors of the two machines are quite different, with the rotor of an induction generator typically consisting of one of two types of arrangements: "squirrel cage" or a "wound rotor." Also, unlike the synchronous generator that has to be "synchronized" with the electrical grid before it can generate power, induction generator can be connected directly to the utility grid and driven directly by the turbines rotor blades at variable wind or running water speeds. Induction motor is an economical and reliable choice as generator in many wind and running water power turbines where its rotational speed, performance, and efficiency can be increased by coupling a mechanical gearbox.

Being cheap, reliable, and readily available in a wide power range from fractional horse power to multi-megawatt capacities leads squirrel cage induction motor type machines to be used in both domestic and commercial renewable energy/running water power applications. The features that make this motor desirable make also the induction generator desirable over other types of generators. Generally, induction generators are constructed based on the squirrel cage induction motor type.

However, the induction generator may provide the necessary power directly to the mains utility grid, but it also needs reactive power to its supply which is provided by the utility grid. Stand-alone (off-grid) operation of the induction generator is also possible but the disadvantage here is that the generator requires additional capacitors connected to its windings for self-excitation.

Three-phase induction machines are very well-suited for wind power and even hydroelectric (running water) generation. Induction machines, when functioning as generators, have a fixed stator and a rotational rotor, the same as that for the synchronous generator. However, excitation (creation of a magnetic field) of the rotor is performed differently and typical designs of the rotors are the squirrel-cage structure where conducting bars are embedded within the rotors' body and connected together at their ends by shorting rings and the wound (slip-ring) rotor structure that carries a normal 3-ph winding, connected in star or delta and terminated on three slip-rings, which are short- circuited when the machine is in normal operation.

Induction machines are also known as asynchronous machines, that is, they rotate below synchronous speed when used as a motor and above its synchronous speed by some prime mover when used as a generator. The prime mover may be a turbine, an engine, a windmill, or anything that is capable of supplying the torque and speed needed to drive the motor into the overspeed condition. So when rotated faster than its normal operating or no-load speed, induction generator produces AC electricity. In this position, the speed is hypersynchronous, the slip is negative (and usually small), the rotor e.m.f.s and currents have such direction as to demand active power output from the stator terminals. But magnetization is still dependent on the stator winding accepting reactive power for this purpose from the electrical source, so that the induction generator can only operate when connected to a live and synchronous AC system. If a lagging reactive power input is equated with a leading reactive power output, then the generator can be described as operating with a lending power factor. The torque acts in a direction opposite to that of the rotating field, requiring a mechanical drive at the shaft. Because an induction generator synchronizes directly with the main utility grid– that is, produces electricity at the same frequency and voltage – no rectifiers or inverters are required. A major advantage of the induction generator is frequency regulation. The output frequency and volts are regulated by the power system in the induction generators and are independent of speed variations. The self-regulation effect minimizes control system complexity. But the performance characteristics as a generator will vary slightly from those as a motor. In general, the slip rpm and power factor will be lower and the efficiency will be higher. The differences may be so insignificant as to be undetectable by normal field measuring methods. On the other hand, alongside advantages mentioned above of the squirrel-cage and wound rotor induction generators whose rotor windings are short- circuited, there are some disadvantages [2, 3]. For example, the active and reactive power of generators cannot be controlled independently and stably. A computer and a cycloconverter in rotor circuit are needed to accomplish this task.

In recent years, there has been an increased attention toward wind power generation. Conventionally, grid-connected cage rotor induction machines are used as wind generators at medium power level. When connected to the constant frequency network, the induction generator runs at near-synchronous speed drawing the magnetizing current from the mains, thereby resulting in constant-speed constant-frequency (CSCF) operation. However, the power capture due to fluctuating wind speed can be substantially improved if there is flexibility in varying the shaft speed. In such variable-speed constant-frequency (VSCF) application, rotor-side control of grid-connected wound rotor induction machine is an attractive solution [5]. A doubly-fed wound rotor induction generator can produce constant stator frequency even though rotor speed varies. This system can be controlled by a small-capacity converter compared with the generator capacity, when the control range speed is narrow. Because of these features, this system is currently considered to be adaptable to power systems for hydroelectric and wind-mill-type power plants [6-8].

When adapted to the power system, it is important to examine the effects of this system on the power system. In the system under consideration, the stator is directly connected to the three-phase grid and the rotor of the doubly-fed induction machine is excited by three-phase low-frequency AC currents, which are supplied via slip-rings by either a cycloconverter or a voltage–fed PWM rectifier-inverter. The AC excitation on the basis of a rotor-position feedback loop makes it possible to achieve stable variable-speed operation. Adjusting the rotor speed makes the induction machine either release the electric power to the utility grid or absorb it from the utility grid [9]. The concept of power control was applied to reactive power compensator applications some 20 years ago, but the application to electrical machine control is new [10].

To control induction generator, several methods are used: electrically [vector control [11], active and reactive power control [12], direct torque control [13], direct power control [14], variable structure or sliding mode control [15], passivity control [16]], and mechanical (pitch, stall, and active stall control [17], yaw control [18], flywheel storage [9]). More information can be found in [19].

In literature, two kinds of approach are proposed for independent control of active and reactive powers. One of them is stator flux oriented vector control with rotor position sensors. The other is position sensorless vector control method. The control with rotor position sensors is the conventional approach and the performance of the system depends on the accuracy of computation of the stator flux and the accuracy of the rotor position information derived from the position encoder. Alignment of the position sensor is, moreover, difficult in a doubly-fed wound rotor machine [5].

Position sensorless vector control methods have been proposed by several research groups in the recent past [20-23]. A dynamic torque angle controller is proposed. This method uses integration of the PWM rotor voltage to compute the rotor flux; hence, satisfactory performance can not be achieved at or near synchronous speed. Most of the other methods proposed make use of the measured rotor current and use coordinate transformations for estimating the rotor position [21-23]. Varying degree of dependence on machine parameters is observed in all these strategies.

Alternative approaches to field-oriented control such as direct self control (DSC) and direct torque control (DTC) have been proposed for cage rotor induction machines. In these strategies, two hysteresis controllers, namely a torque controller and a flux controller, are used to determine the instantaneous switching state for the inverter. These methods of control are computationally very simple and do not require rotor position information. However, the application of such techniques to the control of wound rotor induction machine has not been considered so far. A recently developed algorithm for independent control of active and reactive powers with high dynamic response in case of a wound rotor induction machine is direct power control. In direct power control, the directly controlled quantities are the stator active and reactive powers. The proposed algorithm as direct power control also differs from DTC in that it does not use integration of PWM voltages. Hence, it can work stably even at zero rotor frequency. The method is inherently position sensorless and does not depend on machine parameters like stator/rotor resistance. It can be applied to VSCF applications like wind power generation as well as high-power drives [5].

Little literature has been published on control strategy and dynamic performance of doubly-fed induction machines [22; 24-27]. Leonhard (1985) describes a control strategy for an

http://dx.doi.org/10.5772/61130

adjustable-speed doubly-fed induction machine intended for independent control of the active and reactive power. The control strategy provides two kinds of current controllers: inner feedback loops of the rotor currents on the d-q coordinates and outer feedback loops of the stator currents on the M-T coordinates. However, it is not clarified theoretically why the control strategy requires the two kinds of current controllers.

This chapter describes the power control characteristics on the rotating reference frame fixed on the air-gap flux of a doubly-fed wound rotor induction generator and proposes a new approach to control with rotor position sensor. The proposed approach is the enhanced version of a previous study [28]. Thus, in this chapter, a new power control system that has been developed by using the computer and driver for a wound rotor induction generator takes place. This system is a new theoretical approach and this power control system has applied a control method using a rotating reference frame fixed on the air-gap flux of the generator. By using this control system, the active and reactive power of generator can be controlled independently and stably. Therefore, to achieve this purpose, firstly the complex power expression (and thus the active and reactive power expressions) for an induction machine in space vector notation and in two-axis system has been gotten. Then, power and current control, which are fundamental subjects, have been analyzed and as a result a computer- and driver-aided circuit is given to achieve the power and current control.

2. Analysis of the control system

For the stable control of the active and reactive power, it is necessary to independently control them. As known, the active power control is the control of torque produced by the machine and the reactive power control is the control of flux. The stator active and reactive power of doubly-fed wound rotor induction generator is controlled by regulating the current and voltage of the rotor windings. Therefore, to achieve independent control, the current and voltage of the rotor windings must be divided into components related to stator active and reactive power. It is well-known that an induction machine can be modeled as a voltage behind a total leakage inductance. Therefore, after a three-phase to two-phase power variant transformation, the induction machine model becomes that of Fig. 1 [10].

Approximate vector diagram of an induction machine is shown in Fig. 2. In this section, the analysis of the doubly-fed wound rotor induction generator on the rotating frame fixed on the air-gap flux (M-T frame) is carried out.

The M-axis is fixed in the air-gap flux and the T-axis is fixed in the quadrature with the M-axis. The relations of stator α_1-β_1 axis, rotor α_2-β_2 axis, and M-T axis are shown in Fig. 3.

2.1. Complex power expression

Assuming that the voltage vector is used as the reference for the determination of lagging and leading, we can write the complex stator power expression for a machine in space vector notation as:

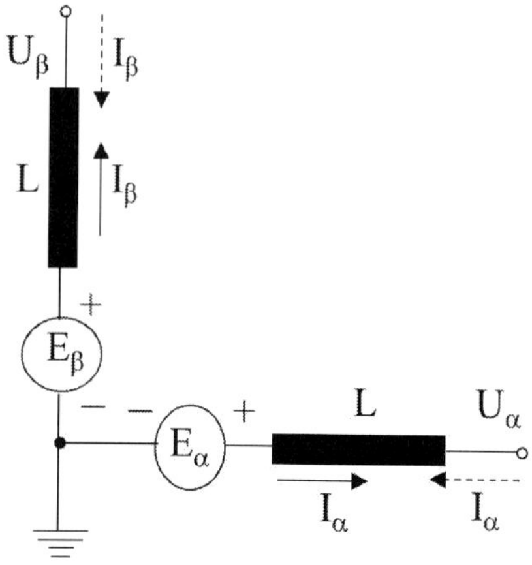

Figure 1. Two-phase representation of a three-phase induction machine (→generating; ⇢ motoring)

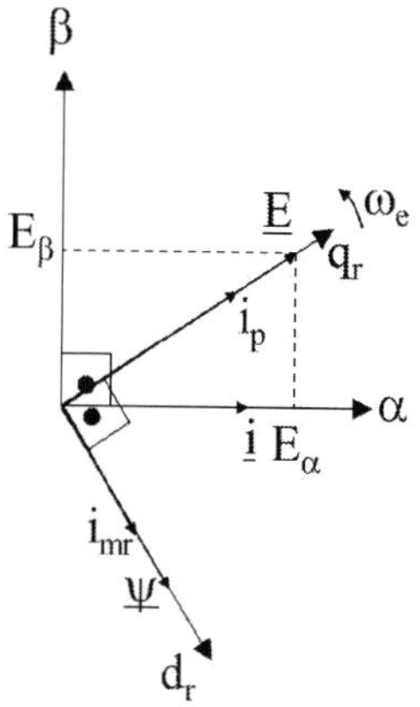

Figure 2. Approximate vector diagram of an induction machine

$$S_1 = I_1.U_1^* \tag{1}$$

which can be written in two-phase stationary frame variables as:

$$S_1 = \left[I_{M1,T1}\right]\left[U_{M1,T1}\right]^* \tag{2}$$

where

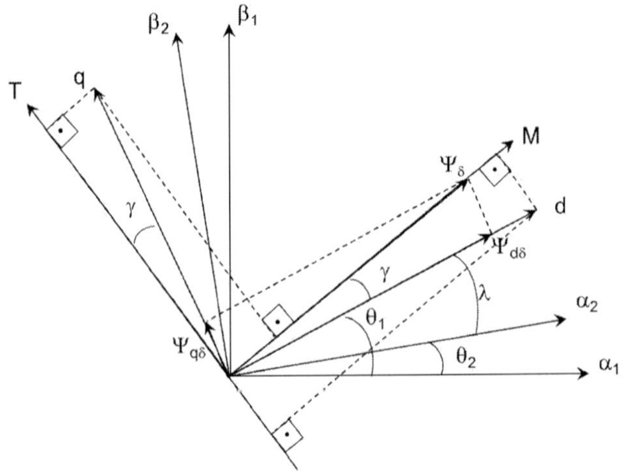

Figure 3. Vector diagram of M-T frame

$$\left[I_{M1,T1}\right] = I_{M1} + jI_{T1}; \left[U_{M1,T1}\right]^{*} = U_{M1} - jU_{T1}$$

Expanding Equation (2) we get the expressions for the stator active and reactive power as defined in [29]:

$$S_1 = U_{M1}I_{M1} + U_{T1}I_{T1} + j\left(U_{M1}I_{T1} - U_{T1}I_{M1}\right) \tag{3}$$

Therefore,

$$P_1 = U_{M1}I_{M1} + U_{T1}I_{T1} \tag{4}$$

$$Q_1 = U_{M1}I_{T1} - U_{T1}I_{M1} \tag{5}$$

where P_1 is the stator active power, and Q_1 is the stator reactive power, I_{M1} and I_{T1} are the M and T axis stator currents, and U_{M1} and U_{T1} are the M and T axis stator voltages.

2.2. Power control

In this section, the relationship between stator power and rotor current is analyzed. In Equations (4) and (5), the stator active and reactive power was expressed by the stator current based on the M-T frame. The relationships between the rotor and stator currents are:

$$I_{T1} = I_{q1}Cos\gamma - I_{d1}Sin\gamma \tag{6}$$

$$I_{T2} = I_{q2}Cos\gamma - I_{d2}Sin\gamma \tag{7}$$

$$I_{T1} + I_{T2} = \left(I_{q1} + I_{q2}\right)Cos\gamma - \left(I_{d1} + I_{d2}\right)Sin\gamma \tag{8}$$

$$I_{d1} = I_{M1}Cos\gamma;\ I_{d2} = I_{M2}Cos\gamma \tag{9}$$

$$I_{q1} = I_{M1}Sin\gamma;\ I_{q2} = I_{M1}Sin\gamma \tag{10}$$

$$I_{T1} + I_{T2} = 0; I_{T1} = -I_{T2} \tag{11}$$

By using Equations (9) and (10), the term of ($I_{T1} + I_{T2}$) is expressed in terms of ψ_δ and L_M:

$$I_{T1} + I_{T2} = \left(\frac{\psi_\delta}{L_M}\right)Sin\gamma Cos\gamma - \left(\frac{\psi_\delta}{L_M}\right)Cos\gamma Sin\gamma = 0 \tag{12}$$

$$\psi_\delta = L_M\left(I_{M1} + I_{M2}\right) \tag{13}$$

where ψ_δ is the air-gap flux, I_{M2} and I_{T2} are the M and T axis rotor currents, and L_M is the mutual inductance.

By using Equations (12) and (13), the stator active and reactive power is expressed in terms of ψ_δ and L_M:

$$\begin{gathered} P_1 = U_{M1}\left[\left(L_M\left(I_{M1} + I_{M2}\right)/L_M\right) - I_{M2}\right] + U_{T1}\left(-I_{T2}\right) \\ P_1 = \left(\psi_\delta / L_M\right)U_{M1} - U_{M1}I_{M2} - U_{T1}I_{T2} \end{gathered} \tag{14}$$

$$\begin{gathered} Q_1 = U_{M1}\left(-I_{T2}\right) - U_{T1}\left[\left(L_M\left(I_{M1} + I_{M2}\right)/L_M\right) - I_{M2}\right] \\ Q_1 = U_{T1}I_{M2} - \left(\psi_\delta / L_M\right)U_{T1} - U_{M1}I_{T2} \end{gathered} \tag{15}$$

In this system, the stator winding is directly connected to the power system. The conditions $U_{M1} \approx 0$, $U_{T1} \approx$ constant, $\psi_\delta \approx$ constant are derived from this feature [25]. By using these relationships and Equations (12), (13), (14), and (15), the stator active and reactive power is expressed in terms of rotor current. Equations (4) and (5) are rewritten as follows:

$$P_1 \approx -U_{T1} I_{T2} \tag{16}$$

$$Q_1 \approx U_{T1} I_{M2} - (\psi_\delta / L_M) U_{T1} \tag{17}$$

Equation (16) shows that the stator active power (P_1) is expressed by the terms proportional to the rotor current I_{T2}. Equation (17) shows that the stator reactive power (Q_1) is expressed by the terms proportional to the rotor current I_{M2} and constant value $(\psi_\delta/L_M).U_{T1}$. From the above relationships, the rotor current is divided into the active power (P_1) and the reactive power (Q_1) components. That is, the independent control of the stator active and reactive power can be actualized by regulating rotor currents I_{M2} and I_{T2}.

2.3. Current control

In this section, the relation between the rotor currents and the rotor voltages is analyzed. The equations for wound rotor induction machine based on M-T frame are shown as follows:

$$U_{M1} = R_1 I_{M1} + \frac{L_{1\sigma} dI_{M1}}{dt} - (\omega_1 + \omega_\delta) L_{1\sigma} I_{T1} + \frac{d\psi_{M1\delta}}{dt} - (\omega_1 + \omega_\delta) \psi_{T1\delta} \tag{18}$$

$$U_{T1} = R_1 I_{T1} + \frac{L_{1\sigma} dI_{T1}}{dt} + (\omega_1 + \omega_\delta) L_{1\sigma} I_{M1} + \frac{d\psi_{T1\delta}}{dt} + (\omega_1 + \omega_\delta) \psi_{M1\delta} \tag{19}$$

$$U_{M2} = R_2 I_{M2} + \frac{L_{2\sigma} dI_{M1}}{dt} - (\omega_S + \omega_\delta) L_{2\sigma} I_{T2} + \frac{d\psi_{M2\delta}}{dt} - (\omega_S + \omega_\delta) \psi_{T2\delta} \tag{20}$$

$$U_{T2} = R_2 I_{T2} + \frac{L_{2\sigma} dI_{T2}}{dt} + (\omega_S + \omega_\delta) L_{2\sigma} I_{M2} + \frac{d\psi_{T2\delta}}{dt} + (\omega_S + \omega_\delta) \psi_{M2\delta} \tag{21}$$

where $\psi_{M1\delta}$ and $\psi_{T1\delta}$ are the M and T axis stator air-gap flux, R_1 and R_2 are the stator and rotor resistance, ω_1 is the stator angular speed, ω_δ is the angular speed of the air-gap flux, $\psi_{M2\delta}$ and $\psi_{T2\delta}$ are the M and T axis rotor air-gap flux, $L_{1\sigma}$ and $L_{2\sigma}$ are the stator and rotor leakage inductance, ω_S is the slip angular speed.

These equations are transformed by using the relationships $\psi_{T1\delta} = \psi_{T2\delta}$, $\psi_{M1\delta} = \psi_{M2\delta}$, and $\omega_S = \omega_1 - \omega_2$, and the following expressions are derived:

$$U_{M2} = R_2 I_{M2} + \frac{L_{2\sigma} dI_{M2}}{dt} - (\omega_S + \omega_\delta) L_{2\sigma} I_{T2} + \frac{d\psi_{M1\delta}}{dt} - (\omega_S + \omega_\delta)\psi_{T2\delta} \tag{22}$$

$$U_{T2} = R_2 I_{T2} + \frac{L_{2\sigma} dI_{T2}}{dt} + (\omega_S + \omega_\delta) L_{2\sigma} I_{M2} + \frac{d\psi_{T1\delta}}{dt} + (\omega_S + \omega_\delta)\psi_{M2\delta} \tag{23}$$

$$U_{M2} = R_2 I_{M2} + \frac{L_{2\sigma} dI_{M2}}{dt} - (\omega_S + \omega_\delta) L_{2\sigma} I_{T2} + U_{M1\delta} + \omega_2 \psi_{T2\delta} \tag{24}$$

$$U_{T2} = R_2 I_{T2} + \frac{L_{2\sigma} dI_{T2}}{dt} + (\omega_S + \omega_\delta) L_{2\sigma} I_{M2} + U_{T1\delta} - \omega_2 \psi_{M2\delta} \tag{25}$$

where ω_2 is the rotor angular speed.

When the Equations (22), (23), (24), and (25) are transformed by the rotor current I_{M2} and I_{T2}, Equations (26) and (27) are given:

$$I_{M2} = \left(U_{M2} + (\omega_S + \omega_\delta) L_{2\sigma} I_{T2} - U_{M1\delta}\right) / \left(R_2 + pL_{2\sigma}\right) \tag{26}$$

$$I_{T2} = \left(U_{T2} - (\omega_S + \omega_\delta) L_{2\sigma} I_{M2} - U_{T1\delta} + \omega_2 \psi_\delta\right) / \left(R_2 + pL_{2\sigma}\right) \tag{27}$$

where p is the differential operator.

The stator winding is directly connected to the power system. Therefore, stator voltage becomes constant in the normal state, leading to the conditions $U_{M1\delta} \approx 0$ and $U_{T1\delta} \approx$ constant. And $L_{2\sigma}$ is negligible because it is generally small [25].

Thus, Equations (26) and (27) become as follows:

$$I_{M2} = U_{M2} / R_2 \tag{28}$$

$$I_{T2} = \left(U_{T2} - U_{T1\delta} + \omega_2 \psi_\delta\right) / R_2 \tag{29}$$

Equations (28) and (29) show that rotor voltages along the M and T axes, respectively depend only on the rotor currents along the M and T axes. In other words, the relationships between the currents and voltages along the M and T axes are linear. Consequently, the rotor currents I_{M2} and I_{T2} can be controlled independently by regulating the rotor voltages U_{M2} and U_{T2}.

2.4. Composition of the control system

Figure 4 illustrates the control system diagram. Considering the analysis of control system, it can be interpreted that M-T frame can be used to describe the composition of the active and the reactive power. Mentioned control system has six parts: (1) power control loop (regulation of the rotor current references from the deviation between detection and reference values for both active and reactive power); (2) current PI regulator for rotor currents I_{M2} and I_{T2} (same application with the first part for regulation of the voltage); (3) air-gap flux calculator (using the stator currents, voltages, and the signals of the position sensor); (4) P_1 and Q_1 detector (calculation of the stator active and reactive power); (5) detector of the rotor current (vector values of M-T axis, Equation (30)); (6) coordinate transformer (three-phase voltage references, Equation (31)) [25].

$$\begin{bmatrix} I_{M2} \\ I_{T2} \end{bmatrix} = \begin{bmatrix} Cos(\lambda+\gamma) & Sin(\lambda+\gamma) \\ Sin(\lambda+\gamma) & Cos(\lambda+\gamma) \end{bmatrix} \begin{bmatrix} 1 & -1/2 & -1/2 \\ 0 & \sqrt{3}/2 & -\sqrt{3}/2 \end{bmatrix} \begin{bmatrix} I_{a2} \\ I_{b2} \\ I_{c2} \end{bmatrix} \tag{30}$$

$$\begin{bmatrix} U_{a2} \\ U_{b2} \\ U_{c2} \end{bmatrix} = \begin{bmatrix} 1 & 0 \\ -1/2 & \sqrt{3}/2 \\ -1/2 & -\sqrt{3}/2 \end{bmatrix} \begin{bmatrix} Cos(\lambda+\gamma) & -Sin(\lambda+\gamma) \\ Sin(\lambda+\gamma) & Cos(\lambda+\gamma) \end{bmatrix} \begin{bmatrix} U_{M2} \\ U_{T2} \end{bmatrix} \tag{31}$$

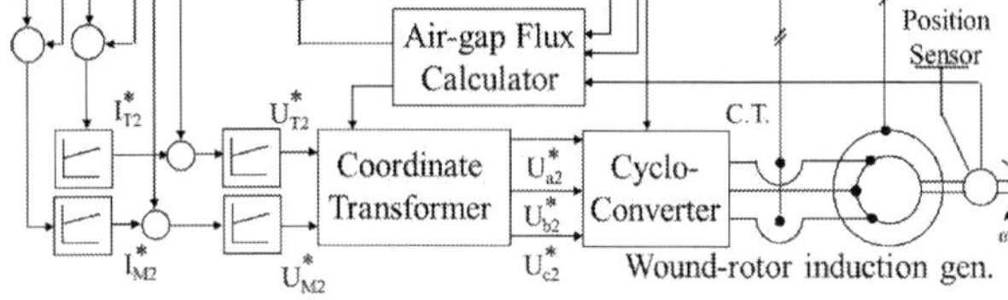

Figure 4. Block diagram of the control system

2.5. Power reference generation

Power Reference Generation (PRG) is one of the key modules of the algorithm. Strangely, it is the only module that depends on the parameter values. Due to direct connection to the torque, active power reference is the simplest way to find out the torque value desired to produce. On the other hand, there are some difficulties related with the active power reference.

Normally, the following equation is enough to calculate active power reference [10]:

$$P_{ref} = T_{ref}\omega_{m2} = T_{ref}\left(2\pi n_2\right) \tag{32}$$

where T_{ref} is supplied from some outer loop such as a speed control loop, and n_2 is the actual mechanical shaft speed at a particular instant of time.

Equation (32) is reliable under some limits so that calculated value of active power is not adequate to generate the torque T_{ref}. Main reason is the losses (losses in the stator and rotor resistance and iron losses) where all input power can not be transmitted to output. Loss in rotor resistance is influenced by the slip of the machine. Schematic demonstration of the power flows under motoring and generating for an induction machine are shown in Fig. 5.

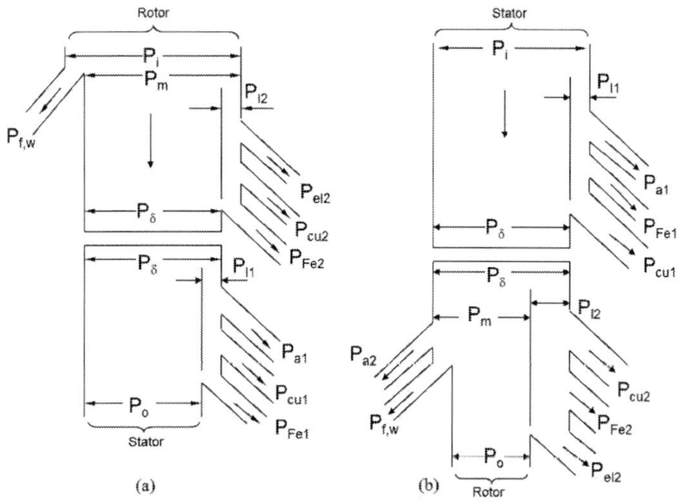

Figure 5. Schematic demonstration of the power flows in an induction machine under (a) motoring and (b) generating (P_i = input power (taken from supply in motor mode, given from shaft in generator mode), P_m = mechanical power, P_δ = air-gap power (transferred from stator to rotor under motoring and from rotor to stator under generating), P_l = power loss, P_{fe} = iron loss, P_{cu} = copper loss, P_a = additional losses produced by harmonics, $P_{f,w}$ = friction and windage losses, P_{el} = effective electrical power taken from rotor circuit, P_o = output power (shaft power in motor mode, electrical power in generation mode) (1 subscript means stator, 2 subscript means rotor) [28]

As seen in literature, whilst some researchers can assume that the stator resistance power and the iron losses can be ignored (under many practical situations), the power in the rotor resistance cannot be ignored, especially under heavy load conditions when the slip of the machine can be large [10].

As the diagram shows, under the motoring condition the input power separates into two parts: one part is for the losses (stator, iron, and harmonics) and the other part is related with the rotating field power (air-gap power).

Also, under generating the input shaft power separates into two parts: the losses (friction, windage) and mechanical power. All these parts should be balanced to reach the goal power value. To manage this aim, we should be able to calculate the slip and afterward we could calculate required power expressions as follows for the motoring/generating situation (if the friction and windage losses are neglected):

$$P_{ref} = \frac{P_{shaft}}{(1-s)} = \frac{T_{ref}\omega_{m2}}{(1-s)} \tag{33}$$

where P_{shaft} is the desired shaft power, and P_{ref} is the terminal reference power as defined previously.

The reactive power reference generation is linked with the slip. So, first we should deal with it. Figure 2 gives an approximate vector diagram of the voltages, currents, and fluxes of an induction machine. Reactive power can be calculated as the multiplication of the emf voltage E in one axis (e.g. q_r) of the machine by the current in the other axis (e.g. d_r):

$$|Q| = I_{mr}E \tag{34}$$

$$|Q| = I_{mr}\omega_1\psi_m \tag{35}$$

where ψ_m is the flux magnitude, I_{mr} is the magnetising current, and ω_1 is the stator angular speed related with electrical frequency.

Rearranging (35) one can write:

$$\omega_{sref} = -\left(\frac{|Q_{ref}|}{(I_{mrref}\psi_{mref})}\right) - \omega_2 \tag{36}$$

realizing:

$f_2 = p_p.n_2$

$\omega_2 = p_p.\omega_{m2}$

$\omega_1 = \omega_2 + \omega_s$

ω_s = slip angular speed related with slip frequency

$\omega_{s\ ref}$ = the desired slip angular speed related with slip frequency

$I_{mr\ ref}$ = the desired magnetizing current

p_p = the pole pairs

$| Q_{ref} |$ = the desired reactive power

The negative sign in Equation (36) results from the sign of Q_{ref}, the reference reactive power.

Given Equation (36), we can now write the expression for the slip:

$$s = \frac{\omega_s}{(\omega_2 + \omega_s)} = 1 + \left(\omega_2 I_{mrref} \psi_{mref} / Q_{ref}\right) = 1 + \left(\omega_2 L_m I_{mrref}^2 / Q_{ref}\right) \tag{37}$$

This expression can be substituted into the active power slip compensation term 1/(1-s). The $I_{mr\ ref}$ and $\psi_{m\ ref}$ terms in this expression are reference values. Clearly, I_{mr} and ψ_m for an induction machine are related, that is, $\psi_m = L_m.I_{mr}$. Hence, we have written the numerator of Equation (37) as $L_m.I_{mr\ ref2}$. This requires acquiring the value of magnetizing inductance of the machine.

The reactive power expression can be written as:

$$|Q_{ref}| = I_{mr} \omega_1 \psi_m = I_{mr} \psi_m (\omega_2 + \omega_s) \tag{38}$$

So, it can be concluded that the reactive power is dependent on the slip frequency. The slip frequency is well-related with the torque. In case a rapid change in torque is needed, slip frequency changes as a step and consequently the reactive power value changes. The torque T and ω_s expressions of an induction machine can be written as (using the standard expression from Field Oriented Control):

$$T = 1.5 p_p L_m^2 |I_{mr}| I_p / L_2 \tag{39}$$

$$\omega_s = I_p / \left(\tau_2 |I_{mr}|\right) \tag{40}$$

where $\tau_2 = L_2 / R_2$, L_2 = the rotor inductance, and R_2 = the rotor resistance. Therefore, τ_2 is the rotor time constant.

Rearranging Equation (39), one can write:

$$I_p = (TL_2)/\left(1.5p_p L_m^2 |I_{mr}|\right) \tag{41}$$

Substituting Equation (41) into Equation (40) gives:

$$\omega_s = (TL_2)/\left(1.5p_p L_m^2 |I_{mr}|^2 \tau_2\right) \tag{42}$$

The denominator in Equation (42) can be simplified by assuming that the leakage inductance of the machine is very small in relation to L_2, and hence $L_2 \approx L_m$. Therefore, Equation (42) can be written as:

$$\omega_s \approx \frac{2T}{\left(3p_p L_m |I_{mr}|^2 \tau_2\right)} = \frac{2T}{\left(3p_p \psi_m |I_{mr}| \tau_2\right)} \tag{43}$$

We are now in a position to write an expression for the reference reactive power. Substituting Equation (43) into Equation (38) and simplifying, we get Equation (44):

$$Q_{ref} = -\left[\left(|I_{mr}|_{ref}\, \omega_2 \psi_{mref}\right) + \left(\frac{2T_{ref}}{3p_p\tau_2}\right)\right] = -\left[L_m \left(|I_{mr}|_{ref}\right)^2 \omega_2 + \left(\frac{2T_{ref}}{3p_p\tau_2}\right)\right] \tag{44}$$

3. Experimental studies

An experimental setup using the control system shown in Fig. 4 was established by Yamamoto and Motoyoshi and the characteristics of the control system along the M-T frame have been experimentally examined. Experimental data are taken from their study [25]. The schematic diagram of the experimental system is shown in Fig. 6. Specifications of the generator used in experimental system are shown in Table 1. Equations (16) and (17) show that active and reactive power is proportional to the rotor currents I_{T2} and I_{M2}.

Phase	3	Stator Voltage	270 V
Pole Pairs	2	Rotor Voltage	257 V
Power Capacity	20 kW	Synchronous Speed	1500 rpm

Table 1. Specifications of the generator used in experimental system [25]

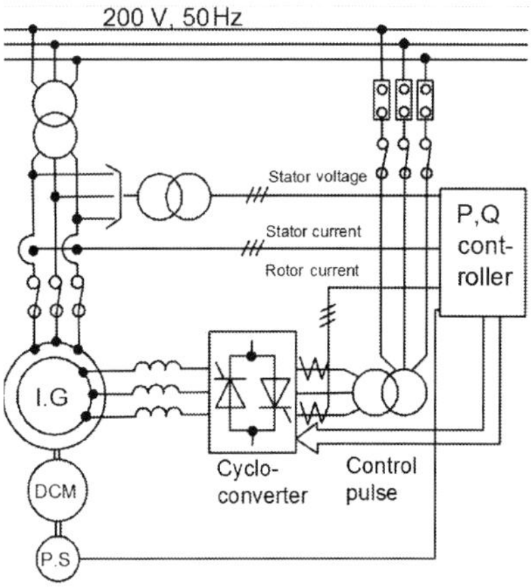

Figure 6. Experimental system [25]

3.1. Experimental results

Figures 7 and 8 show the relationships between stator power and rotor current reference and Figs. 9 and 10 show the relationships between stator power and stator power reference.

Figure 10 shows the step response of the stator reactive power Q_1 and the step response of the rotor current IM2 which is in proportion to the stator reactive power (Q_1). Figure 8 shows the step response of the rotor current I_{T2} which is in proportion to the stator active power (P_1). Figure 9 shows the step response of the stator active power P_1. I_{M2} and I_{T2} respond to stepping of I^*_{M2} and I^*_{T2} in 20 ms without any effect on I_{T2} and I_{M2}. P_1 and Q_1 respond to stepping of P^*_1 and Q^*_1 in 80 ms without any effect on Q_1 and P_1. The effects of this active and reactive power control method have been proved by these experimental results.

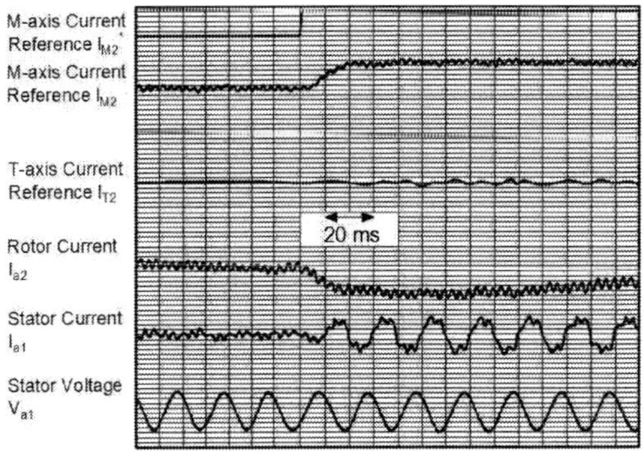

Figure 7. Step response of the M-axis current [25]

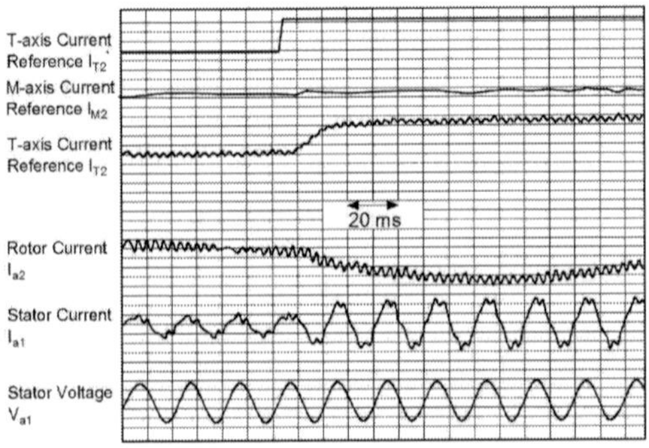

Figure 8. Step response of the T axis current [25]

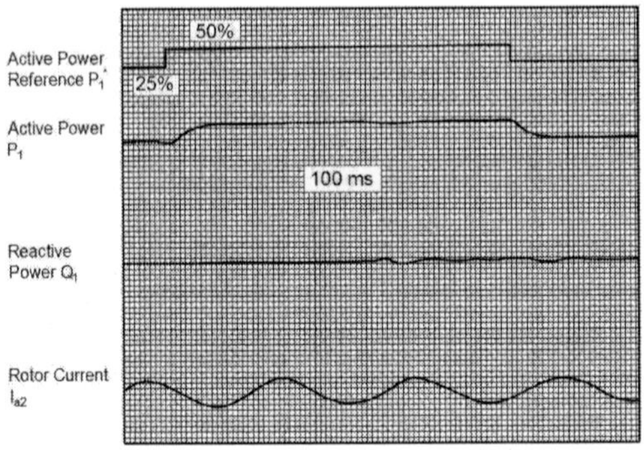

Figure 9. Step response of the active power [25]

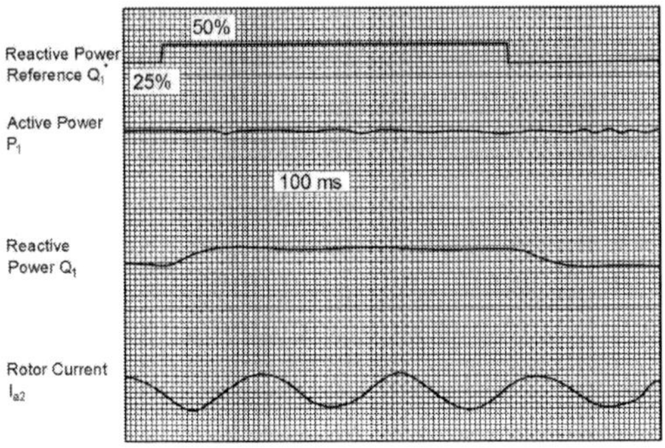

Figure 10. Step response of the reactive power [25]

3.2. Harmonic analysis

For the power converter of this system, the cycloconverter which is suitable for a large capacity system is often used. Generally, when the large-scale converter is applied to the electric power network system, it is a very important item to analyze the harmonic currents of the power converter. As known, the harmonic currents of the power converter are transmitted to the electric power network system through the rotor and stator windings. In this chapter, the characteristics of the transmission of the harmonic currents caused by the cycloconverter are analyzed theoretically.

Before beginning analysis, let us remember the basic principles of cycloconverter and mathematical background (taken from [30]) for the understanding of the subject.

3.2.1. Principle of the cycloconverter

As known, cycloconversion is concerned mostly with direct conversion of energy to a different frequency by synthesizing a low-frequency wave from appropriate sections of a higher-frequency source. As shown in Fig. 11a, a cycloconverter can be considered to be composed of two converters connected back-to-back. The load waveforms of Fig. 11b show that in the general case, the instantaneous power flows in the load fall into one of four periods. The two periods, when the product of load voltage and current is positive, require power flow into the load, dictating a situation where the converter groups rectify, the positive and negative groups conducting respectively during the appropriate positive and negative load-current periods.

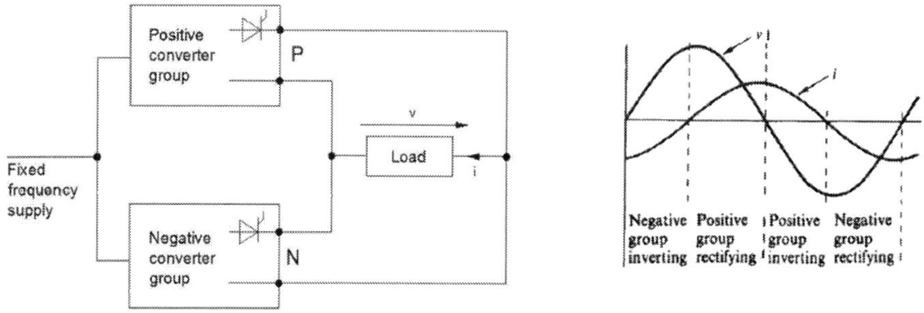

Figure 11. General cycloconverter layout (a) Block diagram representation, (b) Ideal load waveforms [30]

The other two periods represent times when the product of load voltage and current is negative; hence, the power flow is out of the load, demanding that the converters operate in the inverting mode. As shown in Fig. 11a, the principle of the cycloconverter can be demonstrated by using the simplest possible single-phase input to single-phase output with a pure resistance as load. Each converter is a bi-phase half-wave connection, the positive group labeled P and the negative group for reverse current labeled N.

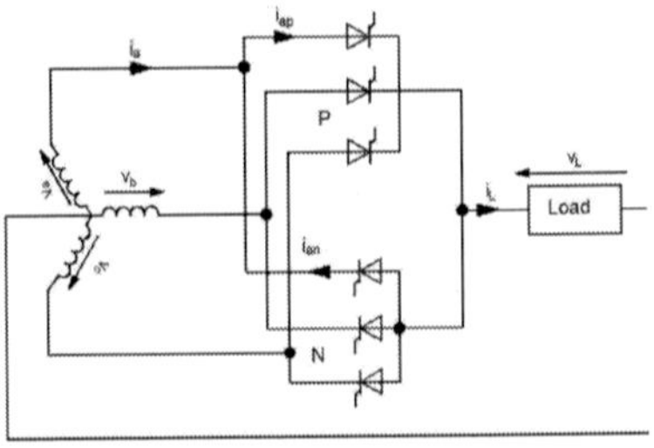

Figure 12. Single-phase load fed from a three-pulse cycloconverter [30]

The operation of the blocked group cycloconverter with various loads can be readily explained by reference to the three-pulse connection shown in Fig. 12, with the associated waveforms for inductive load in Figs. 13 and 14. As known, the wound rotor induction motor is an inductive load. In this chapter, waveforms are drawn for inductive load.

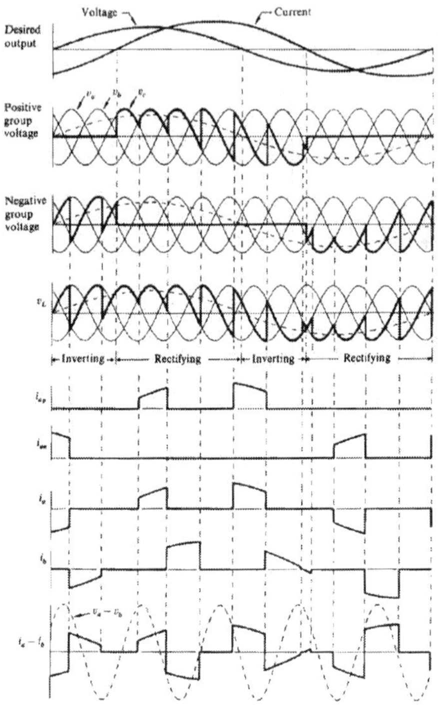

Figure 13. Waveforms with maximum voltage to an inductive load [30]

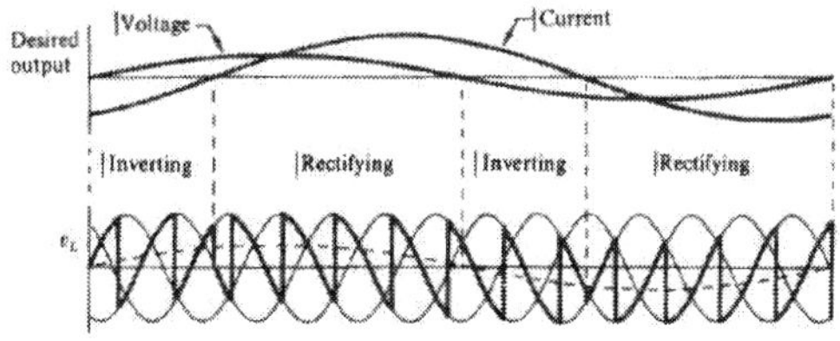

Figure 14. Waveforms when the load voltage is at half maximum (inductive load current continuous) [30]

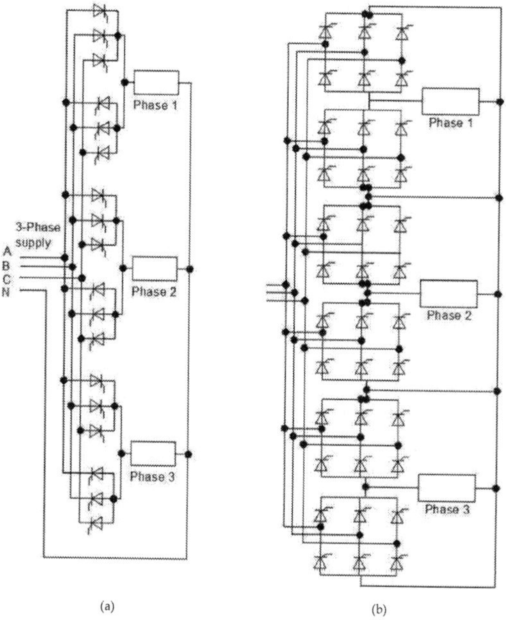

Figure 15. Cycloconverter connections with three-phase output. (a) Three-pulse bridge, (b) Six-pulse bridge [30]

The waveforms are as shown in Fig. 13 when the load is inductive, these being at a condition of maximum voltage. The load current will lag the voltage and, as the load-current direction determines which group is conducting, the group on-periods are delayed relative to the desired output voltage. The group thyristors are fired at such angles to achieve an output as close as possible to a sinewave, but now the lagging load current takes each group into the inverting mode. The group will cease conducting when the load current reverses. The load-voltage waveform is shown as a smooth transfer between groups, but in practice, a short gap would be present to ensure cessation of current in, and the regaining of the blocking state in, the outgoing group, before the incoming group is fired. The waveforms drawn assume the current is continuous within each load half-cycle. The effects of overlap will in practice be present in the waveforms.

In practice, the load-current waveform in Fig. 13 is assumed to be sinusoidal, although it would contain a ripple somewhat smaller than that in the voltage waveform. A light load inductance would result in discontinuous current, giving short zero-voltage periods. Each thyristor will conduct the appropriate block of load current, having the branch currents shown. The current i_a - i_b would represent the transformer input line current if one assumed the supply to be via a delta primary transformer. The input-current waveform shows changes in shape from cycle to cycle but where the input and output frequencies are an exact multiple, the waveform will repeat over each period of output frequency.

As shown in Fig. 14, a reduction in the output voltage can be obtained by firing angle delay. Here firing is delayed, even at the peak of the output voltage, so that control is possible over the magnitude of the output voltage. Comparison of Fig. 14 with Fig. 13 indicates a higher ripple content when the output voltage is reduced.

As shown in Fig. 15a, the three-pulse cycloconverter when feeding a three-phase load can be connected with a total of 18 thyristors. As shown in Fig. 15b, a six-pulse cycloconverter can be based on either six-phase half-wave blocks or the bridge connection when 36 thyristors are required.

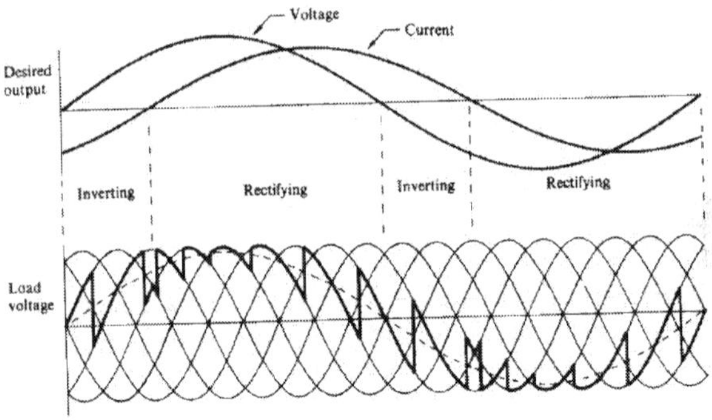

Figure 16. Cycloconverter load-voltage waveform with a lagging power factor load (six-pulse connection) [30]

An example of the cycloconverter output waveforms for the higher-pulse connections is given in Fig. 16, with an output frequency of one-third of the input frequency. It is clear from these waveforms that the higher the pulse-number, the closer is the output waveform to the desired sinusoidal waveform. In general, the output frequency is in general limited to about one-half to one-third of the input frequency, the higher-pulse connections permitting a higher limit.

As in Fig. 15, when the three-pulse cycloconverter feeds a three-phase balanced load, the current loading on the supply is much more evenly balanced. The waveforms to illustrate this are given in Fig. 17 for a frequency ratio of 4/1 with a load of 0.707 power factor lagging. It has been assumed that the load current is sinusoidal, although in practice it must contain ripple. The total load current is not identical from one cycle to the next, obviously contains harmonics, and its fundamental component lags the supply voltage by a larger amount than the load power factor angle.

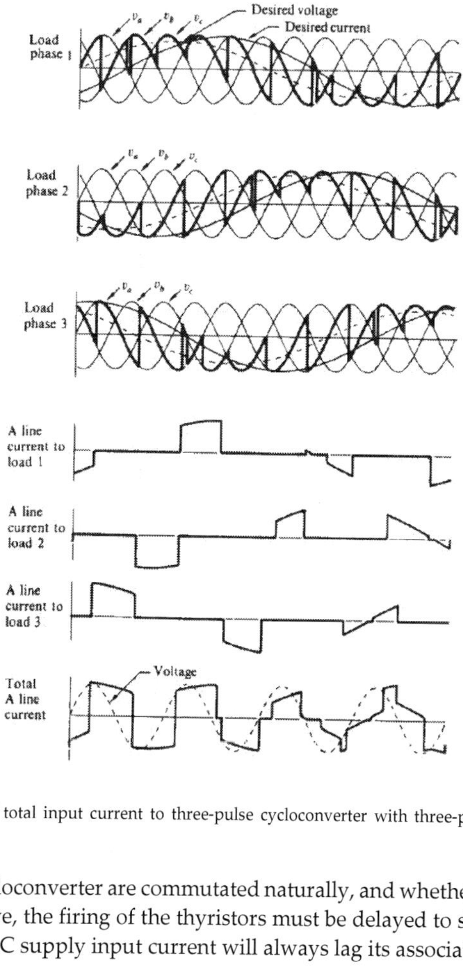

Figure 17. Development of total input current to three-pulse cycloconverter with three-phase lagging power factor load [30]

The thyristors of a cycloconverter are commutated naturally, and whether the load is resistive, inductive, or capacitive, the firing of the thyristors must be delayed to shape the output. The net result is that the AC supply input current will always lag its associated voltage.

3.2.2. Circulating current mode

The previous section specified cycloconverter operation where either the positive or negative groups were conducting, but never together. As shown in Fig. 18, if a center-tapped reactor is connected between the positive group P and negative group N, then both groups can be permitted to conduct. The reactor will limit the circulating current, that is, the value of its inductance to the flow of load current from either group being one quarter of its value to the circulating current, because inductance is proportional to the square of the number of turns.

In Fig. 19, typical waveforms are shown for the three-pulse cycloconverter shown in Fig. 18. Each group conducts continuously, with rectifying and inverting modes as shown. The mean between the two groups will be fed to the load, some of the ripple being cancelled in the combination of the two groups. Both groups synthesize the same fundamental sinewave. The reactor voltage is the instantaneous difference between the two group voltages. The circulating current shown in Fig. 19 can only flow in one direction, the thyristors preventing reverse flow. Hence, the current will build up during the reactor voltage positive periods until in the steady state it is continuous, rising and falling as shown.

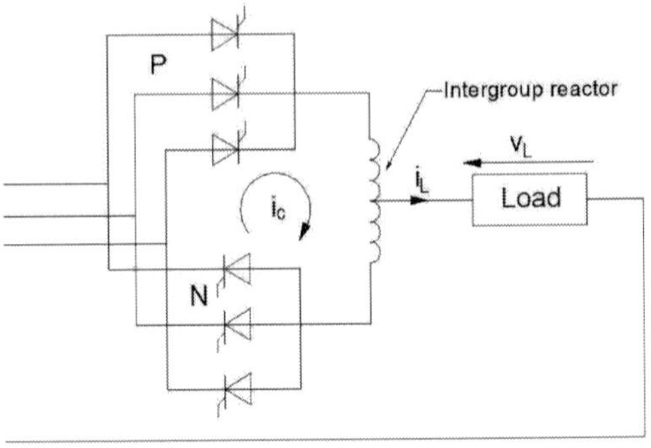

Figure 18. Three-pulse cycloconverter with intergroup reactor [30]

3.2.3. Mathematical analysis

The above subjects have demonstrated that almost all the waveforms associated with power electronic equipment are non-sinusoidal, which contains harmonic components. The purpose of this subject is to analyze the harmonic content of the various waveforms and discuss their effects as regards both supply and load.

Any periodic waveform may be shown to be composed of the superposition of a direct component with a fundamental pure sinewave component, together with pure sinewaves known as harmonics at frequencies which are integral multiples of the fundamental. A non-sinusoidal wave is often referred to as a complex wave [30].

Mathematically, it is more convenient to express the independent variable as x and the dependent variable as y. Then, the series may be expressed as:

$$y = f(x) = a_0 + a_1 \cos(x) + a_2 \cos(2x) + \ldots + a_n \cos(nx) + b_1 \sin(x) + b_2 \sin(2x) + \ldots + b_n \sin(nx) \quad (45)$$

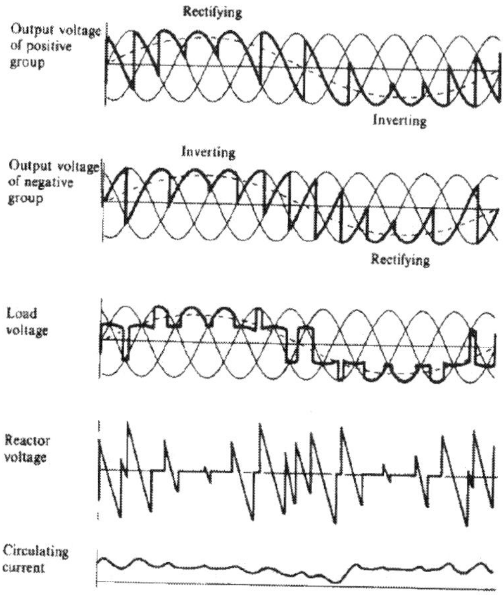

Figure 19. Waveform of a three-pulse cycloconverter with circulating current but without load [30]

Equation (45) is known as a Fourier series, and where f(x) can be expressed mathematically, a Fourier analysis yields that the coefficients are [31]:

$$a_0 = \frac{1}{2\pi}\int_{-\pi}^{\pi} f(x)dx \tag{46}$$

$$a_n = \frac{1}{\pi}\int_{-\pi}^{\pi} f(x)\cos nx dx \tag{47}$$

$$b_n = \frac{1}{\pi}\int_{-\pi}^{\pi} f(x)\sin nx dx \tag{48}$$

Alternatively, the series may be expressed as:

$$y = f(x) = R_0 + R_1\sin(x-\phi_1) + R_2\sin(2x-\phi_2) + R_3\sin(3x-\phi_3) + \ldots + R_n\sin(nx-\phi_n) \tag{49}$$

Equations (49) and (45) are equivalent, with:

http://dx.doi.org/10.5772/61130

$$a_n \cos nx + b_n \sin nx = R_n \sin\left(nx - \phi_n\right) \tag{50}$$

From which the resultant and phase angle:

$$R_n = (a_n^2 + b_n^2)^{1/2} \tag{51}$$

$$\phi_n = arctan\frac{a_n}{b_n} \tag{52}$$

Electrically, it expresses the independent variable as ωt instead of x and the dependent variable as volts or amperes instead of y. Then, the series may be expressed as:

$$v = V_0 + V_1 \sin\left(\omega t - \phi_1\right) + V_2 \sin\left(2\omega t - \phi_2\right) + V_3 \sin\left(3\omega t - \phi_3\right) + ... + V_n \sin\left(n\omega t - \phi_n\right) \tag{53}$$

where

v is the instantaneous value of the voltage at any time t;

V_o is the direct (or mean) value of the voltage;

V_1 is the maximum value of the fundamental component of the voltage;

V_2 is the maximum value of the second harmonic component of the voltage;

V_3 is the maximum value of the third harmonic component of the voltage;

V_n is the maximum value of the nth harmonic component of the voltage;

ϕ defines the relative angular reference; and

$\omega = 2\pi f$, where f is the frequency of the fundamental component, 1/f

defining the time over which the complex wave repeats itself.

The constant term of Equation (46) is the mean value of the function, and is the value found in, for example, the calculation of the direct (mean) voltage output of a rectifier. In the analysis of a complex wave, certain statements and simplifications are possible by inspection of any given waveform. If the areas of the positive and negative half-cycles are equal, then a_0 is zero. If $f(x + \pi) = - f(x)$, then there are no even harmonics, that is, no second, fourth, etc. In plain terms, this means the negative half-cycle is a reflection of the positive half-cycle. If $f(- x) = - f(x)$, then $a_n = 0$; that is, there are no sine terms. If $f(- x) = f(x)$, then $b_n = 0$; that is, there are no cosine terms. Symmetry of the waveform can result in Equations (47) and (48) being taken as twice the value of the integral from 0 to π, or four times the value of the integral from 0 to π / 2, hence simplifying the analysis.

Where it is difficult to put a mathematical expression to f(x), or where an analysis of an experimental or practical waveform obtained from a piece of equipment is required, graphical analysis can be performed [30].

The conditions of analysis are as follows.

1. The frequency of the harmonic current for the three-pulse and six-pulse cycloconverter is 6mfl ± (2n + 1)fs, where m is any integer from 1 to infinity, n is any integer from 0 to infinity, fl is the frequency of the power source, and fs is the output frequency of the cycloconverter [32].
2. The analysis is carried out only on the harmonic components.
3. Zero-phase-sequence current (harmonic currents of multiple of 3) is neglected.
4. The stator is connected to the fundamental frequency voltage source.

Using these conditions, the three-phase rotor current can be defined as follows:

$$I_{a2} = \sum_m \sum_n \begin{matrix} A_{(6n-1)m} Sin\{6m\theta_1 - (6n-1)\theta_s\} + A_{(6n-5)m} Sin\{6m\theta_1 - (6n-5)\theta_s\} \\ +B_{(6n-1)m} Sin\{6m\theta_1 - (6n-1)\theta_s\} + B_{(6n-5)m} Sin\{6m\theta_1 - (6n-5)\theta_s\} \end{matrix} \tag{54}$$

$$I_{b2} = \sum_m \sum_n \begin{matrix} A_{(6n-1)m} Sin\left\{6m\theta_1 - (6n-1)\left(\theta_s - \frac{2\pi}{3}\right)\right\} \\ +A_{(6n-5)m} Sin\left\{6m\theta_1 - (6n-5)\left(\theta_s - \frac{2\pi}{3}\right)\right\} \\ +B_{(6n-1)m} Sin\left\{6m\theta_1 - (6n-1)\left(\theta_s - \frac{2\pi}{3}\right)\right\} \\ +B_{(6n-5)m} Sin\left\{6m0_1 - (6n-5)\left(\theta_s - \frac{2\pi}{3}\right)\right\} \end{matrix} \tag{55}$$

$$I_{c2} = \sum_m \sum_n \begin{matrix} A_{(6n-1)m} Sin\left\{6m\theta_1 - (6n-1)\left(\theta_s + \frac{2\pi}{3}\right)\right\} \\ +A_{(6n-5)m} Sin\left\{6m\theta_1 - (6n-5)\left(\theta_s + \frac{2\pi}{3}\right)\right\} \\ +B_{(6n-1)m} Sin\left\{6m\theta_1 - (6n-1)\left(\theta_s + \frac{2\pi}{3}\right)\right\} \\ +B_{(6n-5)m} Sin\left\{6m\theta_1 - (6n-5)\left(\theta_s + \frac{2\pi}{3}\right)\right\} \end{matrix} \tag{56}$$

http://dx.doi.org/10.5772/61130

where m=1,2,....., n=1,2,......, $\theta_1=\omega_1 t$: ω_1 is the stator angular speed, $\theta_s=\omega_s t$: ω_s is the slip angular speed.

When Equations (54)–(56) are transformed to the d-q axis based on the stator voltage, (57) and (58) are derived:

$$I_{d2}=\sum_m\sum_n \begin{matrix} A_{(6n-1)m}Sin\{6m\theta_1-6n\theta_s\} \\ +A_{(6n-5)m}Sin\{6m\theta_1-(6n-1)\theta_s\} \\ +B_{(6n-1)m}Sin\{6m\theta_1+6n\theta_s\} \\ +B_{(6n-5)m}Sin\{6m\theta_1+(6n-1)\theta_s\} \end{matrix} \tag{57}$$

$$I_{q2}=\sum_m\sum_n \begin{matrix} A_{(6n-1)m}Cos\{6m\theta_1-6n\theta_s\} \\ +A_{(6n-5)m}Cos\{6m\theta_1-(6n-1)\theta_s\} \\ +B_{(6n-1)m}Cos\{6m\theta_1+6n\theta_s\} \\ +B_{(6n-5)m}Cos\{6m\theta_1+(6n-1)\theta_s\} \end{matrix} \tag{58}$$

The characteristics of transmission from the rotor winding to the stator winding are analyzed by substituting into the fundamental equation of a wound rotor induction machine. The analysis uses the symmetrical coordinate method for simplification. In this case, positive phase sequence component value (F component value) and negative phase sequence component value (B component value) have a conjugate relationship. The F component value is used for the analysis in this paper, which gives the following fundamental equation of the wound rotor induction machine:

$$\begin{bmatrix} V_{F1} \\ V_{F2} \end{bmatrix}=\begin{bmatrix} R_1+(p+j\omega_1)(L_1+L_m) & (p+j\omega_1)L_m \\ (p+j\omega_0)L_m & R_2+(p+j\omega_0)(L_2+L_m) \end{bmatrix}\begin{bmatrix} I_{F1} \\ I_{F2} \end{bmatrix} \tag{59}$$

When Equation (59) is transformed relative to I_{F1}, the following equation is obtained:

$$I_{F1}=\frac{(p+j\omega_1)L_m}{R_1+(p+j\omega_1)(L_1+L_m)}I_{F2}+V_{F1} \tag{60}$$

In this analysis, stator voltage is the fundamental voltage source, then, $V_{F1}=0$ is defined. When I_{F1} is expressed by I_{d2} and I_{q2}, using Equation (60) and $I_{F2}=(1/\sqrt{2})(I_{d2}+jI_{q2})$, the following expression is derived:

$$I_{F1} = \frac{(p + j\omega_1)L_m}{R_1 + (p + j\omega_1)(L_1 + L_m)} \frac{1}{\sqrt{2}} (I_{d2} + jI_{q2}) \tag{61}$$

Equation (59) is substituted into Equations (57) and (58). Applying Laplace transformation, the stator current is calculated. A-phase stator current is derived as follows:

$$I_{a1} = \sum_m \sum_n \begin{matrix} C_{(6n-1)m} Cos\{(6m+1)\theta_1 - 6n\theta_s\} \\ +C_{(6n-5)m} Cos\{(6m-1)\theta_1 - (6n-1)\theta_s\} \\ +D_{(6n-1)m} Cos\{(6m-1)\theta_1 + 6n\theta_s\} \\ +D_{(6n-5)m} Cos\{(6m+1)\theta_1 + (6n-1)\theta_s\} \end{matrix} \tag{62}$$

where

$$C_{(6n-1)m} \approx A_{(6n-1)m}\left(\frac{L_m}{(L_m + L_1)}\right), D_{(6n-1)m} \approx B_{(6n-1)m}\left(\frac{L_m}{(L_m + L_1)}\right)$$

$$C_{(6n-5)m} \approx A_{(6n-5)m}\left(\frac{L_m}{(L_m + L_1)}\right), D_{(6n-5)m} \approx B_{(6n-5)m}\left(\frac{L_m}{(L_m + L_1)}\right)$$

The results of this analysis show that the harmonic currents fed to the rotors winding are transmitted to the stator windings by changing its frequency. The rotor harmonic currents at $6m\omega_1 \pm (6n - 5)\omega_s$ change to stator harmonic currents at $(6m \pm 1)\omega_1 \pm 6(n - 1)\omega_s$. And the rotor harmonic currents at $6m\omega_1 \pm (6n - 1)\omega_s$ change to stator harmonic currents at $(6m \pm 1)\omega_1 \pm 6n\omega_s$. This is the effect of the rotating speed of the wound rotor induction generator. Moreover, the ratio between the amplitude of harmonic currents in the rotor and the stator is nearly 1:1.

Analysis results are verified by the experimental system shown in Fig. 6. Table 2 gives an example of experimental results. It is seen that experimental values well match with the theoretical ones. Thus, experiments confirmed that the analysis gives reliable results.

Rotor Harmonic Current		Stator Harmonic Current		
Frequency	Experimental Value	Frequency	Experimental Value	Theoretical Value
296.7 Hz	3.18 A	250 Hz	2.99 A	3.12 A
303.3 Hz	1.65 A	350 Hz	1.53 A	1.61 A
293.3 Hz	0.43 A	340 Hz	0.46 A	0.42 A
316.7 Hz	0.40 A	260 Hz	0.36 A	0.39 A

Table 2. Experimental results (ω_1=50 Hz, ω_s=50 Hz, m=1, N= 1400 r/min, I_1=I_2=25 A) [25]

It has been proved by experiment that this control system can control the active and reactive power independently and stably. In addition, it has been confirmed by analysis and experiment that the harmonic currents fed to the rotor windings of the generator are transmitted to the stator windings changing its frequency [25].

3.3. Proposed experimental setup

By using the computer and driver/buffer, the experimental setups to control active and reactive power of the wound rotor induction generator independently and stably are shown in Figs. 20 and 21 [28]. The terminals in experimental setups shown in Figs. 20 and 21 are numbered considering that the three-pulse cycloconverter with a total of 18 thyristors shown in Fig. 15a will be used.

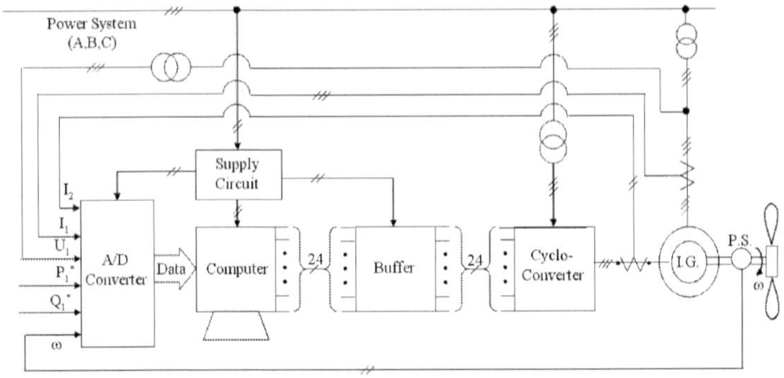

Figure 20. Schematic block diagram of the experimental setup for computer-aided power control of wound rotor induction generator

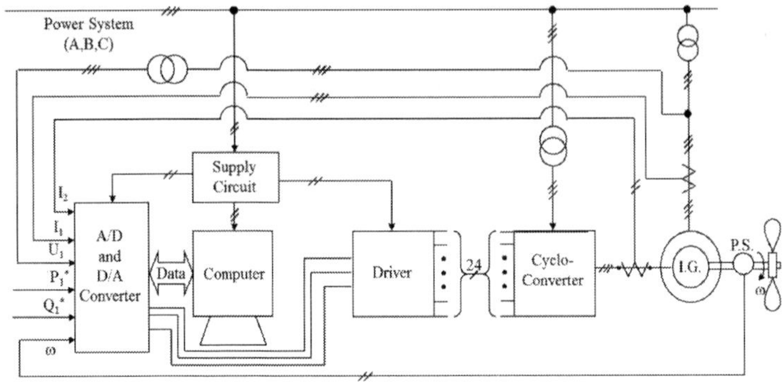

Figure 21. Another schematic block diagram of the experimental setup for computer-aided power control of wound rotor induction generator

4. Conclusion

New configurations for power control system of the doubly-fed wound rotor induction generator have been proposed. These configurations are based on a control method using a rotating reference frame fixed on the air-gap flux of the generator. By using them, the active and reactive power of generator can be controlled independently and stably. To achieve this purpose, power and current control that are fundamental subjects have been analyzed and as a result, a computer-aided circuit is given to achieve the power and current control. Using computers enables application of new technologies for easier control. For example, any new metaheuristic techniques or classification/identification techniques could be applied by just changing the code in the computer.

Author details

Fevzi Kentli*

Address all correspondence to: fkentli@marmara.edu.tr

Marmara University- Faculty of Technology – Department of Mechatronics Engineering Kadıköy-Istanbul, Turkey

References

[1] Boldea I. *The Electric Generators Handbook*. Florida, USA: Taylor & Francis; 2006. p. 552.

[2] Induction Generator [Internet]. [Updated: March 2015]. Available from: http://www.alternative-energy-tutorials.com/wind-energy/induction-generator.html [Accessed: March 2015]

[3] Nidec Motor Corporation. Induction Generator [Internet]. Available from: http://www.usmotors.com/TechDocs/ProFacts/Induction-Generator.aspx [Accessed: March 2015]

[4] Liserre M, Cárdenas R, Molinas M, Rodríguez J. Overview of multi-MW wind turbines and wind parks. *IEEE Transact Indus Electron*. 2011;58(4):1081-1095. DOI: 10.1109/TIE.2010.2103910

[5] Datta R, Ranganathan V T. Direct power control of grid-connected wound rotor induction machine without rotor position sensors. *IEEE Transact Power Electron*. 2001;16(3):390-399. DOI: 10.1109/63.923772

[6] Nakra H L, Dube B. Slip power recovery induction generators for large vertical axis wind turbines. *IEEE Transact Energy Convers*. 1988;3(4):733-737. DOI: 10.1109/60.9346

[7] Herrera J I, Reddoch T W, Lawler J S. Harmonics generated by two variable speed wind generating systems. *IEEE Transact Energy Convers*. 1988;3(2):267-273. DOI: 10.1109/60.4729

[8] Holmes P G, Elsonbaty N A. Cycloconvertor-excited divided-winding doubly-fed machine as a wind-power convertor. *IEE Proc B*. 1984;131(2):61-69. DOI: 10.1049/ip-b. 1984.0010

[9] Akagi H, Sato H. Control and performance of a doubly-fed induction machine intended for a flywheel energy storage system. *IEEE Transact Power Electron*. 2002;17(1): 109-116. DOI: 10.1109/63.988676

[10] Betz R E, Cook B J. Instantaneous power control of induction machines. *J Electric Electron Engin*. 2001;21(1):57-63.

[11] Lei S, Zengqiang M, Yang Y, Tao W, Haifeng T. Active power and reactive power regulation capacity study of DFIG wind turbine. In: International Conference on Sustainable Power Generation and Supply; 6-7 April 2009; Nanjing, China. IEEE; 2009. p. 1-6. DOI: 10.1109/SUPERGEN.2009.5348144

[12] Jiabing H, Heng N, Bin H, Yikang H, Zhu Z Q. Direct active and reactive power regulation of DFIG using sliding-mode control approach. *IEEE Transact Energy Conver*. 2010;25(4):1028–1039. DOI: 10.1109/TEC.2010.2048754

[13] Verij K M, Moradi M, Verij K R. Minimization of powers ripple of direct power controlled DFIG by fuzzy controller and improved discrete space vector modulation. *Electric Power Sys Res*. 2012;89:23–30. DOI: 10.1016/j.epsr.2012.02.008

[14] Arnalte S, Burgos J C, Rodriguez-Amenedo J L. Direct torque control of a doubly-fed induction generator for variable speed wind turbines. *Electric Power Comp Sys*. 2002;30(2):199–216. DOI: 10.1080/153250002753427851

[15] Huang H, Fan Y, Qiu R C, Jiang X D. Quasi-steady-state rotor EMF oriented vector control of doubly fed winding induction generators for wind-energy generation. *Electric Power Comp Sys*. 2006;34(11):1201–1211. DOI: 10.1080/15325000600698597

[16] Sang C L, Kwang H N. Dynamic modelling and passivity-based control of an induction motor powered by doubly fed induction generator. In: Industry Applications Conference: 38th IAS Annual Meeting; 12-16 October 2003; Utah, USA. New Jersey, USA: IEEE; 2003. p. 1970–1975. DOI: 10.1109/IAS.2003.1257837

[17] Muljadi E, Butterfield C P. Pitch-controlled variable-speed wind turbine generation. *IEEE Transact Indust Applic*. 2001;37(1):240–246. DOI: 10.1109/28.903156

[18] Fadaeinedjad R, Moallem M, Moschopoulos G. Simulation of a wind turbine with doubly fed induction generator by fast and simulink. *IEEE Transact Energy Conver.* 2008;23(2):690-700. DOI: 10.1109/TEC.2007.914307

[19] Zin A A B M, Pesaran H A M, Khairuddin A B, Jahanshaloo L, Shariati O. An overview on doubly fed induction generators' controls and contributions to wind based electricity generation. *Renew Sustain Energy Rev.* 2013;27:692–708. DOI: doi:10.1016/j.rser.2013.07.010

[20] Xu L, Cheng W. Torque and reactive power control of a doubly fed induction machine by position sensorless scheme. *IEEE Transact Indust Applic.* 1995;31(3):636 - 642. DOI: 10.1109/28.382126

[21] Datta R, Ranganathan V T. Decoupled control of active and reactive power for a grid-connected doubly-fed wound rotor induction machine without rotor position sensors. In: IEEE Industry Applications Conference: Thirty-Fourth Ias Annual Meeting; 3-7 October 1999; Arizona, USA. IEEE; 1999. p. 2623-2630.

[22] Morel L, Godfroid H, Mirzaian A, Kauffmann J M. Double-fed induction machine: Converter optimisation and field oriented control without position sensor. *IEE Proc B.* 1998;145(4):360-368. DOI: 10.1049/ip-epa:19981982

[23] Bogalecka E. Power control of a double fed induction generator without speed or position sensor. In: Fifth European Conference on Power Electronics and Applications; 13-16 September 1993; Brighton, England. London, England: Institution of Electrical Engineers; 1993. p. 224-228.

[24] Leonhard W. *Control of Electrical Drives*. 1st edn. New York, USA: Springer-Verlag; 1985. p. 346.

[25] Yamamoto M, Motoyoshi O. Active and reactive power control for doubly-fed wound rotor induction generator. *IEEE Transact Power Electron.* 1991;6(4):624-629. DOI: 10.1109/63.97761

[26] Brune C, Spee R, Wallace A K. Experimental evaluation of a variable-speed, doubly-fed wind-power generation system. In: IEEE Industry Applications Conference: 28th IAS annual meeting; 2-8 October 1993; Ontario, Canada. New Jersey, USA: Institute of Electrical & Electronics Engineers; 1993. p. 480-487.

[27] Bhowmik S, Spee R, Enslin J H L. Performance optimization for doubly-fed wind power generation systems. In: Cramer, Don et al. (eds.). Industry Applications Conference, The 1998 IEEE Conference; 12-15 October 1998; Missouri, USA. New York, USA: The Institute of Electrical and Electronic Engineers, Inc; 1998. p. 2387-2394.

[28] Kentli F. Computer aided power control for wound rotor induction generator. *Ozean J Appl Sci.* 2009;2(1):39-48.

[29] Akagi H, Kanazawa Y, Nabae A. Instantaneous reactive power compensators comprising switching devices without energy storage components. *IEEE Transact Indust Applic*. 1984;20(3):625-630. DOI: 10.1109/TIA.1984.4504460

[30] Cyril W. Lander. *Power Electronics*. 3rd ed. England: McGraw-Hill Publishing Company; 1993. p. 480.

[31] Erwin Kreyszig. *Advanced Engineering Mathematics*. 3rd edn. New York: Wiley; 1972. p. 866.

[32] Pelly B R. *Thyristor Phase-Controlled Converter and Cycloconverters*. 1st edn. Wiley-Blackwell; 1971. p. 434.

Chapter 3

Analysis and Methodology for Determining the Parasitic Capacitances in VSI-fed IM Drives Based on PWM Technique

Rudolf Ribeiro Riehl, Fernando de Souza Campos, Alceu Ferreira Alves and Ernesto Ruppert Filho

Additional information is available at the end of the chapter

http://dx.doi.org/10.5772/61544

Abstract

Three-phase induction motors present stray capacitances. The aim of this chapter is to present a methodology to experimentally determine these capacitances and also evaluate the effects of electromagnetic interference on motors in common mode. The proposed procedures for this methodology consist of: a) identifying the motor equivalent electrical circuit parameters through characteristic tests performed in the laboratory; b) setting up configurations between the PWM inverter and the motor for voltage and current measurements: common mode and shaft voltages, leakage and shaft (bearing) currents by using a dedicated measuring circuit; c) calculating the parasitic capacitance values between stator and frame, stator and rotor, rotor and frame and bearings of the motor using the capacitance characteristic equation; d) using the dedicated software Pspice to simulate the system composed by the three-phase induction motor fed by PWM inverter with the equivalent electrical circuit parameters; e) determining the characteristic waveforms involved in the common mode phenomenon.

Keywords: Induction motors, parasitic capacitances, PWM inverter

1. Introduction

The use of inverter controlled by pulse width modulation (PWM Inverter), on drive and control of the three-phase induction motors is increasingly common, especially for the power range of up to 10 Hp.

The recent developments in power semiconductor devices (IGBT, MOSFET, and others) have allowed these drives to achieve switching frequencies up to 20 kHz. In these frequencies, the rise time of PWM voltage becomes very small and responsible for the appearance of phenomena, defined as electromagnetic interference (EMI) in the induction motor [12, 18, 25]. Due to the presence of parasitic capacitances in the motor caused by the free or isolated spaces between metal parts, capacitive couplings occur, which become flow paths of the high-frequency electric current between the motor phases (differential mode coupling) and between phases and the ground (common mode coupling). The higher the switching frequency, the greater are the extensions and consequences of these phenomena.

The differential mode phenomena are responsible for excessive heating of the motor, harming the electrical insulation characteristics, performance, and, consequently, their useful life, and can burn out the motor. This occurs because the PWM inverter does not feed the induction motor with a sinusoidal voltage, but by applying modulated or switched pulses, producing high-frequency harmonics, and high-voltage gradients values (high dV/dt) to the stator windings [3].

The phenomena due to the common mode are responsible for the appearance of circulating currents between motor and ground moving through the frame, bearings, and motor pedestals. As the common mode voltage is different from zero, it raises a shaft voltage between the bearing parts and the ground, which is dependent on both this common mode voltage and the parasitic capacitances that can circulate electric currents capacitive for multiple paths by the motor [10]. One of these paths passes through the bearings, and in this case the motor currents are known as bearing currents, which, due to discharges occurring at the dielectric break, can cause damages at their bearings and if the shaft locks, the motor would be forced to stop and it will cause burning of the windings. Another phenomenon that occurs is electric shock or electric discharge machining (EDM), due to the flow of leakage current motor to the frame when it is not grounded or this ground is not suitable [1, 4, 7].

The conventional method to determine these parasitic capacitances is by measuring the impedance of the induction motor using LCR Bridge [1 to 10]. In this method, the induction motor is disassembled for specific parameters to be measured. For the determination of the parameters between the stator and the motor frame, the rotor is removed and measurements are conducted with each phase bridge between the LCR and the frame and among the three phases (short-circuited in this case) and the frame. After that, the rotor is put back and its axis is short-circuited to the frame. Parameter values between stator and rotor are measured. The LCR Bridge is connected at the common point of the three phases (neutral) and the stator frame.

The parameters of the motor bearings are dynamic and depend on operating conditions (speed) of the three-phase induction motor and also the dielectric characteristics of resistivity of the lubricant temperature and the geometric construction of the motor [1 to 10]. The rotor parameters can only be determined after all previous parameters have been obtained. The LCR Bridge is connected between the rotor shaft and the frame of the induction motor.

In the absence of an LCR Bridge and an appropriate laboratory for this type of test, another methodology [40, 41] is used to determine the parameters of the equivalent circuit of the induction motor in steady state (nominal frequency) [26, 27] and for high frequency [1, 10, 14, 41], through laboratory measurements using features configurations of links between the

PWM inverter and the motor for measuring the following parameters of interest: common mode voltage (V_{CM}) and shaft voltage (V_{SHAFT}); leakage ($I_{LEAKAGE}$) and bearing (I_B) currents, by measurement circuit designed for this purpose; and calculation of the values of the parasitic capacitances between the stator and motor frame (C_{SF}), stator and rotor (C_{SR}), rotor and motor frame (C_{RF}) and bearings (C_B) using characteristic equations as will be shown in this chapter. The computer application PSPICE [36, 40] is used to simulate the three-phase induction motor powered by a PWM inverter system using the high-frequency equivalent circuit of the motor. The characteristic waveforms that represent the common mode phenomena will be obtained to allow comparisons for validation of procedures to determine the capacitance.

In this methodology [40, 41], the three-phase induction motor does not need to be disassembled and reassembled as in the case of measurement with LCR Bridge. Only the bearings are insulated, and the determination of parasitic capacitances is performed while the motor is running. The results of testing for measurement and determination of two 1 Hp induction motor parasitic capacitances and simulations, using the Pspice program, are presented.

2. Equivalent circuit of three-phase induction motor at high frequencies

At high frequencies, the capacitive reactance among the various parts of the three-phase induction motor are shown in Figure 1, which illustrates the equivalent circuit of the three-phase induction motor fed by PWM inverter [1, 6, 15]. The distributed parameters R, L, and C represent the high-frequency coupling between the windings of the stator and the rotor.

Z_{RF} is the impedance between the rotor and the motor frame, also called air gap impedance Z_g; Z_{ER} is the impedance between the windings of the stator and the rotor; Z_{SF} is the impedance between the stator windings and the frame; Z_g is the air gap impedance; and Z_B represents the impedance between the rotor and the bearing. R_W and L_W represent the equivalent impedance of the driver through which circulates the bearing current and R_g is the lead resistance connected between the frame and ground. Using the defined impedances shown by equations (1) through (4), one can obtain the high-frequency simplified circuit of the induction motor presented in Figure 2 [15]:

$$Z_{rf} = Z_g = jXC_g \tag{1}$$

$$Z_{sr} = R_{sr} + jXL_{sr} + jXC_{sr} \tag{2}$$

$$Z_{sf} = R_{sf} + jXL_{sf} + jXC_{sf} \tag{3}$$

$$Z_b = (R_l \, // \, jXC_b) + R_b + R_w + jXL_w \tag{4}$$

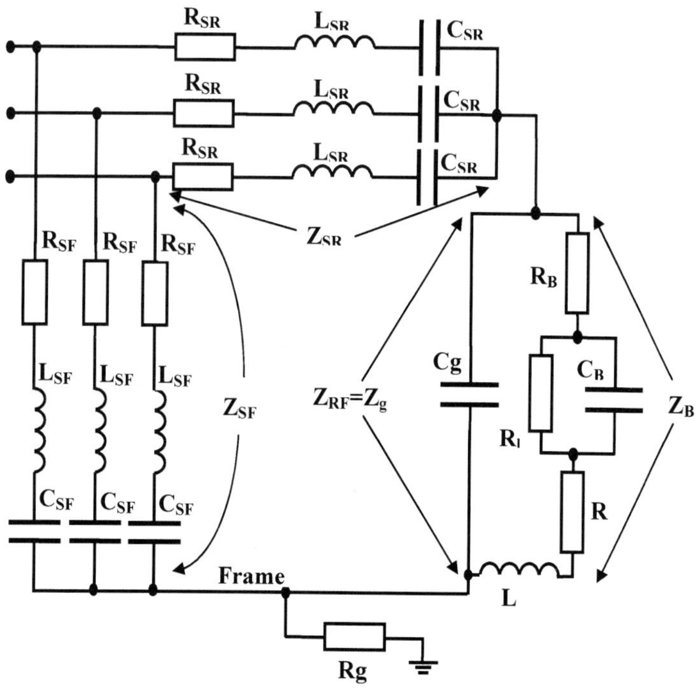

Figure 1. High-frequency equivalent circuit of induction motor

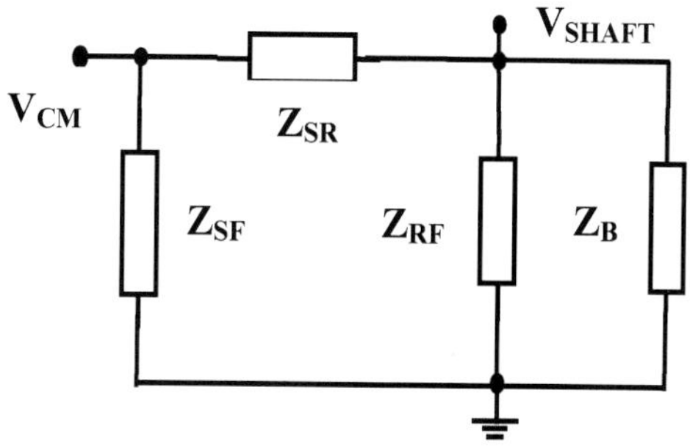

Figure 2. High-frequency simplified circuit for three-phase induction motor according to its impedances

Simplifications of the circuit shown in Figure 2 can be done using the following considerations [1, 3, 6, 9, 10]: a) Z_{RF} is purely capacitive; b) at frequencies lower than 200 kHz, Z_{SR} assumes a capacitive characteristic; and c) Z_{SR} represents a circuit with RC behavior. At frequencies of 4–20 kHz range, which are typical switching frequencies of a PWM inverter, parasitic impedances

http://dx.doi.org/10.5772/61544

of the induction motor assume purely capacitive characteristic [10], according to the example shown in Figure 3.

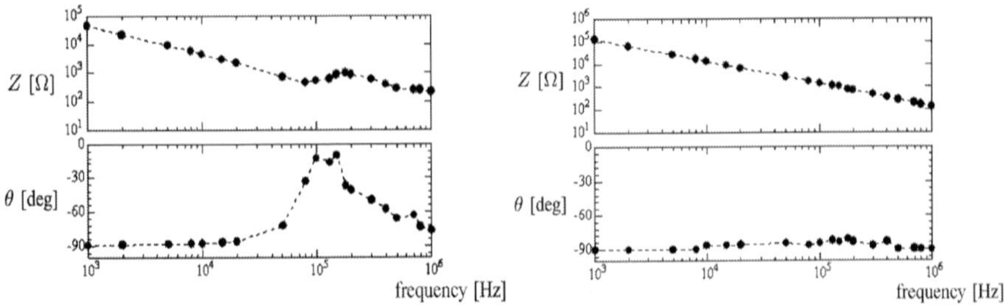

Figure 3. Capacitive impedance characteristic of the three-phase induction motor [10]

Thus, it is possible to simplify the equivalent circuit presented in Figure 2 keeping only the parasitic capacitances at the high frequencies, as shown in Figure 4.

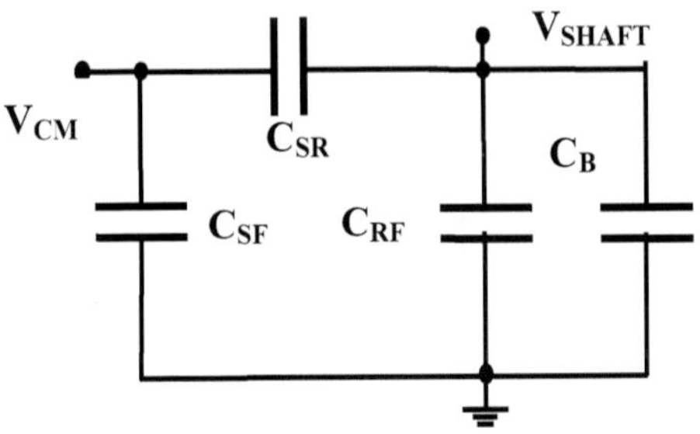

Figure 4. Simplified equivalent circuit of high-frequency induction motor

In Figure 4, V_{CM} is the common mode voltage, C_{SF} is the capacitance between the stator winding per phase and the induction motor frame, C_{SR} is the capacitance among the windings of the stator and the rotor, C_{RF} is the capacitance between the rotor and motor frame, and C_B is the capacitance of the bearing. Using this high-frequency equivalent circuit of the induction motor, the equations of both shaft voltage (V_{SHAFT}) and leakage current ($I_{LEAKAGE}$) are obtained:

$$v_{CM} = \frac{v_a + v_b + v_c}{3} \tag{5}$$

$$V_{SHAFT} = V_{CM} \cdot \left(\frac{C_{SR}}{C_{SR} + C_{RF} + C_B} \right) \tag{6}$$

$$I_{LEAKAGE} = \frac{V_{CM}}{XC_{SF}} + \frac{V_{SHAFT}}{XC_{RF}} + \frac{V_{SHAFT}}{XC_{B}} \quad (7)$$

The capacitances C_{SF}, C_{SR}, C_{RF}, and C_B are defined according to the geometric characteristics of both the stator as the rotor of a three-phase induction motor and its bearings [1, 2, 11, 12, 13, 23]. In [1], these capacitances are set according to the following equations, depending on the geometrical dimensions of the induction motor shown in Figure 5.

$$C_{SF} = \frac{K_{SF}.N_S.\varepsilon_r.\varepsilon_0.(W_d + W_s).L_S}{d} \quad (8)$$

$$C_{SR} = \frac{K_{SR}.N_R.\varepsilon_0.W_r.L_r}{g} \quad (9)$$

$$C_{RF} = \frac{K_{RF}.\pi.\varepsilon_0.L_r}{\ln(R_s / R_r)} \quad (10)$$

$$C_B = \frac{N_b.4.\pi.\varepsilon_0.\varepsilon_r}{(R_b + R_c / R_b - 1)} \quad (11)$$

In the above equations, K_{SF}, K_{SR}, and K_{RF} factors are stacked packages of magnetic stator and rotor, N_S and N_R are the number of slots of the stator winding and the rotor, W_s and W_d are the width and depth of the groove stator, W_r is the width of the rotor slot, L_s and L_r are the lengths of the stator and rotor slots, R_r and R_s are the radii of the stator and rotor, d is the thickness of the insulating dielectric material of the stator channel, g is the gap length, N_b is the number of bearing balls, R_b and R_c are the ball lightning and the raceway, ε_0 and ε_r are the permittivity of the medium (air and insulation). The parasitic capacitances become important when, besides the common mode voltage is different from zero, the frequency of the phase voltages becomes high, resulting in small capacitive reactance and the circuit in parallel with the remaining equivalent.

When the three-phase induction motor is fed by pure sinusoidal voltages, at the power grid frequency, the effect of these capacitances is minimal or nonexistent. If there is an unbalance in phase voltages, the common mode voltage becomes nonzero, establishing current flow through these capacitances (Figure 4) which will be significant, if the amplitudes of phase voltages are high. Thus, assuming that the motor is supplied by balanced phase voltages, the common-mode voltage (5) is zero. When the three-phase induction motor is fed through a PWM inverter, it establishes a "capacitive coupling" created by the modulated phase voltages of the inverter output. These voltages have trapezoidal characteristic value with high dV/dt of

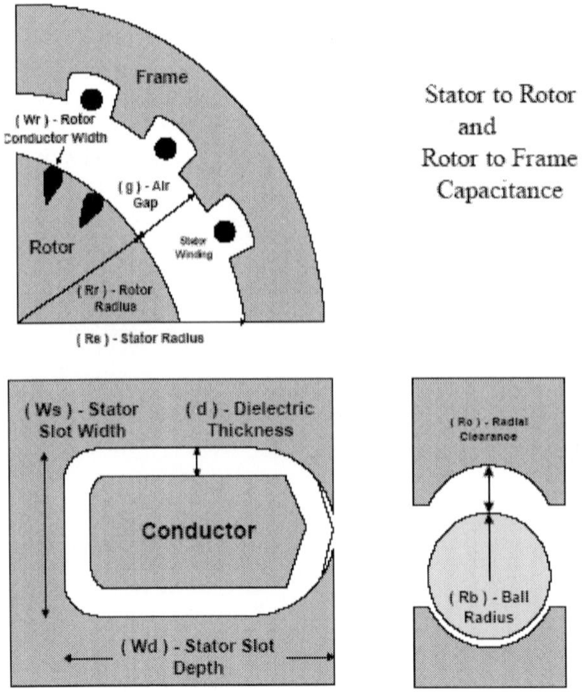

Figure 5. Dimensions – induction motor and bearing [1]

the inverter while the semiconductor switches are turned on or off, as shown in Figure 6 (a, b) [12, 18, 22].

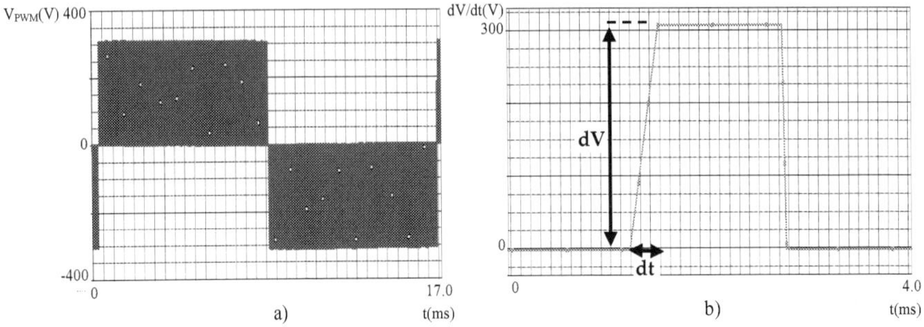

Figure 6. a) Terminal voltage of the PWM inverter, b) dV/dt voltage

As the supply voltages are not sinusoidal, the common mode voltage (V_{CM}) takes nonzero values. Thus, this voltage and also the switching frequency (f_S) of the PWM inverter now have an important role on common mode capacitive currents. For example, Figures 7a and 7b present the waveforms of the output phase voltages of the inverter and the common-mode voltage for a switching frequency (f_S) of 16 kHz.

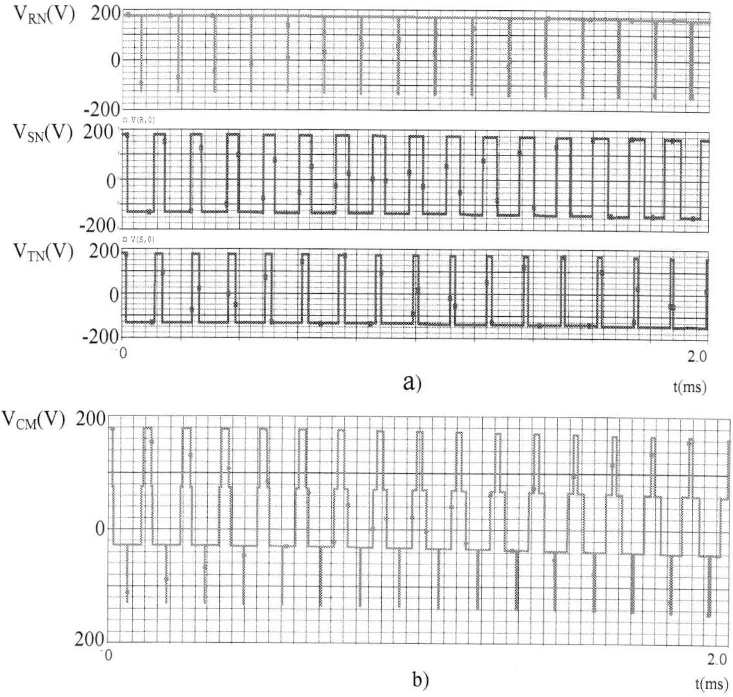

Figure 7. a) Phase voltage, b) a common mode voltage (V_{CM}) - f_S = 16 kHz

The higher the switching frequency, the better is the characteristic of the waveform of the current applied to the induction motor. Besides that, the frequency raise often implies an increase in the switching times of the IGBTs increasingly smaller, providing both increased feature dV/dt, which is directly related to the values of the capacitive currents of common mode current, as the reduction of the capacitive reactance significantly increased the amplitudes thereof. The following equations show these relationships:

$$i = C.\frac{dv}{dt} \tag{12}$$

http://dx.doi.org/10.5772/61544

$$I = 2.\pi.f_S.C.V \tag{13}$$

Observing equations (12) and (13), it is possible to conclude that V and I are effective values of voltage and common mode current. Thus, from the down-movement of the common-mode electrical currents (also called capacitive currents) as shown in Figure 8a [18], one can obtain the equivalent circuit of the motor for high frequency presented in Figure 8b [1 to 10].

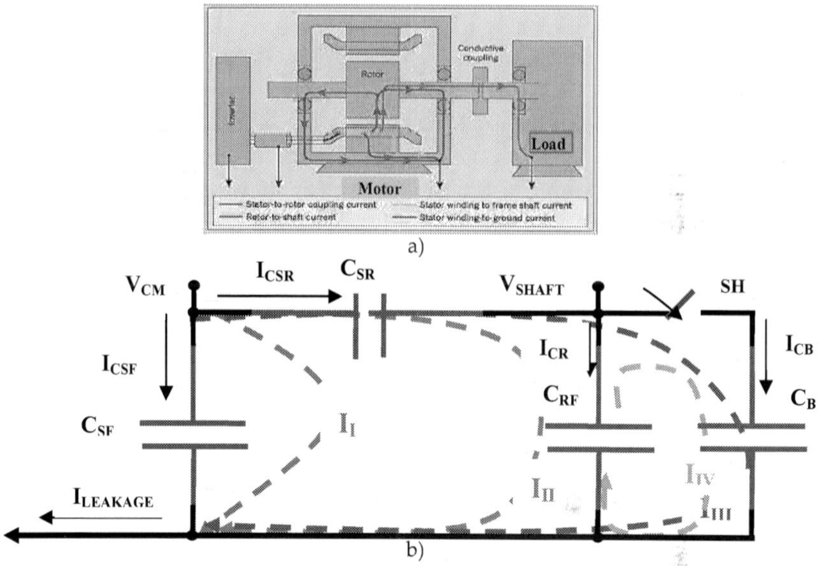

Figure 8. a) Circulating currents in the motor, b) high-frequency equivalent circuit and circulating currents [10]

The electric currents are established as follows: [10] a) II is the current flowing through the common mode voltage point (V_{CM}), which passes through the pump capacitance for the motor frame, ground or neutral and returns to the system to V_{CM} point. It is the largest component of the leakage current ($I_{LEAKAGE}$), compared to other capacitive currents because C_{SF} capacitance is much larger than others. This current is primarily responsible for the electrical discharge of the motor frame to the "ground." If the motor is not grounded to the frame satisfactorily, an important electrical discharge at various parts of it and even at the load may occur and cause an "electric shock" if someone leans on the motor frame. b) III is the current flowing through the common mode voltage point (V_{CM}), which passes through the C_{SR} and capacitances C_{RF}, motor frame, ground or neutral system and returns to the V_{CM} point. c) III is the current flowing through the common mode voltage point (V_{CM}), which passes through the C_{SR} capacitance motor shaft bearing capacitance C_B motor frame, ground or neutral system and returns to the V_{CM} point. d) IV is the current flowing shaft voltage point (V_{SHAFT}); C_{RF} capacitance thus stores energy through the motor shaft, the switch SH, the motor frame, and returns to the C_{RF} capacitance.

The switch SH in Figure 8b is, when closed, the breaking of the bearing insulation dielectric (grease film). When this occurs, and C_{SR} is much smaller than C_{RF}, a new mesh current flow given by IV is then established. The IV mesh is responsible for the electrical discharge in the motor bearings due to the charge stored in the capacitor C_{RF}.

3. Experimental method for determination of parasitic capacitances rotor cage three-phase induction motor [40]

This item presents a methodology [40] for the determination of the parasitic capacitances of the equivalent high-frequency rotor three-phase induction motor circuit cage and the effects of electromagnetic interference caused in the same common mode, when it is driven by a PWM inverter. This is a methodology that uses an electronic circuit to measure variables needed to calculate these parameters.

The procedures proposed for the development of this methodology are: a) determination of the equivalent circuit parameters of the three-phase induction motor in steady state and high frequency and [1, 10, 15, 31] through typical laboratory test; b) establish settings of links between the PWM inverter and the motor for measurements of quantities of interest: common mode (V_{CM}) and shaft (V_{SHAFT}) voltages, leakage ($I_{LEAKAGE}$) and shaft (I_{SHAFT}) currents, developed by measuring circuit for this purpose [13, 14]; c) calculate the values of the parasitic capacitances between stator and frame (C_{SF}), stator and rotor (C_{SR}), rotor and frame (C_{RF}) and bearings (C_B), using their characteristic equations [10, 15]; d) using PSPICE [36] to simulate the system (three-phase induction motor fed by PWM inverter) with the high-frequency equivalent circuit in the same [6, 16]; e) to obtain the waveforms characteristics of the EMI phenomena.

3.1. Methodology determination procedure

The methodology determines the high-frequency equivalent circuit parameters of rotor cage three-phase induction motor, through direct measurement of quantities of interest and, using equation (14), calculate the values of the parasitic capacitances. The quantities of interest are common mode voltage (V_{CM}), shaft voltage (V_{SHAFT}), leakage current ($I_{LEAKAGE}$), and shaft current (I_{SHAFT}).

$$C = \frac{I_C}{2.\pi.f_s.V_C} \tag{14}$$

In equation (14), I_C and V_C represent the current and the effective voltage across the capacitor, respectively, and f_S the switching frequency of the PWM inverter. The schematic diagram of the methodology is shown in Figure 9.

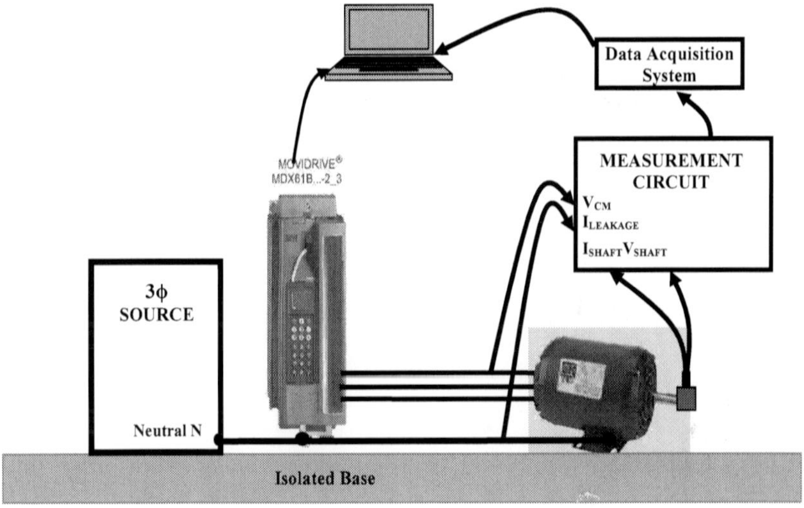

Figure 9. Schematic diagram of the methodology

The structure consists of the following equipment: three-phase power supply 220V/60Hz, 1 Hp three-phase PWM inverter, two three-phase 1Hp induction motors, measuring circuit of the quantities of interest, data acquisition board LabView, notebook PWM inverter management and signal processing through dedicated software LabView 8.5 and MOVITOOLS, base for sustaining fully insulated equipment to allow measurements, especially the currents (as close as possible to an actual situation), connecting cable with the neutral system interconnected to the PWM inverter ground terminals and the induction motor providing a circulating path to the induction motor current.

The magnitudes of the measuring circuit (V_{CM}, V_{SHAFT}, $I_{LEAKAGE}$, and I_{SHAFT}) are shown in Figure 10. The common mode voltage (V_{CM}) is the common point of the measuring voltage (neutral) to the motor frame when it uses star connection (Y). When the same connection uses delta (Δ), it should be equivalent to a star connection using high-value resistors (1MΩ) connected to each phase of the motor with a common point (neutral). In the adjustment of the common mode voltage measurement (V_{CM}) to the acquisition board, a resistive divider (R_1, R_2, R_3, and R_4) is added so that the measured voltage does not exceed the maximum allowed by the acquisition board, that is, +/- 10V, as shown in Figure 10.

For measuring the shaft voltage (V_{SHAFT}), which is the voltage measured between the shaft and the induction motor frame, a copper ring carbon brush system is added to the induction motor shaft.

For leakage current ($I_{LEAKAGE}$) and shaft current (I_{SHAFT}) measurements, a Rogowski coil current sensor, developed for this methodology, was used [40]. The leakage current is the current measured in the connection cable between the induction motor frame and the metal frame of the PWM inverter connected to the neutral system. It is composed of the sum of the currents flowing through the stator-frame capacitances (C_{SF}), rotor-frame (C_{RF}), and bearings (C_B).

The shaft current (I_{SHAFT}) is the current measured in the conductor which is connected to the brush and corresponding contributions from currents flowing through the rotor-frame capacitances (C_{RF}) and bearings (C_B).

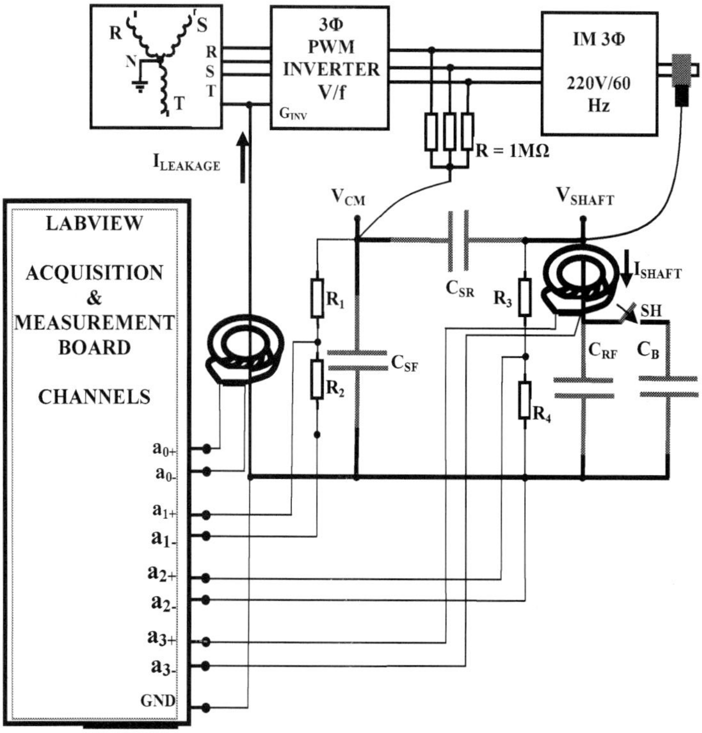

Figure 10. Measuring circuit of the quantities of interest

3.1.1. Induction motor preparation

For the application of the proposed method, the motor rotor three-phase induction cage should be prepared so that the capacitance can be determined [1, 10, 14]. Under normal conditions, the motor bearings are directly connected to both the shaft and the motor frame. It causes the capacitance existing between the rotor and the frame (C_{RF}) to short-circuit.

So the following changes are made. (a) Motor bearings are insulated by a nylon cover (tecnil) high electrical resistance. Thus, there will be no current flow by bearings (IB) and the contri-

http://dx.doi.org/10.5772/61544

bution branch rotor frame can be evaluated, as shown in Figure 11 [10]. (b) A conductor is connected between the external surface of one of the bearings and closest to the motor frame by a switch point. When the switch is open, there is the condition set out in item (a). When the switch is closed, then one has the rolling movement for the current frame and the returning to the bearing [10]. (c) The motor shaft settles in a ring and brush assembly, for measuring the shaft voltage (V_{SHAFT}). This system is also used to measure the shaft current (I_{SHAFT}) [14]. Figure 12 shows the items (b) and (c).

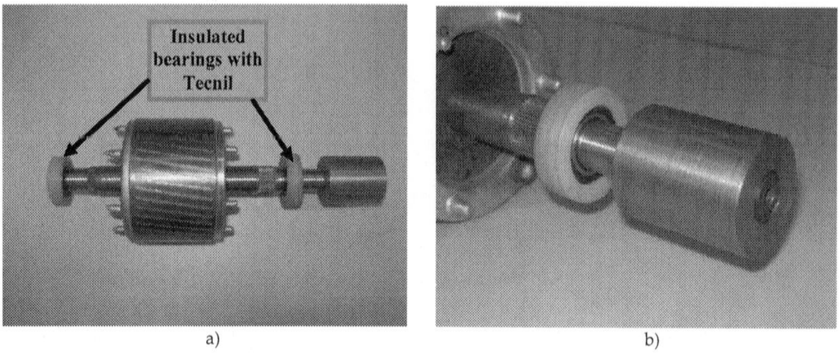

Figure 11. a, b): Insulation of the motor shaft bearings

Figure 12. Switch (b) and ring-brush assembly (c) for measuring

3.1.2. *Data acquisition board and measuring*

In order to acquire the quantities and accomplish the measurements, LabView [32, 33, 34] platform is used. Four channels for the different measurements are defined as follows: (a) a0 channel: leakage current ($I_{LEAKAGE}$); (b) A1 channel: the common mode voltage (V_{CM}), which is set to the voltage of -10V to + 10V limits; (c) a2 channel: shaft voltage (V_{SHAFT}); and (d) a3 channel: shaft current (I_{SHAFT}).

Using dedicated software LabView 8.5, [35] created a block diagram (plant) of the measuring system for the magnitudes of interest. Figure 13 shows the block diagram for leakage current ($I_{LEAKAGE}$). The configuration for each quantity to be measured using channels with differential inputs is shown in Figure 14 to minimize the effects of common-mode voltages (noise).

Observations (Figure 13):

Block A - CORRENTE DE FUGA - Ifuga <=> LEAKAGE CURRENT - Ileakage

Block D - FILTRO RUÍDO <=> NOTCH FILTER

Block E - FILTRO PASSA-BAIXA <=> LOW-PASS FILTER

Block F - ESPECTRO - Ifuga <=> ESPECTRUM - Ileakage

Block G - RMS Ifuga_fs <=> RMS Ileakage_fs

SAÍDA DO FILTRO - Ifuga <=> OUTPUT FILTER – Ileakage

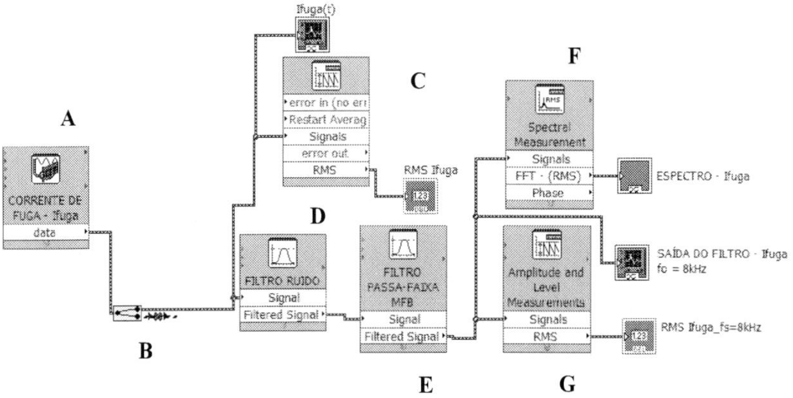

Figure 13. Block diagram for reading the leakage current ($I_{LEAKAGE}$)

Using the DAQ block (A), the following items are defined for each channel: a) type of quantity (voltage or current), b) name of greatness and sample rate, c) type of sampling, d) connection diagram channel and other settings which depend on the desired type of measurement. Can be used up to 16 signals for reading.

http://dx.doi.org/10.5772/61544

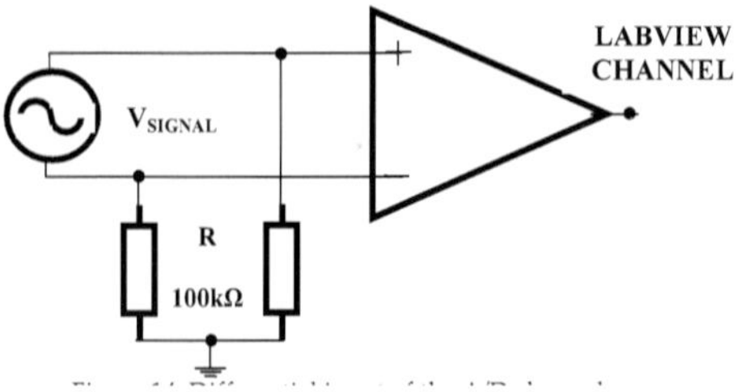

Figure 14. Differential input of the A/D channels

Signals Using Split (B) block quantities of simultaneous readings can be made. In this case, there are 4 outputs. The current $I_{LEAKAGE}$ a function of time is shown on the oscilloscope $I_{LEAKAGE}$ (t).

The Amplitude and Level Measurements block (C) allows for the signs readings DC values (DC) and effective (RMS). The value is shown on the display "RMS $I_{LEAKAGE}$."

The blocks Noise Filter (D) and Pass Filter Range MFB (E) treat the signal so that it can be measured on the switching frequency of the PWM inverter. The filtered signal is displayed on the oscilloscope FILTER OUT – $I_{LEAKAGE}$.

Noise Filter (NOTCH) has the function of not allowing passage of signals of a frequency band or a specific frequency. The Band Pass Filter MFB allows only the frequency band specified signal to pass. This filtering block allows you to define: a) type of filter: low-pass, high-pass, band-pass, band-reject; b) frequency band; c) the type of response: finite impulse response (FIR) or infinite impulse response (IIR); d) topology: Butterworth, Chebyshev, Inverse Chebyshev, Elliptic, and Bessel.

Spectral Measurement block (F) shows the oscilloscope through the spectrum $I_{LEAKAGE}$ already filtered at the desired frequency. And the block (G) similar to block (C) provides effective current at switching frequency f_S, available in display RMS_$I_{LEAKAGE}$_f_S. This entire procedure aims to ensure that measurements are made in order to respect the equation for calculating the capacitance (14).

3.1.3. PWM inverter

The PWM inverter used in this study is a new generation of static power converters that feature improvements in its modular structure, providing better functions in the lower-frequency range and greater overload capacity. Introducing control functions integrated with the possibility of use of communication accessories, this enables applications to AC inverter drive systems requiring high efficiency in a power range 0.55–160 kW.

The Inverter MOVIDRIVE MDX60B/61B [29] has the following main characteristics: a) The MOVIDRIVE MDX61B model allows application for asynchronous motors (induction motor) with or without encoder feedback, synchronous and asynchronous servomotors. b) Control modes: VFC (Voltage Flux Control): to control induction motors. Using encoder feedback, it operates with vector control activated. If there is no feedback loop, it operates with V/f control (scalar) and CFC (Current Flux Control); for controlling synchronous and asynchronous servomotors it always operates with encoder feedback. The inverter model used in this work is the Movidrive MDX61B 0037-2A3-4-0 with the following characteristics: Power 5 Hp (3.7 kw); phase power supply: 220 V/50–60 Hz; rated output current: 15.2 A and PWM switching frequency: 4, 8, 12, and 16 kHz. Using MOVITOOLS [31] dedicated program, both the switching frequency f_S of the inverter induction motor as well as the rotational speed can be changed, allowing a more complete analysis of the behavior of the parasitic capacitances to be realized.

3.2. Tests laboratory procedure

A procedure was adopted within the Electrical Machines Laboratory for the measurement tests so that determining the high-frequency capacitance of the induction motors could be carried out in conditions where there is the least possible interference in the results thereof. These procedures are described as follows: a) Ensure the total isolation of the set: three-phase induction motor being tested and PWM inverter from the base where they are supported in order that there is no current flow to the system from the motor frame and the inverter. b) Verification of the connections between the motor frame and the inverter with the NEUTRAL of three-phase power supply, allowing you to create a single movement path of the motor leakage current and hence its correct measurement. c) The entire measurement system (LABVIEW acquisition board) and the computer (notebook) used in the execution of dedicated programs and measured data storage must be the same reference system NEUTRAL. This ensures reliable measurements. d) Correct positioning and enough brush pressure in the ring, the induction motor shaft, ensuring good contact and, therefore, reliable measurements of shaft voltage and current (V_{SHAFT} and I_{SHAFT}). e) Check all connections of the acquisition system/measurement and inverter with the computer. f) carrying out the tests in air-conditioned environment, to ensure the same conditions for the other tests, and g) if possible, work with a fully isolated power system, not only using an isolating transformer, but also ensuring electrical insulation. This way, voltage and current of the supply system will always be the same in any situation and time, providing measurements and better results. Following these procedures, tests of induction motors to determine the parasitic capacitances can be started.

3.3. Parasitic capacitances determination of the rotor cage three-phase induction motor

The high-frequency equivalent circuit is shown again in Figure 15.

When determining the capacitance the following procedure should be followed: a) Switch SH open: Motor bearings are isolated. So there is no current flowing through the bearings and the C_B capacitance does not contribute to the value of B_{VR} (Bearing Voltage Ratio). b) Switch SH closed: a circulating current starts to flow through the bearings, and the C_B capacitance becomes to influence the value of B_{VR}.

Measurements are performed in the following quantities of interest: common mode voltage (V_{CM}), shaft voltage (V_{SHAFT}), leakage current ($I_{LEAKAGE}$) shaft current with SH off (I_{SHAFT}-$_{OFF}$),

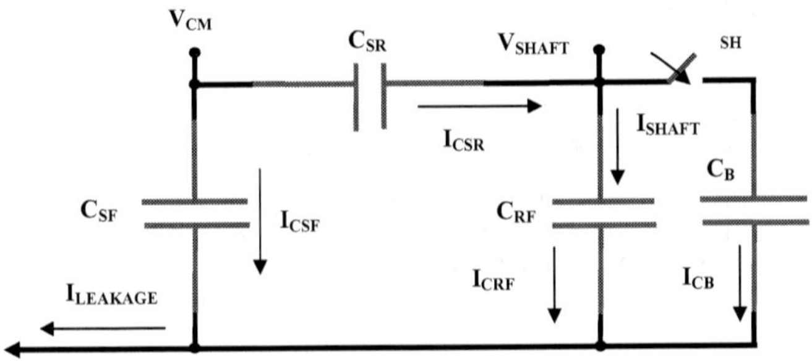

Figure 15. Simplified equivalent circuit of high-frequency induction motor

shaft current with SH on ($I_{SHAFT\text{-}ON}$). The following values are determined from the following streams:

$$I_{C_{SF}} = I_{LEAKAGE} - I_{SHAFT-OFF} \tag{15}$$

$$I_{C_{SR}} = I_{C_{RF}} = I_{SHAFT-OFF} \tag{16}$$

$$I_{C_B} = I_{SHAFT-OFF} - I_{SHAFT-ON} \tag{17}$$

The parasitic capacitances of the induction motor: C_{SF}, C_{RF}, C_{SR}, and C_B are determined through the rewritten equations below:

$$C_{SF} = \frac{I_{C_{SF}}}{2.\pi.f_s.V_{CM}} \tag{18}$$

$$C_{RF} = \frac{I_{C_{RF}}}{2.\pi.f_s.V_{SHAFT}} \tag{19}$$

$$C_{SR} = \frac{I_{C_{SR}}}{2.\pi.f_s.(V_{CM} - V_{SHAFT})} \tag{20}$$

$$C_B = \frac{I_{C_B}}{I_{C_{RC}}} \cdot C_{RC} \tag{21}$$

4. Measurements and simulations

The method described in the previous section was applied in three-phase induction motors (two 1 Hp motors). One of the 1 Hp motors no longer presents its original features because the stator windings were fully replaced without regard to the original design features. This procedure allowed to check if there were changes (or not) in the values of the capacitances for the other 1 Hp motor, which presents original project features, since the objective is to ratify that the parasitic capacitances depending only on geometric-constructive features of the motor. Since the determination of the capacitances depends on the quantities of interest at high frequency, the tests may be performed with or without load on the shaft of the induction motors.

4.1. Test of three-phase induction motors

4.1.1. Determination of low-frequency parameters

Initially, tests for determining the low-frequency equivalent circuit (nominal) parameters of the motors were performed [27, 37]. The parameters were determined from characteristic tests using the two wattmeter method: test empty and locked rotor test with characteristic equations [26, 27]. The equivalent circuit parameters of the two induction motors 1 Hp shown in Table 1 were determined. These parameters are used in the simulation using PSPICE for analysis and comparison with the results obtained in the laboratory (Figure 16).

The two three-phase induction motors used in this test are as follows: a) manufacturer: WEG, b) rated power: 1 Hp, c) nominal voltage: 220V (Δ) / 380V (Y) d) current nominal: 3,8A (Δ) and 2,4a (Y), e) rated speed: 1710 rpm, f) number of poles: 4.

Motor (Hp)	$R_1(\Omega)$	$R_2'(\Omega)$	$L_1 = L_2$ (mH)	L_{mag}(mH)
1 (IM1hp1)	8.55	5.62	16.95	346.31
1 (IM1hp2)	8.70	6.53	18.22	366.40

Table 1. The low-frequency equivalent circuit parameters determined from laboratory

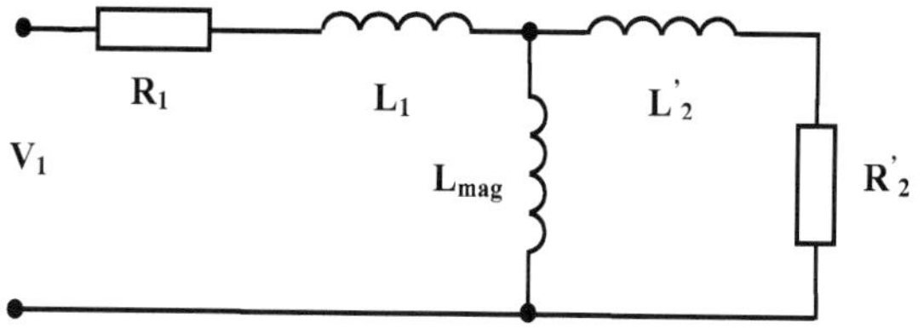

Figure 16. Equivalent circuit of low-frequency induction motor

4.1.2. High-frequency parameters determination

Using the application program MOVITOOLS [31], the PWM inverter is parameterized in terms of induction motor characteristics to be tested (plate data) as well as the drive and operating characteristics. PWM switching frequency and fundamental frequency of the induction motor can be changed in real time.

The application program LabView [35] was set to adjust the frequencies of noise filters and band pass of the quantities of interest to the switching frequency of the PWM inverter.

After all measurements and calculations, the parasitic capacitance curves were constructed: a) capacitance as a function of motor drive frequency (Hz) to a fixed switching frequency, b) capacitance as a function of the switching frequency to drive 60 Hz motor, c) mean value of capacitance as a function of the switching frequency. The waveforms of the quantities of interest were also obtained by measuring with a digital oscilloscope.

4.2. Test of three-phase induction motors of 1 Hp

4.2.1. Motor IM1hp1

The three-phase induction motor 1 Hp (IM1hp1) is connected to stator winding delta (Δ), and 220 V line voltage. A star configuration equivalent to the common mode voltage measurement (V_{CM}) is well utilized. Table 2 shows the measured values of the quantities of interest. Table 3 shows the calculated values of the currents flowing between the stator and the motor frame (I_{CSF}) and bearings (I_{CB}). Table 4 shows the calculated values of the three-phase induction motor parasitic capacitances 1Hp IM1hp1.

Measurements and calculations of quantities of interest of parasitic capacitances of the two induction motors were made to switching frequencies of 12 and 16 kHz. The results for the switching frequencies of 4 and 8 kHz can be observed in [40].

Inverter Switching Frequency (kHz)	Motor Frequency (Hz)	V_{CM} (V)	V_{SHAFT} (V)	$I_{LEAKAGE}$ (mA)	$I_{SHAFT-OFF}$ (mA)	$I_{SHAFT-ON}$ (mA)
12						
	20	93,31	4,36	14,30	0,456	0,382
	30	82,17	3,87	12,45	0,393	0,344
	40	67,42	3,18	10,53	0,343	0,303
	50	50,57	2,37	7,83	0,247	0,218
	60	33,11	1,57	5,23	0,169	0,153
16						
	20	93,91	4,51	19,40	0,544	0,440
	30	83,98	4,08	17,30	0,490	0,412
	40	69,23	3,37	14,20	0,403	0,341

Inverter Switching Frequency (kHz)	Motor Frequency (Hz)	V_{CM} (V)	V_{SHAFT} (V)	$I_{LEAKAGE}$ (mA)	$I_{SHAFT\text{-}OFF}$ (mA)	$I_{SHAFT\text{-}ON}$ (mA)
	50	51,77	2,53	10,50	0,302	0,257
	60	33,71	1,67	7,03	0,202	0,175

Table 2. Measurement of quantities of interest of 1Hp motor IM1hp1

Inverter Switching Frequency (kHz)	Motor Frequency (Hz)	Ic_{SF} (mA)	Ic_B (mA)
12			
	20	13,84	0,07
	30	12,06	0,05
	40	10,19	0,04
	50	7,58	0,03
	60	13,84	0,07
16			
	20	18,86	0,10
	30	16,81	0,08
	40	13,80	0,06
	50	10,20	0,05
	60	6,83	0,03

Table 3. Calculated currents – IM1hp1

Inverter Switching Frequency (kHz)	Motor Frequency (Hz)	C_{SF} (pF)	C_{RF} (pF)	C_{SR} (pF)	C_B (pF)
12					
	20	1967,82	1387,17	67,99	268,72
	30	1946,16	1346,89	66,57	191,85
	40	2004,05	1430,60	70,82	188,86
	50	1988,84	1382,29	67,97	183,88
	60	2027,35	1427,71	71,07	149,30
16					
	20	1997,33	1199,87	60,53	283,61
	30	1991,15	1194,67	61,00	226,18

Inverter Switching Frequency (kHz)	Motor Frequency (Hz)	C_{SF} (pF)	C_{RF} (pF)	C_{SR} (pF)	C_B (pF)
	40	1982,45	1189,56	60,87	216,28
	50	1959,52	1187,41	61,01	207,91
	60	2014,87	1203,23	62,72	185,64

Table 4. Parasitic capacitances motor 1Hp IM1hp1

Figure 17 shows the values of each of the parasitic capacitances on the basis of the motor frequency (speed) specific to the switching frequency of the PWM inverter 12 kHz and 16 kHz.

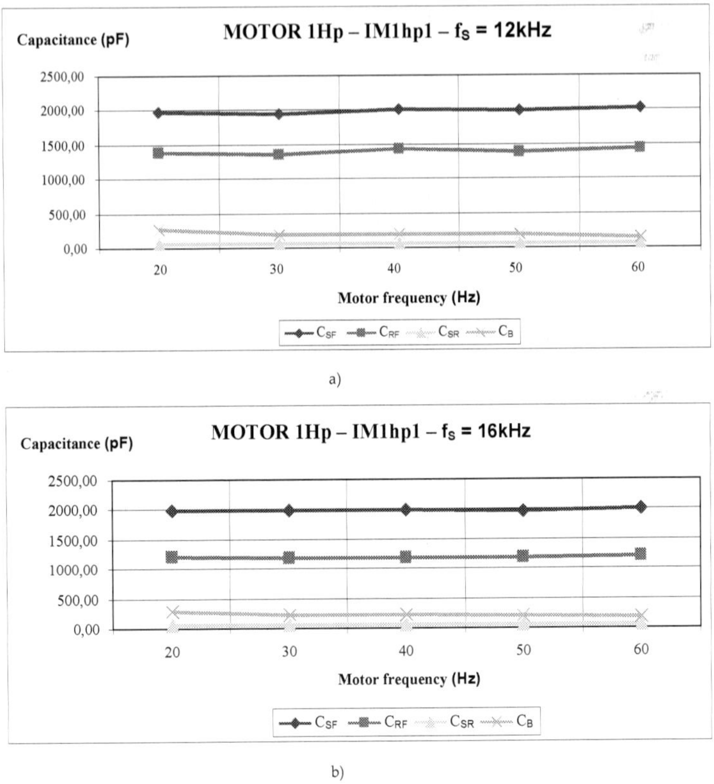

Figure 17. Parasitic capacitances due to motor frequency of 1 Hp (IM1hp1) for 12 kHz (a) and 16 kHz (b) inverter switching frequency

The capacitances C_{SF}, C_{RF}, C_{SR} are almost constant for the variation of the motor frequency and the switching frequency of the PWM inverter, indicating that they depend mainly on the geometric characteristics of the motor [1].

In turn, the C_B capacitance decreases with increasing motor frequency (increase the speed). This is because, besides being dependent on the dimensions of the bearing, this capacitance is also a function of speed [10].

Figure 18 shows the values of the parasitic capacitances on the basis of the switching frequency of the PWM inverter to the motor frequency of 60Hz. The results confirm that the parasitic capacitances depend exclusively on the physical constructive characteristics of the induction motor.

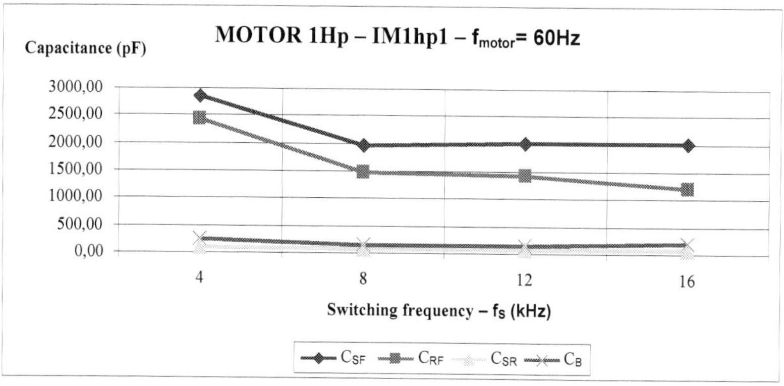

Figure 18. Capacitances due to the motor switching frequency of 1 Hp (IM1hp1) power frequency 60 Hz

Figure 19 presents the average parasitic capacitance values of the motor 1 Hp IM1hp1.

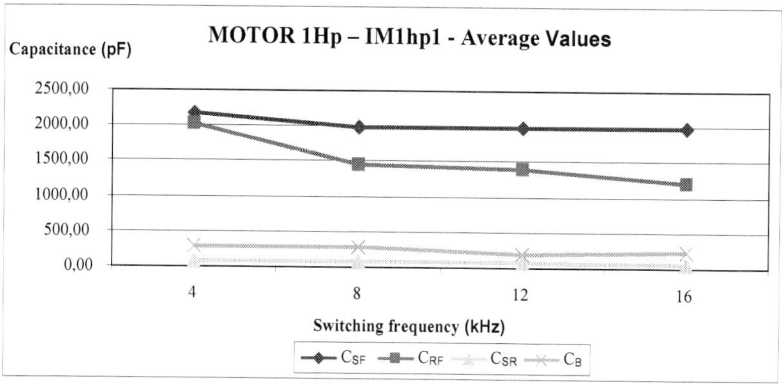

Figure 19. Average values of the capacitances for 1 Hp motor (IM1hp1)

http://dx.doi.org/10.5772/61544

4.2.2. *Motor IM1hp2*

The results for the second three-phase induction motor 1Hp, IM1hp2 are shown below. Table 5 presents the measured values of the quantities of interest.

Table 6 presents the calculated values of the currents flowing between the stator and the motor frame (I_{CSF}) and bearings (I_{CB}).

Table 7 presents the calculated values of the three-phase induction motor parasitic capacitances of 1 Hp IM1hp2.

Inverter Switching Frequency (kHz)	Motor Frequency (Hz)	V_{CM} (V)	V_{SHAFT} (V)	$I_{LEAKAGE}$ (mA)	$I_{SHAFT-OFF}$ (mA)	$I_{SHAFT-ON}$ (mA)
12						
	20	95,42	4,20	14,84	0,393	0,211
	30	85,58	3,74	13,10	0,343	0,222
	40	70,13	3,18	10,93	0,284	0,207
	50	52,68	2,37	8,15	0,222	0,185
	60	34,62	1,53	5,45	0,149	0,133
16						
	20	95,12	4,35	20,00	0,530	0,327
	30	84,88	3,97	18,46	0,478	0,326
	40	69,83	3,31	15,18	0,404	0,313
	50	52,98	2,48	11,10	0,303	0,261
	60	34,31	1,65	7,67	0,212	0,187

Table 5. Measurement of quantities of interest – IM1hp2

Inverter Switching Frequency (kHz)	Motor Frequency (Hz)	I_{CSF} (mA)	I_{CB} (mA)
12			
	20	14,45	0,18
	30	12,76	0,12
	40	10,65	0,08
	50	7,93	0,04

Inverter Switching Frequency (kHz)	Motor Frequency (Hz)	Ic_{SF} (mA)	Ic_B (mA)
	60	5,31	0,02
16			
	20	19,47	0,20
	30	17,98	0,15
	40	14,78	0,09
	50	10,80	0,04
	60	7,46	0,03

Table 6. Calculated currents – IM1hp2

Inverter Switching Frequency (kHz)	Motor Frequency (Hz)	C_{SF} (pF)	C_{RF} (pF)	C_{SR} (pF)	C_B (pF)
12					
	20	2008,24	1239,80	57,09	1062,69
	30	1977,10	1215,68	55,55	661,51
	40	2013,30	1185,36	56,30	442,07
	50	1995,63	1244,62	58,64	253,24
	60	2032,70	1291,66	59,73	159,75
16					
	20	2036,24	1212,65	58,06	749,66
	30	2107,39	1196,70	58,72	554,36
	40	2104,88	1212,93	60,35	349,10
	50	2027,39	1215,36	59,69	196,66
	60	2163,08	1280,51	64,68	172,38

Table 7. Parasitic capacitances of motor 1 Hp IM1hp2

Figure 20 shows the values obtained for the parasitic capacitance depending on the motor frequency, for specific switching frequency of the PWM inverter 12 kHz and 16 kHz.

The same behaviors found in the motor 1Hp IM1hp1 were observed.

Figure 21 shows the values of the parasitic capacitances on the basis of the switching frequency of the PWM inverter to the motor frequency of 60Hz.

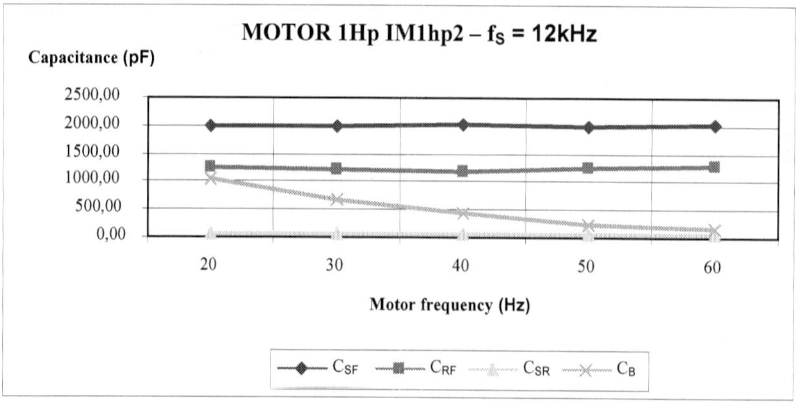

a)

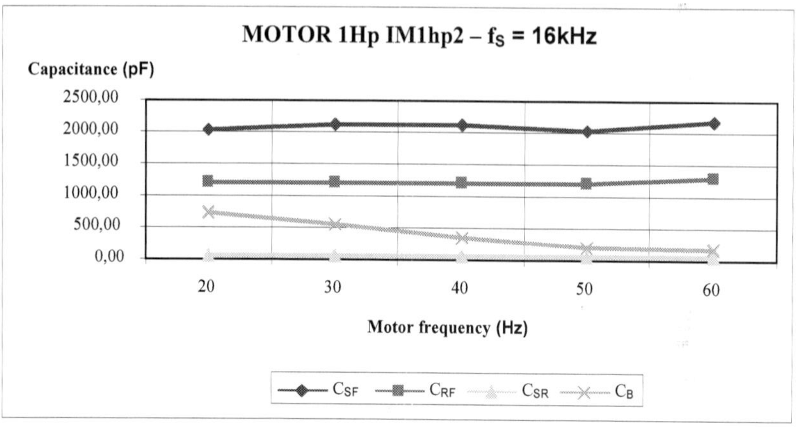

b)

Figure 20. a),b) Parasitic capacitances due to motor frequency for different inverter switching frequency (IM1hp2)

Figure 22 presents the average values of parasitic capacitance of motor 1 Hp IM1hp2.

4.3. Simulations and measurements

The software Pspice [30, 36] was used to simulate the methodology and compare the waveforms obtained in the simulations with actual measurements made using digital oscilloscope. The schematic circuit simulation is shown in Figure 23.

The schematic circuit comprises: a) inverter structure with three-phase supply, three-phase rectifier without control (diodes) with DC bus, and three-phase inverter bridge; b) equivalent

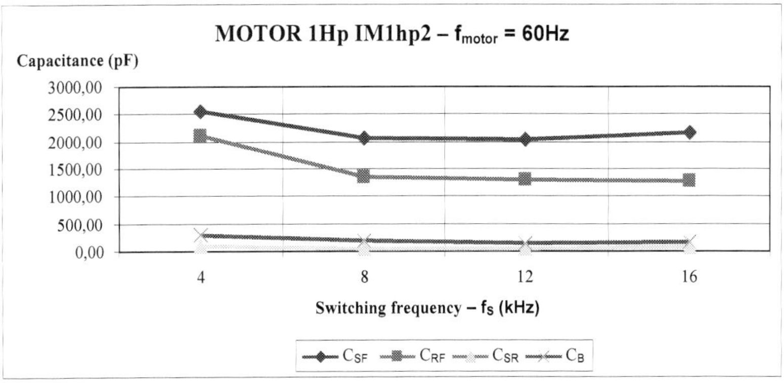

Figure 21. Capacitances due to the motor switching frequency of 1 Hp (IM1hp2) motor frequency 60 Hz

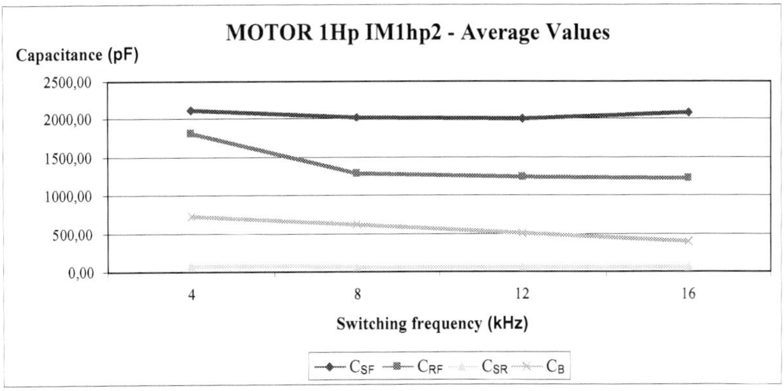

Figure 22. Average values of capacitance for the motor 1 Hp IM1hp2

circuit of low-frequency induction motor; c) equivalent circuit of high-frequency three-phase induction motor; and d) signal generator with pulse width modulation (PWM).

The charts below show waveforms that appear in the named graphs (a) are results of measurements by using digital oscilloscope. The forms of common mode voltage wave (V_{CM}) are attenuated to match the oscilloscope's full-scale capacity. The waveforms that appear in the named graphs (b) are obtained by using PSPICE simulation application. Without loss of generality, in the simulation (Figure 23) are used ideal components of semiconductor switches that facilitate the process of solving differential equations reducing convergence problems.

The waveforms shown below are the quantities of interest with switching frequency of 16 kHz PWM inverter and frequency of 60 Hz motor to three-phase induction motors of 1 Hp (IM1hp1 and IM1hp2), respectively.

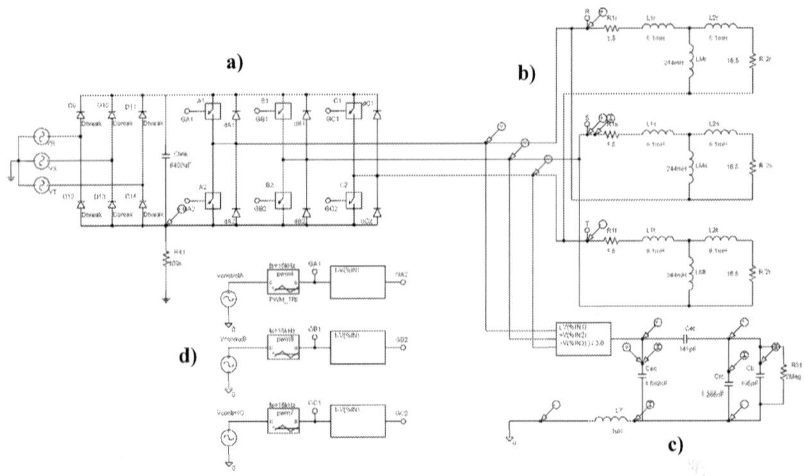

Figure 23. Schematic circuit used for simulation in Pspice [40]

Figures 24 and 25 show the waveforms of various quantities of interest for the three-phase induction motor 1Hp IM1hp1, Figures 26 and 27 show the waveforms for the three-phase induction motor of 1 Hp IM1hp2.

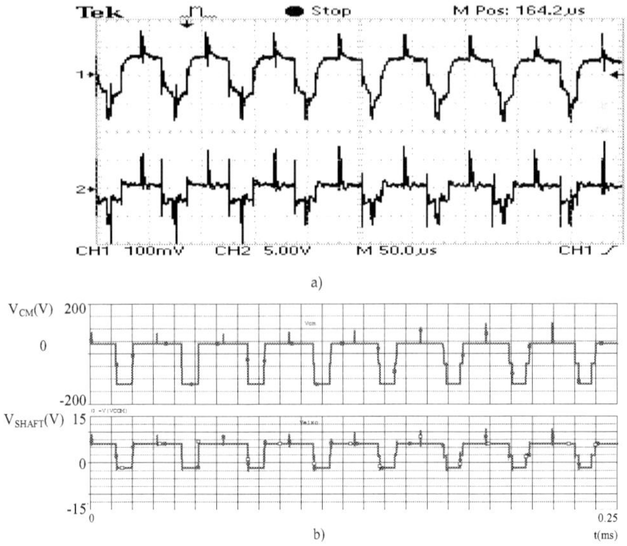

Figure 24. V_{CM} – CH1, V_{SHAFT} – CH2; a) measured, b) simulated (IM1hp1)

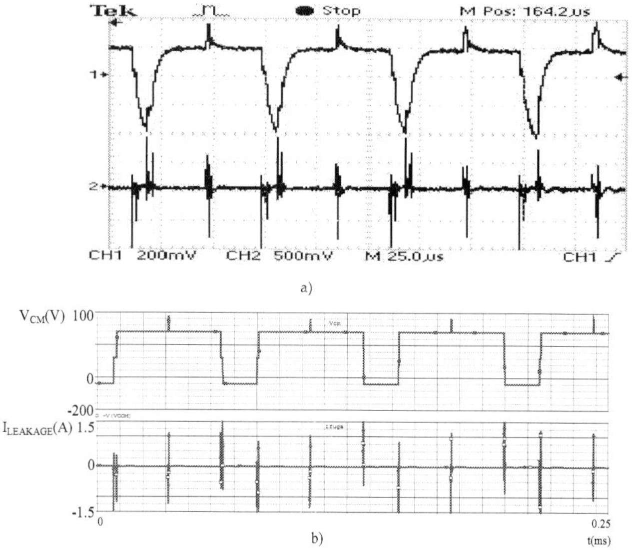

Figure 25. V_{CM} – CH1, $I_{LEAKAGE}$ – CH2; a) measured, b) simulated – (IM1hp1)

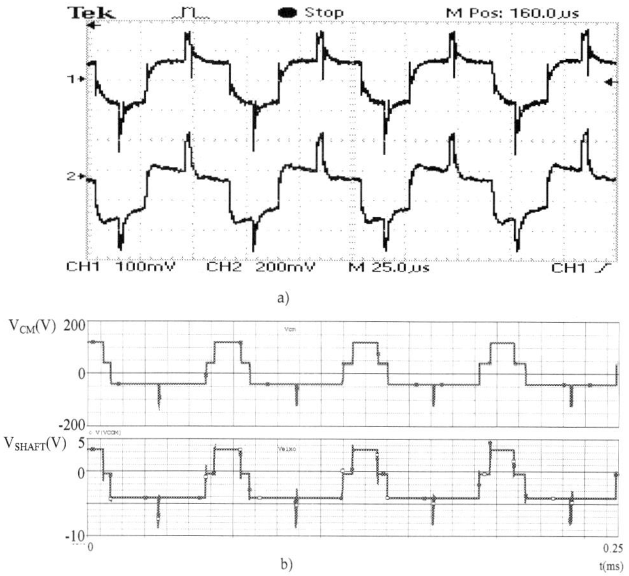

Figure 26. V_{CM} – CH1, V_{SHAFT}– CH2 - a) measured, b) simulated - (IM1hp2)

http://dx.doi.org/10.5772/61544

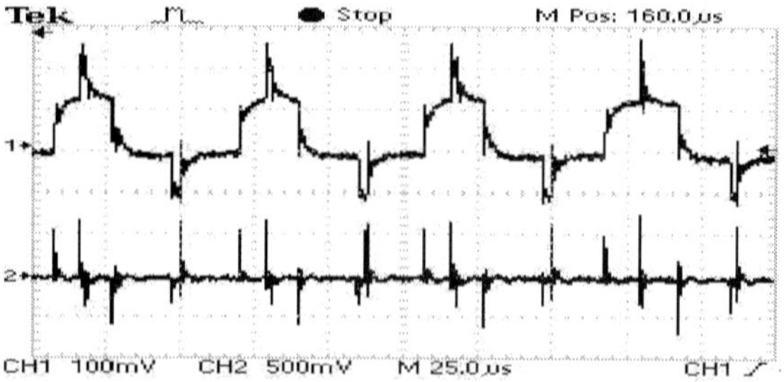

a)

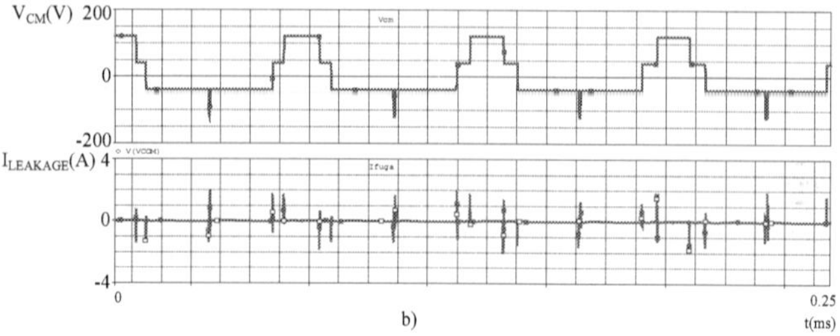

b)

Figure 27. V_{CM} – CH 1, $I_{LEAKAGE}$ – CH 2 - a) measured, b) simulated - (IM1hp2)

5. Results and conclusions

In the comparison of results between the two 1Hp motors (IM1hp1 and IM1hp2), the following considerations are made: (a) Although the motors are of the same power and the same manufacturer, one of them underwent maintenance and the other still has its original components. This results in differences in the values of capacitances C_{SF}, C_{RF}, and C_B. The exchange of maintaining consisted of stator conductors, insulating layer placed in the stator slots, and replacing the original bearings. These changes result in different values of capacitances, although they present very similar values. (b) The use of sensors of both voltage and current characteristics more suitable for this type of experiment such as precision operation range, frequency response, are necessary to avoid problems like those that occurred mainly in the measurements of the motors on the switching frequency of 4 kHz [40]. (c) There was no automation of measurements of quantities of interest according to the equipment available for this purpose. Thus, measurement errors were the same, but it can still be said that they were within a satisfactory tolerance, when comparing these two motors. In Table 8, comparisons are shown between the parasitic capacitances for two 1Hp motors.

The methodology for determining the parasitic capacitance of the induction motor, when driven by a PWM inverter presents consistent and coherent results.

The results of the tests have revealed that the parasitic capacitances of the induction motor are a function only of the geometric-constructive characteristics [40, 41].

Inverter Switching Frequency (kHz)	Motor Frequency (Hz)	C_{SF} (pF)	C_{RF} (pF)	C_{SR} (pF)	C_B (pF)
16					
Motor	20	1997,33	1199,87	60,53	283,61
1Hp	30	1991,15	1194,67	61,00	226,18
IM1hp1	40	1982,45	1189,56	60,87	216,28
	50	1959,52	1187,41	61,01	207,91
	60	2014,87	1203,23	62,72	185,64
16					
Motor	20	2036,24	1212,65	58,06	749,66
1Hp	30	2107,39	1196,70	58,72	554,36
IM1hp2	40	2104,88	1212,93	60,35	349,10
	50	2027,39	1215,36	59,69	196,66
	60	2163,08	1280,51	64,68	172,38

Table 8. Comparison between the parasitic capacitances of 1Hp motors

The switching frequency variations almost do not change the values of the capacitances. In fact, the quantity which have their values changed due the variation of switching frequencies is the reactance of these capacitances. High switching frequency of the PWM inverter, despite improving the characteristic shape of the load current wave, causes the switching time of power electronic devices (IGBT or MOSFET) to be rather low, resulting in increased rates of growth of voltage (dV/dt) [7]. This reflects directly on the currents flowing through the parasitic capacitances and therefore the leakage current.

Also there is an increase of the amplitude of these currents due to the capacitive reactance parasites being substantially diminished in value due to the increase of the switching frequency. So, the effects of electromagnetic interference are increased both on bearings (bearing currents) as on capacitive currents in the motor. It is also noticed that the bearing of capacitance values (C_B) decrease with increasing induction motor speed [10].

Common mode filters are used to minimize the effects of currents flowing through the parasitic capacitances of the induction motor. The accurate determination of these capacitances values, using the methodology proposed in this section, has the main objective of optimizing the design of these filters.

http://dx.doi.org/10.5772/61544

This optimization implies a more accurate and reliable specification of the components used in filter design, thus providing a significant reduction of volume and weight of the filter; the number of components used, and thus cost reduction thereof.

Author details

Rudolf Ribeiro Riehl[1*], Fernando de Souza Campos[1], Alceu Ferreira Alves[1] and Ernesto Ruppert Filho[2]

*Address all correspondence to: rrriehl@feb.unesp.br

1 São Paulo State University, Unesp, Bauru, Brazil

2 State University of Campinas, Unicamp, Campinas, Brazil

References

[1] Busse, D., Erdman, J., Kerkman, R. J., Schlegel, D., and Skibinski, G. System Electrical Parameters and Their Effects on Bearing Currents, *IEEE Trans Indus Applicat*, Vol. 33, no. 2, pp. 577-584, 1997.

[2] Erdman, J., Kerkman, R. J., Schlegel, D., and Skibinski, G. Effect of PWM Inverters on AC Motor Bearing Currents and Shaft Voltages, *IEEE APEC Conf*, Vol. 01, pp. 24-33, 1995.

[3] Akagi, H., Hasegawa, H., and Domouto, T. Design and Performance of a Passive EMI Filter for Use with a Voltage-Source PWM Inverter having Sinusoidal Output Voltage and Zero Common-Mode Voltage, *IEEE Trans Power Electron*, Vol. 19, no. 4, pp. 1069-1076, 2004.

[4] Busse, D., Erdman, J., Kerkman, R. J., Schlegel, D., and Skibinski, G. Bearing Currents and their Relationship to PWM Drives, *IEEE Trans Power Electron*, Vol. 12, no. 2, pp. 243-252, 1997.

[5] Villabona, E. G., Gúrpide, P. S., Sádaba, O. A., Azanza, A. L., and Palomo, L. M. Simplified High-Frequency Model for AC Drives, *IEEE Conf Indus Electron Soc*, Vol. 02, pp. 1144-1149, 2002.

[6] Naik R., Nondahl, T. A. Nondahl, Melfi M. J., Wang, and J. S. Wang. Circuit Model for iShaft Voltage Prediction in Induction Motors Fed by PWM-Based AC Drives, *IEEE Trans Indus Applic*, Vol. 39, no. 5, pp. 1294-1299, 2003.

[7] Charoy, A. and Dunand, P. Bearing Current Induced by a Power Drive, *Automot Power Electronics*, pp. 01-07, Paris, 2007.

[8] Boglietti A., Cavagnino A., and Lazzari M. Experimental High Frequency Parameter Identification of AC Electrical Motors, *IEEE Trans Indus Applic*, Vol. 43, no. 01, pp. 23-29, 2007.

[9] Akagi H. and Doumot T. An Approach to Eliminating High-Frequency Shaft Voltage and Ground Leakage Current from an Inverter-Driven Motor, *IEEE Trans Indus Applic*, Vol. 40, no. 4, pp. 1162-1169, 2004.

[10] Akagi, H. and Tamura, S. A Passive EMI Filter for Eliminating Both Bearing Current and Ground Leakage Current from an Inverter-Driven Motor, *IEEE Trans Power Electron*, Vol. 21, no. 5, pp. 1459-1469, 2006.

[11] Kempski, A. Capacitively Coupled Discharging Currents in Bearings of Induction Motor Fed from PWM (pulsewidth modulation) Inverters, *J Electrostatics*, vol. 51-52, pp. 416-423, 2001.

[12] Adabi, J., Zare, F., Ledwich G., and Ghosh, A. Leakage Current and Common Mode Voltage Issues in Modern AC Drive Systems, *Australasian Universities Power Engineering Conference, 2007 (AUPEC 2007)*, pp. 01-06, 2007.

[13] Muetze, A. and Binder, A. Calculation of Motor Capacitances for Prediction of Discharge Bearing Currents in Machines of Inverter-Based Drive Systems, *IEEE International Conference on Electrical Machines and Drives*, pp. 264-270, 2005.

[14] Chen, L., Lipo, T. A., and Fitzgerald, D. Measurement and Analysis of Induction Motor Bearing Currents in PWM Inverter Drives, *IEEE Trans Indus Applic*, Vol. 32, pp. 1365-1370, 1995.

[15] Esmaeli A., Jiang B., and Sun, L. Modeling and Suppression of PWM Inverter's Adverse Effects, *1st International Symposium on Systems and Control in Aerospace and Astronautics*, EI&IEEE, pp. 1450-1454, China, 2006.

[16] Arnedo, L. and Venkatesan, K. Pspice Simulation for Conducted EMI and Overvoltage Investigations in a PWM Induction Motor Drive System, *IEEE Workshop on Computers in Power Electronics*, pp. 132-137, 2002.

[17] Boglietti, A. and Carpaneto, E. An Accurate High Frequency Model of AC PWM Drive Systems for EMC Analysis, *IEEE Indus Applic Conf*, Vol. 2, pp. 1111-1117, 2001.

[18] Melly, S. New Output Filter Concept for Power Drive Systems, *Product Marketing*, SCHAFFNER, pp. 01-08, 2002.

[19] Lai, Jih-Sheng, Xundong, H., Pepa, H., Chen, S., and Nehl, T. W. Inverter EMI Modeling and Simulation Methodologies, *IEEE Trans Indust Electron*, Vol. 53, no. 3, pp. 736-744, 2006.

[20] Chen, S., Lipo, T. A., and Fitzgerald, D. Source of Induction Motor Bearing Currents Caused by PWM Inverters, *IEEE Trans Energy Conver*, Vol. 11, no. 1, pp. 25-32, 2006.

[21] Bogel, B., Christiansen, T., Friis, C., Pedersen, N., and Valsson, T. Active EMI Filter for Common Mode Noise Suppression in Three-Phase Drives, PED2001, 2001.

[22] Muttaqi, K. M. and Haque, M. E. Electromagnetic Interference Generated from Fast Switching Power Electronic Devices, *Int J Innovat Energy Sys Power*, Vol. 3, no. 1, pp. 19-26, 2008.

[23] Kerszenbaum, I. Shaft Currents in Electric Machines Fed by Solid-State Drives, *IEEE Industrial and Commercial Power Systems Technical Conference*, pp. 71-79, 1992.

[24] Ramachandran, A., Reddy, M. C., and Moodithaya, R. Minimization and Identification of Conducted Emission Bearing Current in Variable Speed Induction Motor Drives Using PWM Inverter, *Sãdhanã*, Vol. 33, no. 5, pp. 615-628, 2008.

[25] Mäki-Otto, P. *Modeling and Reduction of Shaft Voltages in AC Motors Fed by Frequency Converters*, Doctoral Dissertation, Helsinki University of Technology, 2006.

[26] Boldea, I. and Nasar, S. A. *The Induction Machine Handbook*, CRC Press, New York, USA, 2002.

[27] Fitzgerald, A. E., Kingsley, C., and Umans, S. D. *Electric Machinery*, McGraw-Hill, 6th Edition, New York, USA, 2003.

[28] Mohan, N., Undeland, T. M., and Robbins, W. P. *Power Electronics – Converters, Applications and Design*, John Wiley & Sons Inc., USA, 2003.

[29] Movidrive MDX60B/61B Catalog, SEW Eurodrive, 2005.

[30] Rashid, M. H. *Spice for Power Electronics and Electric Power*, Prentice-Hall, NJ, USA, 1993.

[31] MOVTOOLS Software – Motion Studio V.5.60, SEW-Eurodrive, 2009.

[32] Paton, B. LABVIEW – Fundamentals of Digital Electronics, National Instruments Corporations, USA, 1998.

[33] Get Starting with LabVIEW, National Instruments Corporations, USA, 2007.

[34] LabVIEW Fundamentals, National Instruments Corporations, USA, 2007.

[35] LabVIEW 8.5 Software, LabVIEW Profissional Development System, National Instruments Corporations, USA, 2007.

[36] PSPICE Schematics 9.2 Software, Cadence Design Systems, USA, 2000.

[37] Ribeiro Riehl, R. and Ruppert, E. High Frequency Parameters of Small Three-Phase Induction Motors for Operation with PWM Inverters. *Int Electric Mach Drives Conf – IEMDC*, Vol. 01, pp. 1352-1357, Miami, USA, 2009.

[38] Ribeiro Riehl, R. and Ruppert, E. A Simplified Method for Determining the High Frequency Induction Motor Equivalent Electrical Circuit Parameters to be used in EMI Effect – ICEMS 2007, pp. 1244-1248, Seoul, Korea, 2007.

[39] Ribeiro Riehl, R. and Ruppert, E. High Frequency Capacitances Determination in Three-Phase PWM Inverter-Motor Drive System Using an Alternative Method Applied to EMI Effect Studies. 9°. Congresso Brasileiro de Eletrônica de Potência – COBEP, vol. 01, pp.137-141, Blumenau, Brazil, 2007.

[40] Ribeiro Riehl, R. and Ruppert, E. *A Methodology to Determine the Parasitic Capacitances of the Squirrel-Cage Three-Phase Induction Motors*, PhD Thesis, FEEC-UNICAMP, Campinas, SP, Brazil, 2010.

[41] Ribeiro Riehl, R. and Ruppert, E. Experimental Method for Determining the Parasitic Capacitances of Three-Phase Induction Motor Driven by PWM Inverter. *Sba Controle & Automação* [online]. 2012, Vol. 23, no. 2, pp. 153-163. ISSN 0103-1759

[42] Ribeiro Riehl, R., Covolan Ulson, J.A., Andreoli, A.L., and Ferreira Alves, A. A Simplified Approach for Identification of Parasitic Capacitances in Three-Phase Induction Motor Driven by PWM Inverter, *Electrical Machines and Systems (ICEMS), 2014, 17th International Conference on*, Vol., no., pp. 2550,2554, 22-25 Oct. 2014.

[43] Romanenko, A., Ahola, J., Muetze, A., and Niskanen, V. Study of Incipient Bearing Damage Monitoring in Variable-Speed Drive Systems, *Power Electronics and Applications (EPE'14-ECCE Europe), 2014 16th European Conference on*, Vol., no., pp. 1,10, 26-28 Aug. 2014.

[44] Tallam, R.M., Valdez, C.D.R., Kerkman, R.J., Skibinski, G.L., and Lukaszewski, R.A. Common-Mode Voltage Reduction for Regenerative AC Drives, *Energy Conversion Congress and Exposition (ECCE), 2012 IEEE*, Vol., no., pp. 3301,3308, 15-20 Sept. 2012.

Section 2

Fault Diagnostics

Chapter 4

Health Condition Monitoring of Induction Motors

Wilson Wang and Derek Dezhi Li

Additional information is available at the end of the chapter

http://dx.doi.org/10.5772/61110

Abstract

Induction motors (IMs) are commonly used in various industrial applications. A spectrum synch (SS) technique is proposed in this chapter for early IM defect detection using electric current signals; fault detection in this work will focus on defects in rolling element bearings and rotor bars, which together account for more than half of IM imperfections. In bearing fault detection, the proposed SS technique will highlight the peakedness of the fault frequency components distributed over several fault related local bands. These bands are synchronized to form a fault information spectrum to accentuate fault features. A central kurtosis indicator is proposed to extract representative features from the fault information spectrum and formulate a fault index for incipient IM fault diagnosis. The effectiveness of the developed SS technique is tested on IMs with broken rotor bars and with damaged bearings.

Keywords: Induction motors, Bearing fault detection, Broken rotor bars, Current signal, Spectrum synch analysis

1. Introduction

Induction motors (IMs) are the workhorse of many industries such as manufacturing, mining, and transportation; and more importantly, they consume up to 50% of the generated electrical energy in the world [1]. Due to these facts, a series of R&D activities have been directed, for decades, to improve the performance and efficiency of IMs. For example, in industrial applications, an effective and reliable condition monitoring system is very valuable in the detection of an IM fault at its earliest stage in order to prevent performance reduction and malfunction of the driven machinery. It could also be utilized to schedule predictive maintenance operations without periodically shutting down machines for manual inspections. Maintenance costs can be further reduced (especially for large expensive motors) by quickly identifying the faulty component(s) without inspecting all components in the IM.

As illustrated in Figure 1, a typical IM consists of a stator, a rotor, a shaft, rolling element bearings, and the related supplementary components. IM components could be damaged during operations due to reasons such as impacts, fatigue, insufficient lubrication, aging, and so on. Investigations have revealed that bearing faults account for approximately 75% of small and medium-sized motor defects and 41% of large motor imperfections in domestic and industrial applications [2]. Other IM defects include broken rotor bars (up to 10%), stator winding faults, shaft imbalance, and phase imperfection.

Figure 1. Structure of induction motors: 1-shaft, 2-bearings, 3-rotor, 4-stator.

The traditional IM fault diagnostic method, which is still widely practiced by maintenance crews in industry, relies on human diagnosticians for periodic inspections based on warning signs such as smell, temperature increase, excessive vibration, and increased acoustic noise level. However, these physical symptoms are prone to being contaminated with noise from other sources. The alternative is the use of signal processing techniques for fault detection. Signal processing is a process to extract representative features from the collected signals. Traditional machinery fault detection is based on thermal signals [3], acoustic signals, and vibration signals [4,5]. The local or bulk temperature can be used to diagnose IM defects, however the heat accumulation and progression are slow, which may not be suitable for incipient fault detection. The acoustic noise can indicate IM faults, especially for severe and distributed defects; however the acoustic signal is prone to contamination by background noise such as noise from other machines in the vicinity. Vibration signals can be collected by the use of the related vibration sensors mounted in the vicinity of the IM support bearings. Although vibration signals have relatively high signal-to-noise ratio, the vibration sensors are expensive and require a high degree of installation accuracy. The alternative is to use the stator current signal for analysis, which is non-invasive to the IM structure. In addition, electric current sensors are inexpensive and easy to install [6]. Thus, the proposed research in this work will focus on IM fault diagnosis using stator current signals.

Several motor current signature analysis techniques have been proposed in the literature for fault detection in IMs, mainly for rotors and bearings, which are briefly summarized next.

1.1. Fault detection of IM rotors

Broken rotor bars are common rotor defects that will render asymmetries of an IM rotor. The rotor bar failures can be caused by several factors, such as overheating due to frequent starts under loading, unbalanced thermal load due to air gap eccentricity, manufacturing defects, and corrosion of rotor material caused by chemicals or moisture [7].

Because of the aforementioned reasons, the rotor bar(s) may be fully or partially damaged, which will cause the rotor cage asymmetry and result in asymmetrical distribution of the rotor currents. When a crack forms in a rotor bar, the cracked bar will overheat and tend to break. Then the adjacent bars have to carry higher currents and consequently they become prone to damage, leading to multiple bar fractures. Moreover, the broken parts from the faulty bars may hit the end winding of the motor and cause serious mechanical damage to the IM [8].

The Fast Fourier Transform (FFT) spectral analysis is a commonly used method for rotor bar breakage detection, by examining the characteristic frequency components in the spectrum. For example, Elkasabgy et al. [9] used spectral analysis of IM current signals to detect rotor bar breakage. It has been reported that the IM current signal becomes non-stationary if rotor bars are damaged. However, the FFT is useful for stationary signal analysis only, which lacks the capability of capturing the transitory characteristics such as drifts, abrupt changes, and frequency trends in non-stationary signals. To solve the problem, time-frequency methods, such as short time Fourier transform (STFT), can be used to process small segments of non-stationary signals for broken rotor bar defect detection. For example, Arabaci and Bilgin [10] applied the STFT to detect IM rotor bar faults. However, the STFT cannot provide the information corresponding to different time resolutions and frequency resolutions due to its fixed length windowing functions [11]. To solve this problem, the wavelet transform (WT) can be employed to explore the information associated with different frequency resolutions. For example, Daviu et al. [12] used discrete WT to detect IM broken rotor bars fault. The wavelet packet decomposition (WPD) was used to explore the whole frequency band with high resolution. For example, Sadeghian et al. [13] used WPD to extract features and applied neural networks to diagnose IM rotor bar breakage. Pineda-Sanchez et al. [14] employed polynomial-phase transform to diagnose broken rotor bar fault in time-varying condition. Riera-Guasp et al. [15] extracted broken rotor bar fault features from transient state of IM using Gabor analysis. Although the WPD can explore details of the signal for some advanced signal analysis, it is usually difficult to recognize the real representative features from the map with redundant and misleading information. Akin et al. [16] performed real-time fault detection using the reference frame theory. Soualhi et al. [17] diagnosed broken rotor bar fault through the classification of selected fault features using the improved artificial ant clustering method. Gunal et al. [18] conducted IM broken rotor bar fault diagnosis by using fault indices in the time domain. Nevertheless, the aforementioned techniques only focus on limited fault information, thus their performance may be degraded.

1.2. Fault detection of IM bearings

Rolling element bearings are commonly used not only in electric motors, but also in various types of rotating machinery facilities. As illustrated in Figure 2, a rolling element bearing is a system consisting of an outer ring (usually the fixed ring), an inner ring (usually the rotating ring), a number of rolling elements, and a cage.

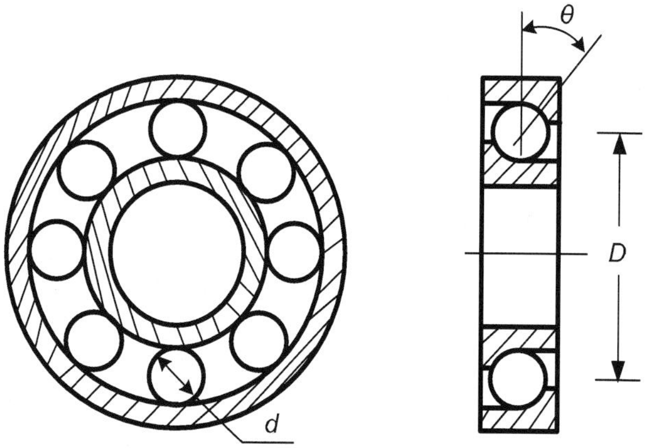

Figure 2. Structure of a rolling element bearing.

Since bearing materials are subjected to dynamic loading, fatigue pitting is the most common defect in bearing components. The bearing defects can occur on the outer race, inner race, and rolling elements. Under normal operating conditions, after the load cycles exceed some threshold, fatigue pitting may occur on the fixed ring race first, and then on the rotating race and rolling elements. Pitting defects not only deteriorate transmission accuracy, but also generate excessive vibration and noise. Other bearing defects, such as scoring and severe wear [7], can be generated by several external causes such as impacts, overloading and overheating, inadequate lubrication, contamination and corrosion from abrasive particles or acid, and improper installation of a bearing, which will introduce excessive misalignment errors.

When a bearing component is damaged, the corresponding characteristic frequencies will be associated with the bearing geometry, rotation speed, and defect location. Suppose the outer race of a bearing is fixed and the inner race rotates with the shaft, which is common case in most applications. The outer race defect characteristic frequency f_{od} inner race defect characteristic frequency f_{id} and rolling element defect characteristic frequency f_{rd} are determined by

$$f_{od} = \frac{N}{2}\left[f_i\left(1 - \frac{d\cos(\theta)}{D}\right)\right] \tag{1}$$

$$f_{id} = \frac{N}{2}\left[f_i\left(1 + \frac{d\cos(\theta)}{D}\right)\right] \tag{2}$$

$$f_{rd} = \frac{D}{2d} f_i\left[1 - \left(\frac{d\cos\theta}{D}\right)^2\right] \tag{3}$$

where f_i is the inner race rotating speed or shaft speed in Hz; d is the diameter of the rolling element; D is the pitch diameter; $\theta$is the contact angle.

When bearing defects occur, these bearing characteristic vibration frequencies f_v (i.e., f_{od} , f_{id}, and f_{rd}) will be modulated with the power supply frequency f_p in the spectrum of stator current signals, because of the air gap eccentricity and load torque variations. Thus, the characteristic stator current frequencies f_c in terms of characteristic vibration frequencies f_v will be calculated by [19]:

$$f_c = \left|f_p \pm mf_v\right|, m = 1, 2, 3,... \tag{4}$$

For IM bearing fault detection, the characteristic stator current frequency components can be used as frequency domain indicators in spectrum analysis [20]. Several techniques have been proposed in the literature for IM bearing fault detection using stator current signals. For example, Devaney and Eren [21] applied IM stator current spectrum analysis for bearing defect detection. FFT can be used to conduct spectrum analysis, so as to detect IM bearing faults under deterministic motor conditions. Similar to the previous discussion regarding broken rotor bar analysis, the WT can be used to catch the transitory characteristics of the signal for IM bearing fault detection. For example, Konar and Vhattopadhyay [22] employed discrete WT to detect IM bearing faults. The WPD can also be employed to explore transient fault information for IM bearing fault detection [23]. Nevertheless, the WPD generates massive non-fault-related information that may mask the fault features in the map, and increase the difficulties in fault detection. Frosini and Bassi [24] used features from stator current signals and IM efficiency for bearing fault detection. Zhou et al. [25] utilized the Wiener filter for noise reduction, so as to detect IM bearing defect. Romero-Troncoso et al. [26] conducted online IM fault detection using information entropy and fuzzy inference. Pineda-Sanchez et al. [27] employed Teager-Kaiser energy operator to enhance fault features to detect IM bearing defect. Nevertheless, these available techniques conduct IM bearing fault detection based on limited fault information rather than comprehensively explore fault features from the time domain, the frequency domain, and the time-frequency domain. Therefore their performance may be degraded.

Typically, the onset of IM faults begins with small imperfections, and propagates to a severe stage as the operation progresses. The severe IM faults will cause machinery malfunction, and even catastrophic failures. Therefore, the detection of IM faults at their earliest stage is of great importance in IM condition monitoring. The IM fault features from stator current signals would be associated with fault size, motor type, supply frequency, load condition, and so on. To date, fault feature extraction from IM current signals, especially associated with bearing defects, still remains a challenging task due to the complex transmission path and environmental noise.

To tackle the aforementioned difficulties, a spectrum synch (SS) technique is proposed in this work to gather fault-related information and generate representative features of IM faults, such as broken rotor bar fault and outer race defect in a bearing. The SS will examine characteristic frequency components as well as their features over their neighborhood local bands, in order to comprehensively highlight fault features, and mitigate the effects of high amplitude outliers. The specific approaches in the proposed SS technique include the following: (1) a synch technique is proposed to form fault information spectrum (FIS) by synchronizing several fault-related local bands, so as to accentuate the fault features and improve the SNR; (2) a central kurtosis technique is suggested to extract fault information from the resulting FIS and generate a fault indicator for incipient IM fault detection. The effectiveness of the proposed SS technique is verified by IM broken rotor bar fault detection and IM bearing fault detection.

The remainder of this chapter is organized as follows: the developed SS technique is described in Section 2. The effectiveness of the proposed diagnostic tool is examined in Section 3 by using two common types of IM fault conditions; finally, some concluding remarks of this study are summarized in Section 4.

2. The spectrum synch technique for IM health condition monitoring

2.1. Theory of spectrum synch analysis

The proposed SS technique is composed of two procedures: local band synch and central kurtosis analysis. The local band synch is used to form the fault information spectrum (FIS) and accentuate fault features. The central kurtosis is suggested to generate fault indices for IM health condition monitoring.

2.1.1. Local band synch

The IM fault characteristic frequency components are distributed over the spectrum, which, however, are usually difficult to recognize due to their low amplitude. To highlight fault features in the spectrum, the FIS is used to enhance the local peakedness of the fault frequency components. Firstly, to mitigate the noise effect in the IM current signal, the spectrum averaging of J data sets φ_j, $j = 1, 2, \ldots, J$, is applied to improve the signal-to-noise ratio (SNR), computed by

$$\Phi = \frac{1}{J}\sum_{j=1}^{J}\log\left(P\left(\varphi_j\right)\right) \tag{5}$$

where Φ is the averaged spectrum over J spectra; $P(\varphi_j)$represents the nonparametric power spectral density (PSD) estimate of the data set [28], given by

$$P\left(\varphi_j\right) = \frac{2}{f_s N}\sum_{i=1}^{N/2+1}\left|F_j\left(i\right)\right|^2 \tag{6}$$

where F_j is the spectrum of φ_j using the Fourier transform (FT); N is the length of φ_j ; and f_s is the sampling frequency.

The fault features are related to fault characteristic frequencies, most of which are masked over the local bands by some other higher level frequency components considered as noise. To tackle this problem, the local bands containing the fault characteristic frequencies are synchronized to reduce the noise effect and protrude the fault frequency components. In each selected local band, the fault frequency component f_c is located in the center of the window, and the width of the local band is selected to properly reveal the peakedness of f_c.

To synchronize the corresponding bands at different locations (frequencies) of the spectrum, the spectrum is transformed from the frequency domain $\Phi(f)$ to discrete point representation $\Phi(d)$. Each frequency f can be represented by its nearest discrete point d. Then, fault characteristic frequency f_c (k) is transformed into a discrete point, $d_c(k)$, whose corresponding frequency is the one closest to f_c, where $k = 1, 2, \ldots, K$, and K is the total number of fault characteristic frequencies considered. Thus, K local bands will be used for this synch operation. The widths of local bands are identical in this work to facilitate the synch operation. Given the bandwidth in frequency f_w the length of the local band in discrete point representation, d_w will be

$$d_w = 2R\left\langle \frac{1}{2} f_w \frac{D_s}{f_s} \right\rangle \tag{7}$$

where f_s is the sampling frequency in Hz, D_s is the discrete point representing f_s, and $R\langle \cdot \rangle$ represents round-off operation. The kth local band ψ_k in the discrete point representation can be determined by

$$\psi_k = \{\Phi(i)\}_{i = d_c(k) - \frac{1}{2}d_w, \ldots, d_c(k) + \frac{1}{2}d_w} - \frac{1}{d_w + 1} \sum_{i = d_c(k) - \frac{1}{2}d_w}^{d_c(k) + \frac{1}{2}d_w} \Phi(i) \tag{8}$$

The ith discrete point in the kth local band ψ_k is denoted as $\psi_{i,k}$, $i = 1, 2,..., d_w+1$; $k = 1, 2,..., K$. The ith discrete points over K local bands $\{ \psi_{i,k} \}$ are sorted in a descending order in terms of their values to generate $\pi_{i,k}$ $k = 1, 2,..., K$; the synchronized band FIS will be

$$g_i = \begin{cases} \frac{2}{(K-1)} \sum_{j=1}^{(K-1)/2} \pi_{i,j} & K \text{ is odd} \\ \frac{2}{K} \sum_{j=1}^{K/2} \pi_{i,j} & K \text{ is even} \end{cases}, i = \frac{1}{2} d_w + 1 \tag{9}$$

$$g_i = \xi \{\pi_{i,j}\}_{j=1,\ 2,\ \ldots,\ K}, i = 1,\ 2, \ldots, \frac{1}{2}d_w, \frac{1}{2}d_w + 2, \ldots, d_w + 1 \tag{10}$$

where ξ $\{\cdot\}$ represents the computation of median value. The top 50% high amplitude center frequency components in local bands are averaged in Equation (9) to enhance the fault feature. The median value calculation in Equation (10) will suppress other frequency components in local bands and reduce the amplitude of outliers. The processing procedures of the proposed FIS formation are illustrated in Figure 3, where the frequency resolution $\Delta f = 0.5$ Hz.

2.2. Central kurtosis analysis

The classic kurtosis is a measure of the peakedness of a signal, computed as $\chi = \frac{\mu_4}{\sigma^4}$, where σ and μ_4 are the standard deviation and the fourth moment of the signal distribution, respectively. The classic kurtosis measures all peaked frequency components of the FIS, which may not properly reveal the fault information. In this work, the fault detection aims to evaluate the peakedness of the center frequency component in the FIS. Therefore, a central kurtosis indicator is proposed to facilitate fault detection. Given the FIS $g(i)$; $i = 1, 2,..., d_w+1$, the relative amplitude of the center frequency components can be determined by

$$v_s = g_s - \xi\{g\} \tag{11}$$

where $g_s = \{g_i\}_{i=d_w/2+1}$ is the center discrete point in the FIS. The amplitude of fault frequency component over synchronized local bands (i.e., FIS), rather than the entire spectrum as in the classical methods, is used to examine fault information.

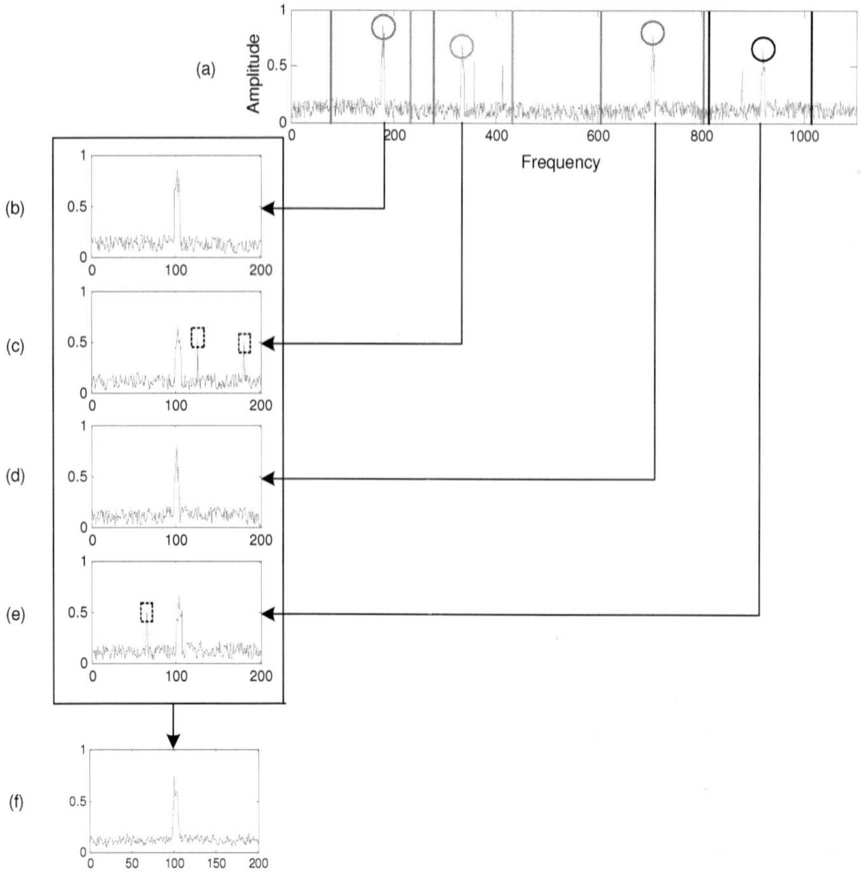

Figure 3. The formulation of FIS: (a) is the original spectrum; (b)-(e) are respective extracted local bands corresponding to the four circled fault frequency components (red, green, pink, and black); (f) is the formulated FIS. The dotted lines in graph (a) represent the boundaries of the local bands; the dashed rectangular boxes represent outliers in the local bands.

The variation of the FIS excluding center frequency component g_s can be evaluated by

$$\sigma_s = E\left\{\left(\tilde{g} - \xi\{\tilde{g}\}\right)^2\right\}^{\frac{1}{2}} \tag{12}$$

where $E\{\cdot\}$ represents the expectation function, and $\tilde{g}$ = $\{g_i, i=1, 2, ..., d_w/2, d_w/2+2, ..., d_w+1\}$.

Then the peakedness of the fault frequency component in the FIS can be measured by the central kurtosis, determined by

$$\chi_s = \begin{cases} v_s^4 / \sigma_s^4 & \text{if } v_s > 0 \\ 0 & \text{if } v_s \le 0 \end{cases} \tag{13}$$

2.3. Implementation of Spectrum Synch Technique

To recapitulate, the proposed SS technique is implemented for IM defect detection in the following steps:

a. Collect J electric current data sets φ_j, $j = 1, 2,..., J$, with the same time delay.

b. Determine the spectrum average Φ over J spectra. Then extract characteristic local bands using Equation (8). Synchronize the local bands to form the FIS using Equations (9) and (10), in order to reduce the noise effect and highlight fault features.

c. Compute the center frequency representative feature using Equation (11), and the variation of the FIS using Equation (12). The fault diagnosis can be performed by analyzing the central kurtosis computed from Equation (13).

3. Performance evaluation

To evaluate the effectiveness of the proposed SS technique for IM fault detection, a series of tests have been conducted for the two common types of IM defects, IM broken rotor bar fault and IM bearing defect, using stator current signals. In rolling element bearings, defect occurs on the race of the fixed ring first since fixed ring material over the load zone experiences more cycles of fatigue loading than other bearing components (i.e., the rotating ring and rolling elements). Correspondingly, this test focuses on incipient bearing defect, or fault on the outer race (fixed ring in this case). The tests are conducted for two power supply frequencies f_p: f_p = 35 Hz and 50 Hz.

3.1. Experimental setup

Figure 4 shows the experimental setup employed in the current work. The speed of the tested IM is controlled by a VFD-B AC speed controller (from Delta Electronics) with output frequency 0.1~400 Hz. A magnetic particle clutch (PHC-50 from Placid Industries) is used as a dynamometer for external loading. Its torque range is from 1 to 30 lb ft (1.356-40.675 N m). The motor used for this research is made by Marathon Electric, and its specifications are summarized in Table 1. The gearbox (Boston Gear 800) is used to adjust the speed ratio of the dynamometer. The current sensors (102-1052-ND) are used to measure different phase currents. A rotary encoder (NSN-1024) is used to measure the shaft speed with the resolution

http://dx.doi.org/10.5772/61110

of 1024 pulses per revolution. Stator current signals are collected using a Quanser Q4 data acquisition board, which are then fed to a computer for further processing.

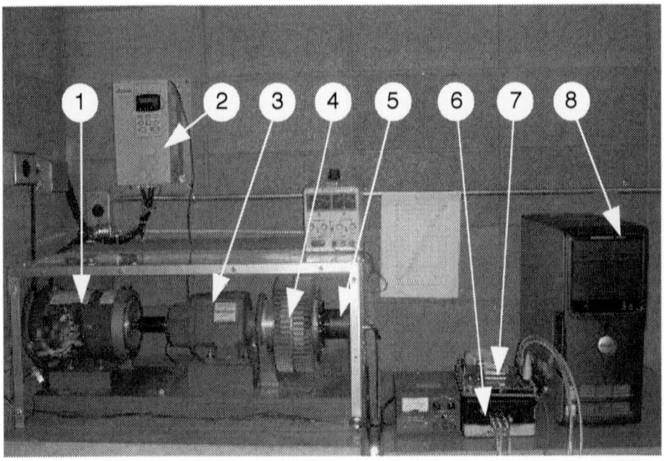

Figure 4. IM experimental setup: (1) tested IM, (2) speed controller, (3) gearbox, (4) load system, (5) encoder, (6) current sensors, (7) data acquisition system, and (8) computer.

Phase	3	Connection	Y
Poles	2	RPM	2850
HP	1/3	Rotor Bars	34
HZ	50	Stator Slots	24

Table 1. Motor specifications.

3.2. Broken rotor bar fault detection

The fault detection of IM broken rotor bar defect is generally based on spectral analysis by inspecting fault-related sideband components in the spectrum:

$$f_{bl} = (1 - 2ks) f_p \tag{14}$$

$$f_{br} = (1 + 2ks) f_p \tag{15}$$

where f_{bl} and f_{br} are the respective left sideband and right sideband of the IM broken rotor bar fault, k = 1, 2,... ; f_p is the power supply frequency of the IM; $s = \frac{n_s - n_a}{n_s} \times 100\%$ is the slip of the IM. n_s (rpm) is the speed of rotating magnetic field, and n_a (rpm) is the shaft rotating speed. In the following tests, an IM containing three broken rotor bars is used to evaluate the proposed SS technique.

To examine the effectiveness of the proposed SS technique, the power spectral density (PSD) based fault detection and the envelope analysis based fault detection are used for comparison. The PSD explores the energy distribution of the data over the spectrum; the envelope analysis performs amplitude demodulation to reveal fault features. In the PSD-based fault detection, the fault index can be represented as,

$$\chi_p = \begin{cases} v_p^4 / \sigma_p^4 & if \;\; v_p > 0 \\ 0 & if \;\; v_p \le 0 \end{cases} \tag{16}$$

The fault index of envelope analysis is given as

$$\chi_e = \begin{cases} v_e^4 / \sigma_e^4 & if \;\; v_e > 0 \\ 0 & if \;\; v_e \le 0 \end{cases} \tag{17}$$

where v_p and v_e represent the averages of the top 50% high amplitude fault frequency components from PSD and envelope analysis respectively; $\sigma_p$and σ_e represent the standard deviations of the entire spectrum band of interests from PSD and envelope analysis respectively; $\chi_p$and χ_e are the respective fault indices from PSD and envelope analysis. Therefore, these two techniques can be used to compare the local band synch method in the proposed SS technique, and the corresponding central kurtosis index.

3.2.1. 35 Hz supply frequency

The first test aims to detect the IM with three adjacent broken rotor bars, 35 Hz power supply frequency, and a half load (50% of rated power). To reduce the noise effect in the spectrum, twenty data sets are collected for spectrum averaging (i.e., J = 20). Other settings are f_s = 65,500 Hz and f_w = 10 Hz. Since 1,024 low-to-high voltage transitions represent one shaft revolution in the encoder signal, the high sampling frequency f_s is chosen to properly capture the encoder signal, so as to accurately estimate shaft speed (i.e., rotor speed). The frequency band [25 Hz, 45 Hz] is used to detect broken rotor fault, because the amplitudes of high order (i.e., k in Equations (14) and (15)) characteristic frequencies are not prominent in the spectrum. Figure 5 illustrates the PSD of a healthy IM (Figure 5a), the PSD of an IM with broken rotor bars (Figure 5b), the envelope analysis of a healthy IM (Figure 5c), and the envelope analysis of an IM with broken rotor bars (Figure 5d), respectively. From Figures 5(b) and5(d), it is seen that the broken rotor bar fault frequency components, although visible, do not prominently protrude in the spectrum. Therefore, a better fault detection technique is needed to extract useful information from multiple characteristic frequency components in the spectrum to generate a more reliable fault index.

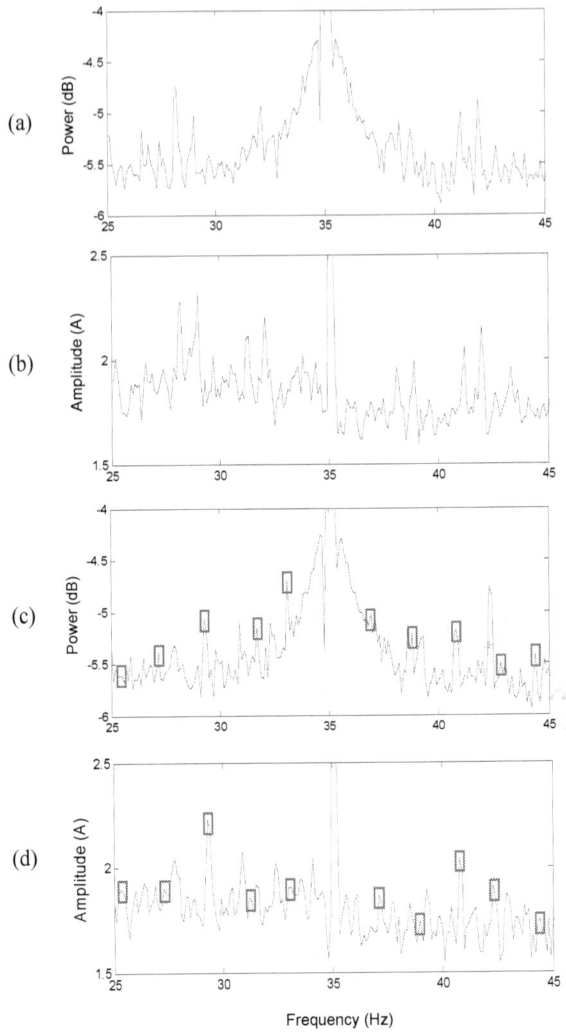

Figure 5. The spectrum average Φ corresponding to: (a) a healthy IM using PSD; (b) an IM with broken rotor bar fault using PSD; (c) a healthy IM using envelope analysis; and (d) an IM with broken rotor bar fault using envelope analysis, at 35 Hz supply frequency and medium-load condition. The red solid rectangular boxes in (b) and (d) highlight fault frequency components.

The FIS, corresponding to a healthy IM (Figure 5a) and a broken rotor bar faulted IM (Figure 5b), are given in Figures 6(a) and 6(b),respectively. The unit of amplitude of the FIS is dB because the local bands are extracted from the PSD logarithmic spectrum. It is seen from Figure 6 that the center frequency component (i.e., synchronized broken rotor bar fault frequency components) in Figure 6(b) has higher relative amplitude than that in Figure 6(a), which is evaluated by the index ν_s in the proposed SS technique. Figure 6(b) has similar spectrum variation as in Figure 6(a), which is examined by the value σ_s in the SS technique. Therefore, the fault information in the FIS can be characterized by the index χ_s using the proposed SS technique.

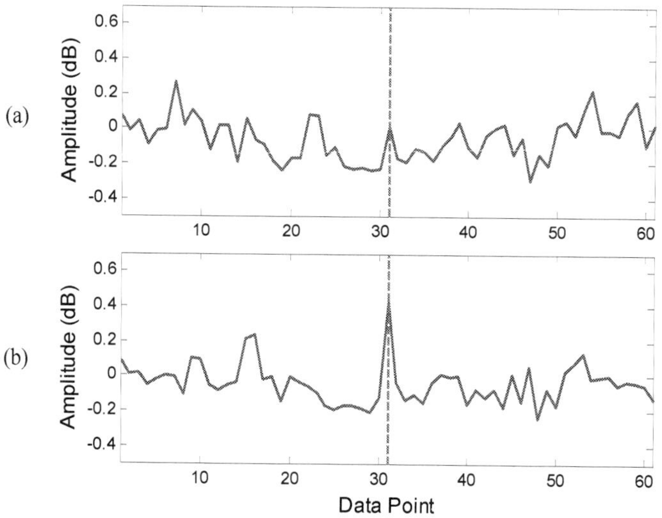

Figure 6. The FIS generated by the SS technique at 35 Hz and a half load condition: (a) from a healthy IM, (b) from an IM with broken rotor bar fault.

The values of IM speed f_r (Hz) and indices corresponding to PSD, envelope analysis and the proposed SS are summarized in Table 2. It is seen from Table 2 that it is difficult to differentiate the IM broken rotor bar faulted condition from the IM healthy condition using envelope analysis, because the values of χ_e corresponding to these two IM conditions are similar. The PSD has a relatively large difference of χ_p of different IM conditions in this case; however, the PSD suffers from interference of non-fault-related high amplitude frequency components and its χ_p values are too small to be relied on. The IM with broken rotor bar defect has considerably larger value of χ_s than that of the healthy IM using the proposed SS technique. Consequently, the proposed SS technique associated with its index χ_s can be used as a fault index for IM broken rotor bar fault detection in the stator current spectrum.

Methods	PSD		Envelope analysis		SS	
Conditions	Healthy	Faulty	Healthy	Faulty	Healthy	Faulty
f_r	34.333	34.343	34.333	34.343	34.333	34.343
$v_p; v_e; v_s$	0.012	0.289	0.132	0.159	0.048	0.472
$\sigma_p; \sigma_e; \sigma_s$	0.751	0.764	0.176	0.161	0.122	0.098
$\chi_p; \chi_e; \chi_s$	6.519e-8	0.021	0.316	0.951	0.024	538

Table 2. Comparisons of central kurtosis indices for IM broken rotor bar fault detection.

3.2.2. 50 Hz supply frequency

The proposed SS technique is then used for IM broken rotor bar fault detection with 50 Hz supply frequency and a half load condition (50% of rated power). Other settings remain the same as in the previous tests. The spectrum of frequency band [35, 65] Hz is used for fault diagnosis. The selected band is shown in Figure 7 using PSD in Figure 7(a) and the envelope analysis in Figure 7(b), respectively. It is seen from Figure 7 that most of the fault frequency components are masked by noise, which cannot be used effectively for reliable fault diagnosis.

The FIS, corresponding to a healthy IM and an IM with broken rotor bar fault are given in Figures 8(a) and 8(b), respectively. The related IM condition indices are summarized in Table 3. It is seen that the relative amplitude of fault frequency component in Figure 8(b) is greater than that in Figure 8(a). From Table 3, the IM with broken rotor bar fault has a larger value of χ_s than that of a healthy IM, which indicates a broken rotor bar fault. The IM health condition with broken rotor bars cannot be differentiated from healthy condition using envelope analysis, associated with its fault index χ_e. Although the PSD index χ_p has a relatively large difference corresponding to different IM conditions, the performance of PSD may be degraded by the interference of non-fault-related high amplitude frequency components and its χ_p values are too small to be relied on. Hence, the proposed SS technique associated with its fault index χ_s can accurately discern the health condition of IMs with broken rotor bar fault under different supply frequencies, when compared to the related two classical methods.

3.3. Incipient bearing defect detection

As mentioned earlier, bearing defects are the most common faults in IMs, which also represent the most challenging task in IM health condition monitoring, especially when using stator current signals [29]. A small dent of diameter approximately 1/16-inch was introduced on the outer race of the bearing to simulate fatigue pitting defect. Whenever a rolling element rolls over the damaged region, impulses are generated, which then excite the resonance frequencies of the IM structures. The vibration-related outer race bearing defect characteristic frequency f_v is given in Equation (1). The corresponding characteristic current frequency f_c can be calculated using Equation (4).

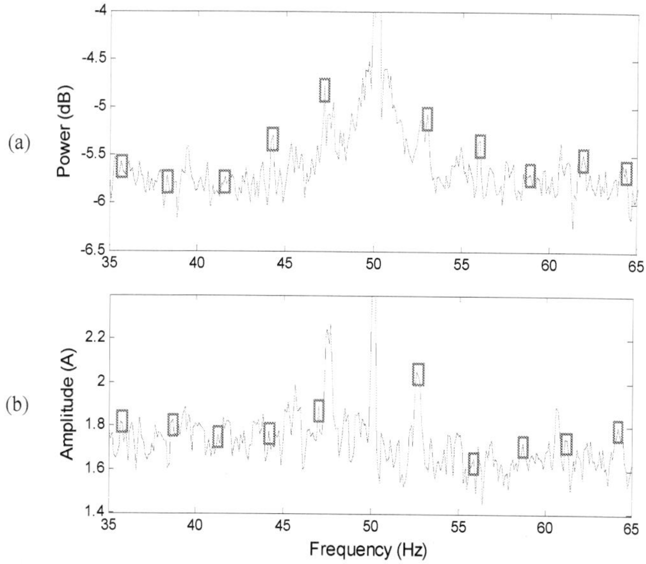

Figure 7. The spectrum Φ for an IM with broken rotor bar fault, 50 Hz supply frequency and half-load condition, using: (a) PSD; and (b) envelope analysis. The red solid rectangular boxes highlight fault frequency components.

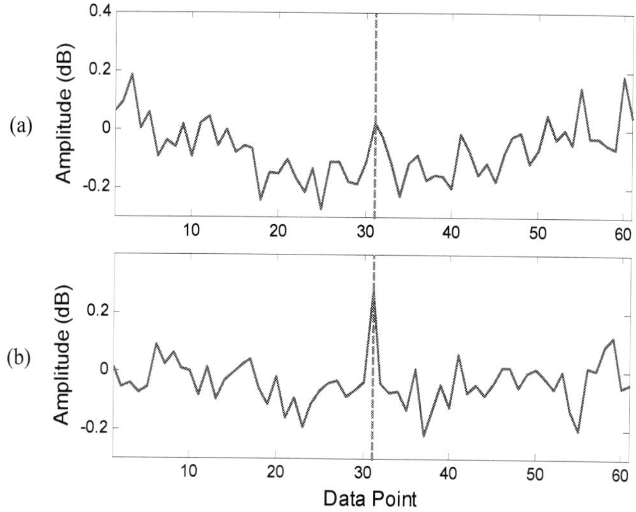

Figure 8. The FIS generated by the SS technique at 50 Hz supply frequency and a half-load level (a) from a healthy IM; (b) from an IM with broken rotor bars.

Methods	PSD		Envelope analysis		SS	
Conditions	Healthy	Faulty	Healthy	Faulty	Healthy	Faulty
f_r	49.180	49.089	49.180	49.089	49.180	49.089
$v_p;v_e;v_s$	0.103	0.195	0.083	0.119	0.092	0.325
$\sigma_p;\sigma_e;\sigma_s$	0.634	0.653	0.149	0.146	0.098	0.068
$\chi_p;\chi_e;\chi_s$	6.966e-4	0.008	0.096	0.441	0.777	522

Table 3. Comparisons of central kurtosis indices for IM broken rotor bar fault detection.

3.3.1. 35 Hz supply frequency

The proposed SS technique is first tested with stator current signals collected from an IM with the outer race defect, 35 Hz power supply frequency, and a light-load (20% of rated power). The settings for the proposed SS technique are selected as $J = 20$, $f_s = 65{,}500$ Hz, $f_w = 10$ Hz, and $f_v = 3.066 f_r$. The high sampling frequency is used to accurately estimate the IM shaft speed. To obtain representative fault features, the frequency band [1000, 2000] Hz is selected for bearing fault detection.

To have a clear view of fault frequency components, the frequency band [1090, 1360] Hz from an IM with outer race bearing defect is shown in Figure 9 using PSD (Figure 9) and envelope analysis (Figure 9b). It is seen that the bearing fault frequency components are difficult to recognize due to the modulation of the signals with other IM frequency components.

The FIS, corresponding to a healthy IM and an IM with the outer race defect, are given in Figures 10(a) and 10(b), respectively. The values of indices corresponding to these three fault detection techniques are summarized in Table 4. It is seen from Figure 10 that the fault frequency component in Figure 10(b) protrudes more significantly than that in Figure 10(a). In Table 4, the fault index χ_s of the IM with faulty bearing is greater than that of a healthy IM using the proposed SS technique, whereas the envelope analysis, associated with its index χ_e, cannot recognize different IM health conditions. The PSD index χ_p generates small values that cannot be relied on. Therefore, the SS technique can be used effectively for IM outer race bearing fault detection in this case, when compared to PSD and envelope analysis.

3.3.2. 50 Hz supply frequency

In this test, the IM supply frequency is set as 50 Hz. The other settings remain the same as in previous test. The frequency band [1000, 2000] Hz of an IM with an outer race bearing defect is used for testing. The band [1400, 1750] Hz is shown in Figure 11, using PSD (Figure 11a) and envelope analysis (Figure 11b), respectively. From Figure 11, the bearing fault frequency components in the spectrum are masked by higher amplitude frequency components unrelated to the bearings, which will degrade the fault detection reliability.

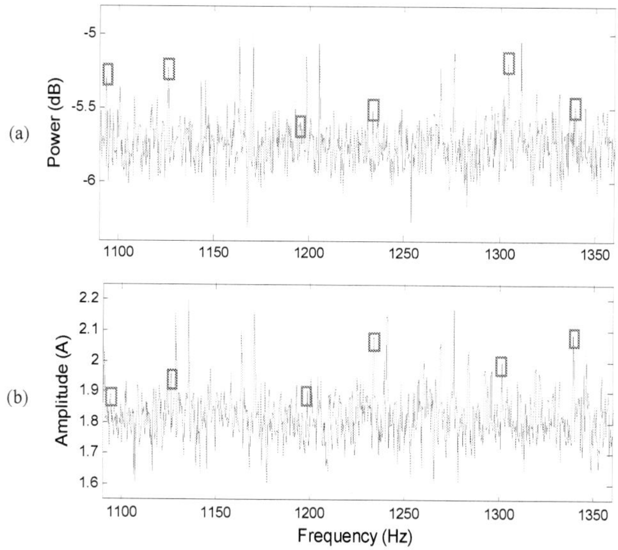

Figure 9. The spectrum average Φ for an IM with outer race bearing defects, 35 Hz supply frequency, and light-load condition, using (a) PSD; and (b) envelope analysis. The rectangular boxes indicate bearing fault frequency components.

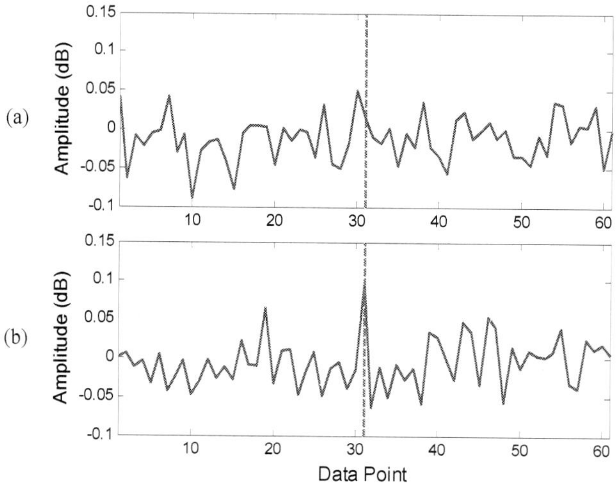

Figure 10. The FIS generated by the SS technique at 35 Hz supply frequency and with a light-load condition (a) from a healthy IM; (b) from an IM with outer race bearing defect.

Methods	PSD		Envelope analysis		SS	
Conditions	Healthy	Faulty	Healthy	Faulty	Healthy	Faulty
f_r	34.746	34.716	34.746	34.716	34.746	34.716
$v_p; v_e; v_s$	0	0.006	0	0	0.023	0.107
$\sigma_p; \sigma_e; \sigma_s$	0.203	0.161	0.102	0.081	0.029	0.029
$\chi_p; \chi_e; \chi_s$	0	1.929e-6	0	0	0.396	185

Table 4. Comparisons of central kurtosis indices for IM outer race bearing fault detection.

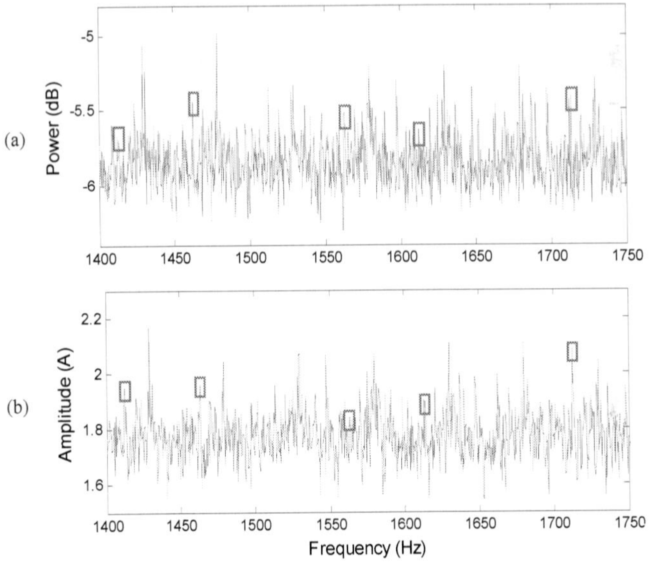

Figure 11. The spectrum average Φ for an IM with outer race bearing defects, 50 Hz supply frequency, and light-load condition, using (a) PSD; and (b) envelope analysis.

The FIS of a healthy IM and an IM with outer race bearing defect are shown in Figure 12. The values of the shaft speed f_r and indices of the PSD, the envelope analysis and the proposed SS technique are listed in Table 5. It is seen that peaked center frequency component can be highlighted in Figure 12(b) than in Figure 12(a). From Table 5, it is seen that the values of χ_s in the proposed SS technique are much greater than that from a healthy IM. Thus, the SS technique and its index χ_s can be used for IM outer race bearing defect detection at different supply frequencies.

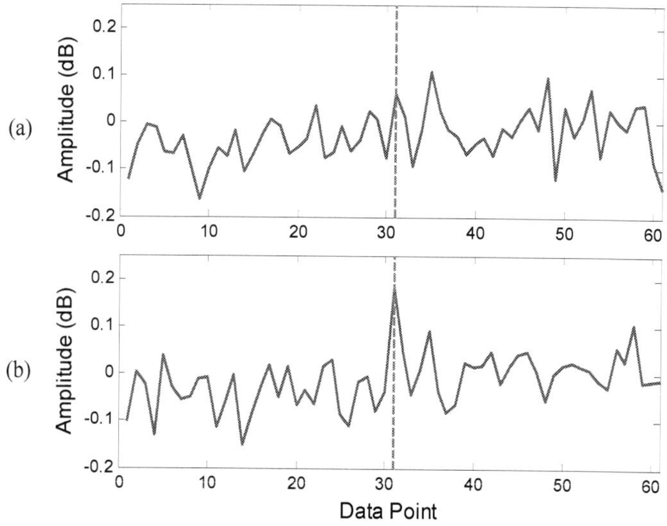

Figure 12. The FIS generated by the SS technique at 50 Hz supply frequency and light-load condition (a) from a healthy IM; (b) from an IM with an outer race bearing defect.

Methods	PSD		Envelope analysis		SS	
Conditions	Healthy	Faulty	Healthy	Faulty	Healthy	Faulty
f_r	49.757	49.705	49.757	49.705	49.757	49.705
$\nu_p;\nu_e;\nu_s$	0	0.041	0.005	0.007	0.090	0.196
$\sigma_p;\sigma_e;\sigma_s$	0.176	0.173	0.087	0.086	0.055	0.052
$\chi_p;\chi_e;\chi_s$	0	0.003	1.091e-5	4.389e-5	7.170	202

Table 5. Comparisons of central kurtosis indices for IM outer race bearing fault detection.

4. Conclusion

A spectrum synch, SS, technique has been proposed in this work for IM fault detection using electric current signals. This research focuses on broken rotor bar fault and outer race bearing fault detection. The local band synch technique is employed to synthesize bearing fault related features to form an FIS to enhance IM defect-related features. A central kurtosis analysis method is proposed to extract some features from the FIS, which are then used to formulate a fault indicator. The effectiveness of the proposed IM fault detection technique is verified using IMs with the bearing defect and the broken rotor bar fault, under different operating conditions. Test results have shown that the proposed SS technique and the related central kurtosis indicator can capture IM defect features effectively and can provide more accurate IM health condition monitoring information. Further research is underway to improve its robustness of the SS technique and adopt it for fault detection in other IM components such as bearings with defects on inner races and rolling elements.

Author details

Wilson Wang[1*] and Derek Dezhi Li[2]

*Address all correspondence to: Wilson.Wang@Lakeheadu.ca

1 Dept. of Mechanical Engineering, Lakehead University, Canada

2 Dept. of Mechanical and Mechatronics Engineering, University of Waterloo, Waterloo, Ontario, Canada

References

[1] Fei R., Fuchs E, Huang H: Comparison of two optimization techniques as applied to three-phase induction motor design. IEEE Transaction on Energy Conversion. 1989; 4:651-660.

[2] Tavner P: Review of condition monitoring of rotating electrical machines. IET Electrical Power Applications. 2008; 2:215-247.

[3] Okoro O: Steady and transient states thermal analysis of a 7.5-kW squirrel-cage induction machine at rated-load operation. IEEE Transactions on Energy Conversion. 2005; 20:730-736.

[4] Widdle R, Krousgrill C, Sudhoff S: An induction motor model for high-frequency torsional vibration analysis. Journal of Sound and Vibration. 2006; 290:865-881.

[5] Tandon N, Yadava G, Ramakrishna K: A comparison of some condition monitoring techniques for the detection of defect in induction motor ball bearings. Mechanical Systems and Signal Processing. 2007; 21:244-256.

[6] Benbouzid M: A review of induction motors signature analysis as a medium for faults detection. IEEE Transactions on Industrial Electronics. 2000; 47:984-993.

[7] Nandi S, Toliyat H, Li X: Condition monitoring and fault diagnosis of electrical motors – a review. IEEE Transactions on Energy Conversion. 2005; 20:719-729.

[8] Thomson W, Fenger M: Current signature analysis to detect induction motor faults. IEEE Industry Application Magazine. 2001; 7:26-34.

[9] Elkasabgy N, Eastham A, Dawson G: Detection of broken bars in the cage rotor on an induction machines. IEEE Transactions on Industrial Applications. 1992; 22:165-171.

[10] Arabaci H, Bilgin O: The detection of rotor faults by using short time Fourier transform. IEEE 15th Signal Processing and Communications Applications. 2007; 1:1-4.

[11] Ambardar A: Analog and Digital Signal Processing. Brooks/Cole Publishing Company; 1999.

[12] Daviu J, Guasp M, Llinares J, Park J, Lee S, Yoo J, Kral C: Detection of broken outer-cage bars for double-Cage induction motors under the startup transient. IEEE Transactions on Industry Applications. 2012; 48:1539-1548.

[13] Sadeghian A, Ye Z, Wu B: Online detection of broken rotor bars in induction motors by wavelet packet decomposition and artificial neural networks. IEEE Transactions on Instrumentation and Measurement. 2009; 8:2253-2263.

[14] Pineda-Sanchez M, Riera-Guasp M, Roger-Folch J, Antonino-Daviu J, Perez-Cruz J, Puche-Panadero R: Diagnosis of induction motor faults in time-varying conditions using the polynomial-phase transform of the current. IEEE Transactions on Industrial Electronics. 2011; 58:1428-1439.

[15] Riera-Guasp M, Pineda-Sanchez M, Perez-Cruz J, Puche-Panadero R, Roger-Folch J, Antonino-Daviu J: Diagnosis of induction motor faults via gabor analysis of the current in transient regime. IEEE Transactions on Instrumentation and Measurement. 2012; 61:1583-1596.

[16] Akin A, Choi S, Orguner U, Toliyat H: A simple real-time fault signature monitoring tool for motor-drive-embedded fault diagnosis systems. IEEE Transactions on Industrial Electronics. 2011; 58:1990-2001.

[17] Soualhi A, Clerc G, Razik H: Detection and diagnosis of faults in induction motor using an improved artificial ant clustering technique. IEEE Transactions on Industrial Electronics. 2013; 60:4053-4062.

[18] Gunal S, Ece D, Gerek O: Induction machine condition monitoring using notch-filtered motor current. Mechanical Systems and Signal Processing. 2009; 23:2658-2670.

[19] Blodt M, Granjon P, Raison B, Rostaing G: Models for bearing damage detection in induction motors using stator current monitoring. IEEE Transactions on Industrial Electronics. 2008; 55:1813-1822.

[20] Tran V, Thobiani F, Ball A, Choi B: An application to transient current signal based induction motor fault diagnosis of Fourier-Bessel expansion and simplified fuzzy ARTMAP. Expert Systems with Applications. 2003; 4:5372-5384.

http://dx.doi.org/10.5772/61110

[21] Devaney M, Eren L: Detecting motor bearing faults. IEEE Instrumentation and Measurement Magazine. 2004; 7:30-50.

[22] Konar P, Vhattopadhyay P: Bearing fault detection of induction motor using wavelet and support vector machines. Applied Soft Computing. 2011; 11:4203-4211.

[23] Lau E, Ngan H: Detection of motor bearing outer raceway defect by wavelet packet transformed motor current signature analysis. IEEE Transactions on Instrumentation and Measurement. 2010; 59:2683-2690.

[24] Frosini L, Bassi E: Stator current and motor efficiency as indicators for different types of bearing faults in induction motors. IEEE Transactions on Industrial Electronics. 2010; 57:244-251.

[25] Zhou W, Lu B, Habetler T, Harley R: Incipient bearing fault detection via motor stator current noise cancellation using Wiener filter. IEEE Transactions on Industry Applications. 2009; 45:1309-1317.

[26] Romero-Troncoso R., et al.: FPGA-based online detection of multiple combined faults in induction motors through information entropy and fuzzy inference. IEEE Transactions on Industrial Electronics. 2011; 58:5263-5270.

[27] Pineda-Sanchez M., et al.: Application of the Teager-Kaiser energy operator to the fault diagnosis of induction motors. IEEE Transactions on Energy Conversion. 2013; 28:1036-1044.

[28] Press W, Teukolsky S, Vetterling W, Flannery B: Numerical Recipes 3rd Edition: The Art of Scientific Computing. Cambridge: Cambridge University Press; 2007.

[29] Ilonen J, Kamarainen J, Lindh T, Ahola J, Kalviainen H, Partanen J: Diagnosis tool for motor condition monitoring. IEEE Transactions on Industry Applications. 2005; 41:963-971.

Chapter 5

Failure Diagnosis of Squirrel-Cage Induction Motor with Broken Rotor Bars and End Rings

Takeo Ishikawa

Additional information is available at the end of the chapter

http://dx.doi.org/10.5772/60964

Abstract

This chapter investigates the diagnosis of not only broken bar but also broken end ring faults in an induction motor. The difference between the broken bars and broken end ring segments is experimentally clarified by the Fourier analysis of the stator current. This difference is verified by two-dimensional finite element (FE) analysis that takes into consideration the voltage equation and the end ring. The electromagnetic field in the undamaged motor and the motor with broken bars and broken end ring segments is analyzed. The effect of the number of broken bars and broken end ring segments on the motor performance is clarified. Moreover, transient response is analyzed by the wavelet analysis.

Keywords: Failure diagnosis, finite element method, motor current signature analysis, wavelet analysis

1. Introduction

Squirrel-cage induction motors are widely used in many industrial applications because they are cost effective and mechanically robust. However, production will stop if these motors fail. Therefore, early detection of motor faults is highly desirable. Induction motor faults are summarized in [1] and [2], and rotor failures account for approximately 10% of the total induction motor failures. Several studies have carried out diagnosis of induction machines using motor current signature analysis (MCSA). For example, Davio et al. proposed a method to diagnose rotor bar failures in induction machines based on the analysis of the stator current during start-up using the discrete wavelet transform (DWT) [3]. Moreno et al. developed an automatic online diagnosis algorithm for broken-rotor-bar detection, which was optimized for single low-cost field-programmable gate array implementation [4]. Guasp et al. proposed a method based on the identification of characteristic patterns introduced by fault components in the wavelet signals obtained from the discrete wavelet transformation of transient stator currents [5]. Kia et al. proposed a time-scale method based on DWT to make the broken-bar fault diagnosis slip independent [6]. Gritli et al. carried out diagnosis of induction machines using DWT under a time-varying condition [7]. However, most of the literature has studied only broken-bar faults, and broken end ring faults have been marginally dealt with. For example, Bouzida et al. dealt with the fault diagnosis of induction machines with broken rotor bars and end ring segment and loss of stator phase during operation using DWT [8]. Concerning the FE analysis of rotor failures in induction motors, several papers have been presented. For example, Mohammed et al. studied the broken rotor bar and stator faults using FE and discrete wavelet analyses [9]. Weili et al. analyzed the flux distribution in the air gap of an induction motor with one and two broken rotor bars [10]. Faiz et al. analyzed the stator current under different numbers of broken bars and different loads of an induction motor [11]. They dealt with broken rotor bars but not a broken end ring.

This chapter addresses not only broken bar but also broken end ring faults. First, we manufacture some rotors with broken bars or end rings [12]. Next, the difference between the broken end ring segments and broken bars is verified by MCSA [12]. The electromagnetic field in the rotor is analyzed to clarify the effect of the number of broken bars and broken end ring segments on the motor performance [13]. Moreover, the stator voltage and current waveforms in a transient response are analyzed by the wavelet analysis.

2. Induction motor with broken rotor and experimental system

Figure 1 shows the photographs of a rotor with broken bars and a broken end ring segments that we have manufactured [13]. Figure 1(a) shows a rotor with one broken bar drilled at the center of the rotor, and Fig. 1(b) shows a rotor with two broken bars drilled at the adjacent aluminum bars. Figure 1(c) shows a rotor with a broken segment in the end ring, which is made by cutting aluminum. Figure 1(d) shows a rotor with two broken segments in the end ring, which are separated by two rotor bars. Figure 1(e) shows a rotor with two broken segments in the end ring whose distance is 45 degrees, that is, 90 electrical degrees. The experimental motor shown in Fig. 2 has the following specifications: 50 Hz, 200 V, 400 W, four poles, and 1,400 min^{-1} speed.

Figure 3 shows the experimental system for the failure diagnosis, which is composed of a 200-V 1.1-kVA 3-A inverter, an induction motor, a torque meter, and a servo motor used as load. Figure 4 shows the developed measurement system using NI cDAQ and Lab VIEW [14]. Lab VIEW is a system design platform and development environment for visual programming language and can be easily used for data acquisition in Microsoft Windows. Figure 4(a) shows the interface part, which is composed of a channel selector, x and y scales, trigger, and two

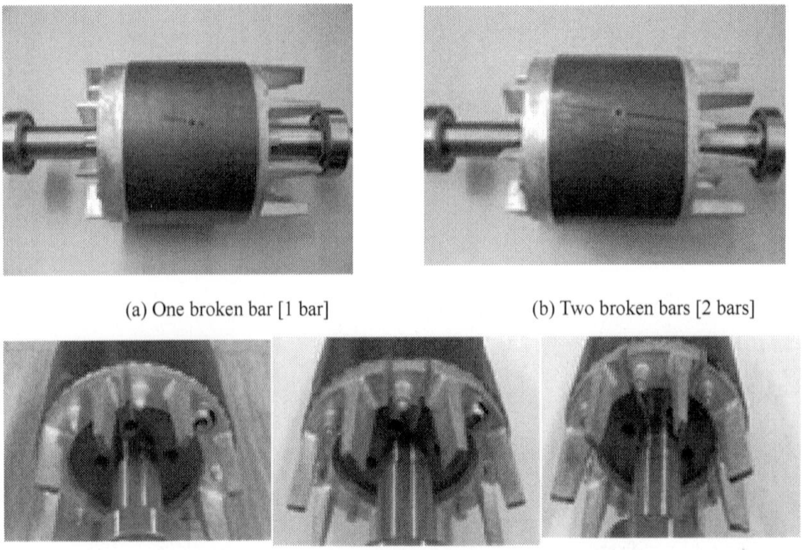

(a) One broken bar [1 bar]

(b) Two broken bars [2 bars]

(c) One broken segment [1 ring]

(d) Two broken segments [2 rings]

(e) Two broken segments (90) [2 rings (2)]

Figure 1. Rotors with broken bars and end rings.

Figure 2. Experimental induction motor.

displays that look like a multi-channel oscilloscope. Figure 4(b) shows a Lab VIEW block, which is composed of blocks for the acquisition and correction of data, trigger detection, and saving of data. Several physical variables, including the stator voltage v, stator current i, torque, and speed are measured by the NI cDAQ and Lab VIEW and are then analyzed by Fourier analysis.

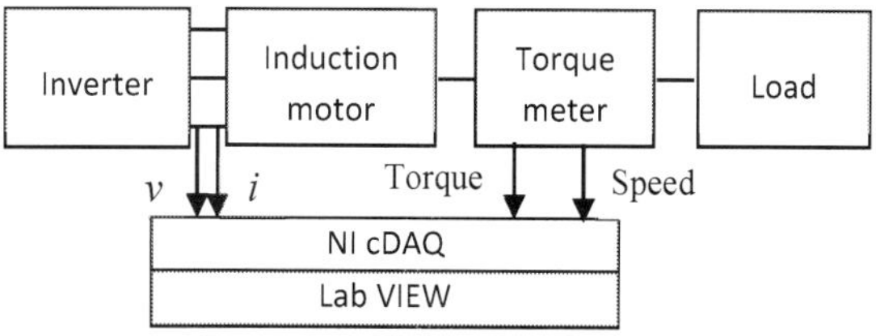

Figure 3. Experimental measurement system.

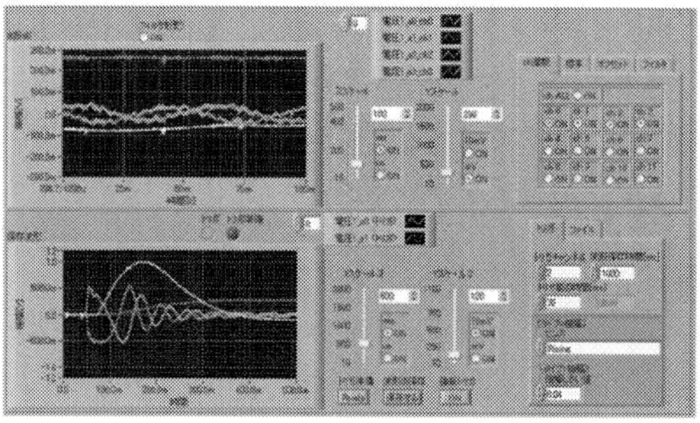

(a) Interface.

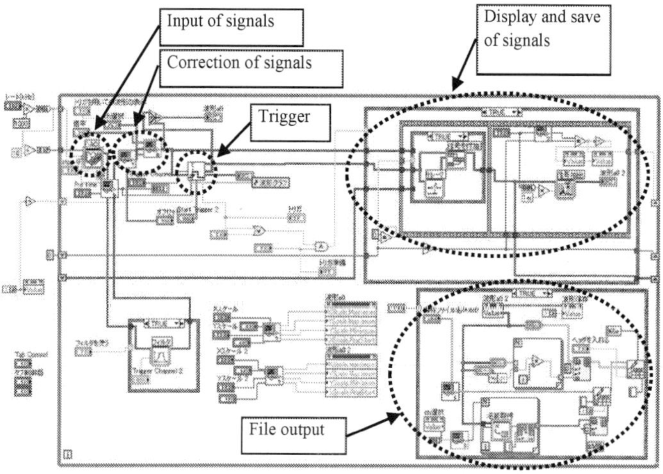

(b) Block diagram using Lab VIEW

Figure 4. Developed measurement system [14].

3. Fourier analysis of the measured data

Figure 5 shows the Fourier analysis of the stator current at a rated speed of 1,400 min^{-1}, where the Fourier component of several rotors is shown at every 0.33 Hz to easily compare the broken situations. Here, 1 bar, 2 bars, 1 ring, 2 rings, and 2 rings (2) mean one broken bar [see Fig. 1(a)], two adjacent broken bars [Fig. 1(b)], end ring broken at one position [Fig. 1(c)], end ring broken at two positions separated by two rotor bars [Fig. 1(d)], and end ring broken at two positions separated by five rotor bars, that is, 90 electrical degrees [Fig. 1(e)], respectively.

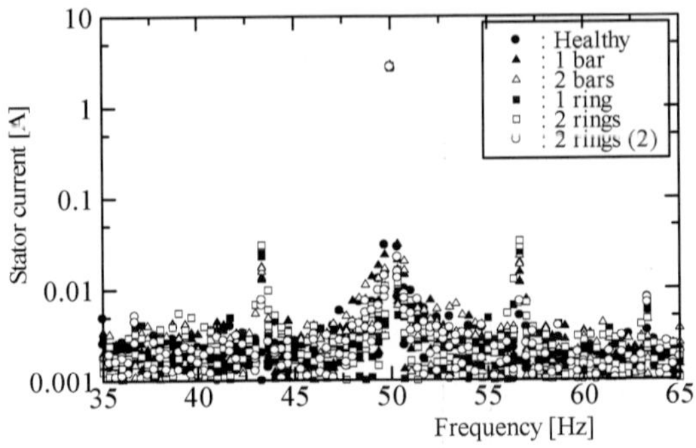

(a) Fourier analysis.

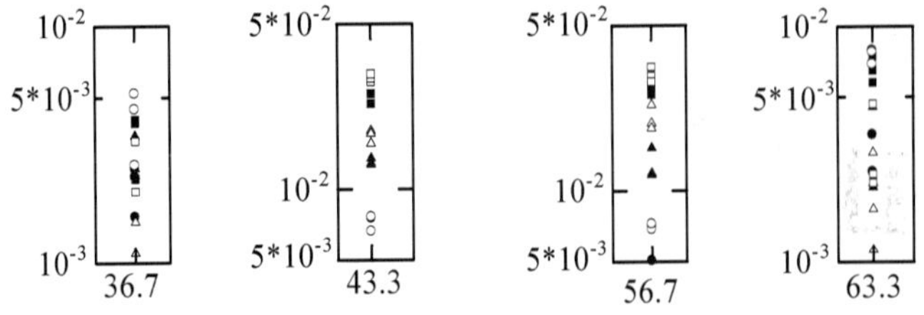

(b) Enlarged around 36.7 Hz, (c) 43.3 Hz, (d) 56.7 Hz, (e) 63.3 Hz

Figure 5. Fourier analysis of the measured stator current at the rated speed.

In Fig. 3, the inverter rating is 1.1 kVA, which is a sufficient capacity for the 400-W experimental induction motor. Therefore, the Fourier analysis of the stator voltage did not include the $(1 \pm 2s)f$ Hz components, where s is the slip and f is the fundamental supply frequency. Because the servo motor is a synchronous motor with surface permanent magnets, the torque ripple is very small. Therefore, the effect of the inverter source and the load equipment on this measurement was very small. However, because the measured results are sensitive to the experimental setup, the experiment was performed three times. We expected a slight difference in the mechanical loss because the connection of the motor, the torque meter, and the servo motor could not be perfect even under the same situation. To take this slight difference into account, the experimental system was set up for every measurement. Therefore, three sets of measured results were slightly different.

We found 50 ± 6.67 Hz components, that is, $(1 \pm 2s)f$ Hz as expected [15]. Let us explain the frequency of $(1 \pm 2s)f$ Hz. When the rotor is running at slip s and the supply frequency is f, the frequency of the forward current in the rotor is sf. If the rotor has a defect, such as a broken bar or broken end ring, a backward current frequency of $-sf$ can flow in the rotor. The forward current frequency sf in the rotor is considered as frequency f at the stator. Because the rotor is running at $(1 - s)f$, the frequency recognized at the stator is $sf + (1 - s)f = f$. In contrast, the backward current frequency of $-sf$ is considered as a stator frequency of $(1 - 2s)f$ because $-sf + (1 - s)f = (1 - 2s)f$. Then, a torque with a frequency of $2sf$ is developed because the stator current has two frequencies f and $(1 - 2s)f$. The torque produces a vibration of $2sf$, and thus, the rotor speed becomes $(1 - s)f \pm 2sf$. This speed vibration and the forward current produce a stator current frequency of $(1 - s)f \pm 2sf + sf = (1 \pm 2s)f$.

In each fault, the components at 50 - 6.67 Hz, that is, $(1 - 2s)f$ Hz, are approximately the same as those at 50 + 6.67 Hz, that is, $(1 + 2s)f$ Hz. Table 1 shows the average components at $(1 \pm 2s)f$ Hz. The order of the components for the different faults is as follows: [2 rings] > [1 ring] > [2 bars] > [1 bar] > [2 rings (2)] > [healthy]. Therefore, we have experimentally clarified that we can detect the fault of the rotor end ring segments and rotor bars from the $(1 \pm 2s)f$ Hz component of the stator current at the rated speed. In contrast, it is difficult to detect the failure type from the Fourier analysis of the stator current at $(1 \pm 2s)f$ Hz because no order of the components relative to the failure type is available.

	43.33 Hz	50 Hz	56.67 Hz
Healthy	0.0010 A	2.818 A	0.0009 A
1 bar	0.0135 A	2.755 A	0.0119 A
2 bars	0.0176 A	2.890 A	0.0193 A
1 ring	0.0286 A	2.872 A	0.0258 A
2 rings	0.0306 A	2.725 A	0.0333 A
2 rings (2)	0.0076 A	2.760 A	0.0072 A

Table 1. Fourier components of the measured stator current.

Figure 6 shows the Fourier analysis of the stator current under a no-load condition. The rotating speed of 1,495 min^{-1} was almost the same. We also find 50 ± 0.33 Hz components, that is, $(1 \pm 2s)f$ Hz components. However, because no order of the components relative to the failure types is available, it is difficult to detect the failure type by Fourier analysis of the stator current under the no-load condition.

Figure 7 shows the Fourier analysis of the torque at the rated speed of 1,400 min^{-1}. We find 6.67 Hz components, that is, $2sf$ Hz. The order of the components for the $2sf$ for different faults was not clear. For the end ring fault, the order of the components is as follows: [2 rings] > [1 ring]

http://dx.doi.org/10.5772/60964

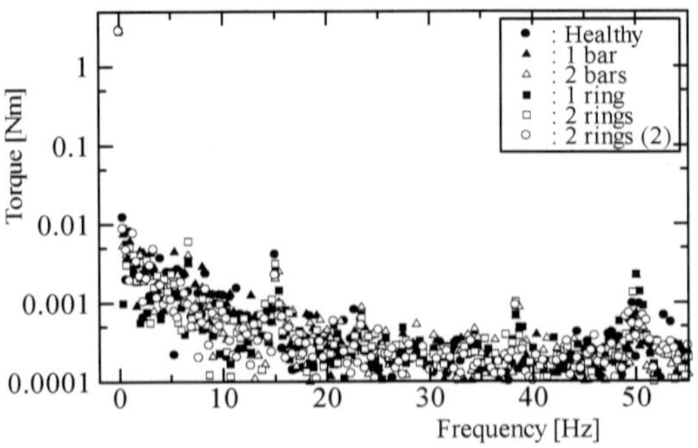

(a) Fourier analysis

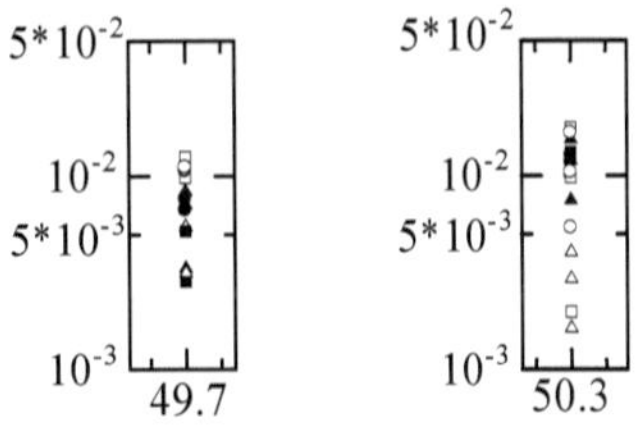

(b) Enlarged around 49.7 Hz (b) Enlarged around 50.3 Hz

Figure 6. Fourier analysis of the measured stator current at no load.

> [2 rings (2)]. However, the component of [2 rings (2)] is approximately the same as that of the healthy motor. For the bar fault, the component of [2 bars] was approximately the same as that of [1 bar]. Therefore, it is difficult to detect the difference in the faults of the rotor end rings and rotor bars by Fourier analysis of the torque even at the rated speed.

4. Simulation of induction motor with broken rotor bars and broken end ring segments

4.1. Analysis method

The experimental motor has rotor skew of one slot pitch. Although a three-dimensional FE analysis is necessary to consider the rotor skew, it is very time consuming. This study calculates the electromagnetic field in the motor using a two-dimensional FE method, which considers the voltage equation and the rotor end ring. The cross section of the motor is shown in Fig. 8. The stator has 36 slots, and the rotor has 44 slots. The following assumptions have been made:

1. Two-dimensional analysis is employed, and the skew in the rotor is ignored.

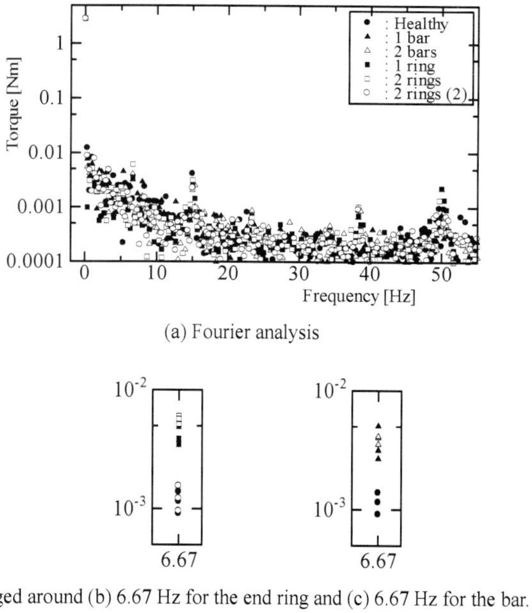

Figure 7. Fourier analysis of the measured torque at the rated speed.

2. Rotor bars and end ring are insulated from the rotor core, and no current flows from the rotor bars to the rotor core.

3. The rotating speed is constant.

4. The supply voltage is assumed to be sinusoidal.

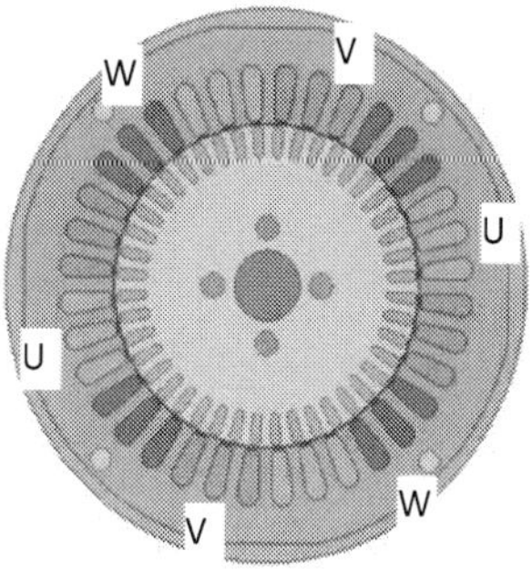

Figure 8. Cross section of the experimental induction motor.

http://dx.doi.org/10.5772/60964

Although the motor is fed by a pulse width modulation (PWM) inverter, the Fourier components of the measured stator current around the switching frequency differ very slightly among the rotor fault types. Therefore, the PWM inverter does not affect the harmonic components of the stator current.

Figure 9 shows the FE analysis region and the connection of the end ring segments in the rotor where 44 bars are included in the FE analysis. Because the end ring is connected to each rotor bar, it is represented by 44 conductor segments whose resistance is R. For example, $R_{1\text{-}2a}$ represents the resistance of the end ring segment between bars 1 and 2, and its value is expressed by

$$R = \rho \frac{l}{S} \tag{1}$$

where of aluminum is. l is the length of conductor segment in the circumferential direction, and S is the area of cross section. Fins connected to the end ring for cooling are neglected. When rotor bar 1 is broken, its conductivity is set to zero. When part of the end ring is broken between bars 1 and 2, resistance $R_{1\text{-}2a}$ is set to infinity. The FE analysis was conducted for 3 s with 6,750 steps, where the number of nodes and elements were 34,909 and 60,866, respectively.

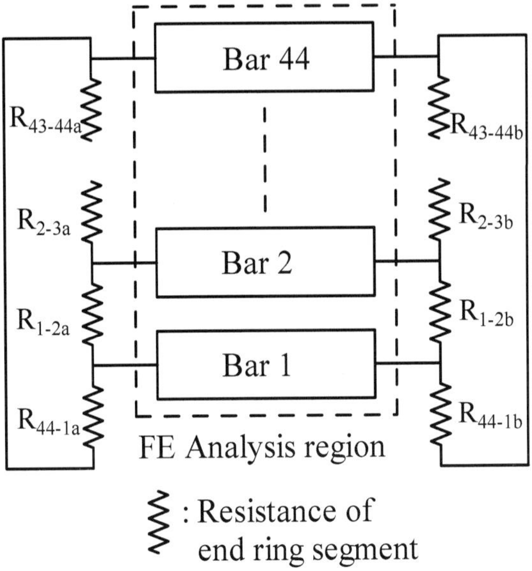

Figure 9. FE analysis region and end ring segments.

4.2. Analysis results

Figure 10 shows the Fourier analysis of the calculated stator current at the rated speed of 1,400 min^{-1}. We find the $(1 \pm 2s)f$ and $(1 \pm 4s)f$ components. Table 2 shows the calculated results of the Fourier analysis at the fundamental and $(1 \pm 2s)f$ components. By comparing these results with those in Table 1, we find that the calculated fundamental components are slightly smaller than the measured ones, and the calculated $(1 \pm 2s)f$ components are larger than the measured ones. We believe that these differences are attributed to the assumption where the skew is not taken into account in the calculation. We also verify from the calculated results that the $(1 \pm 2s)f$ components of the end ring faults are larger than those of the bar faults and that the $(1 \pm 2s)f$ component of the end ring broken at two positions separated by electrical radian is smaller than that of the other faults.

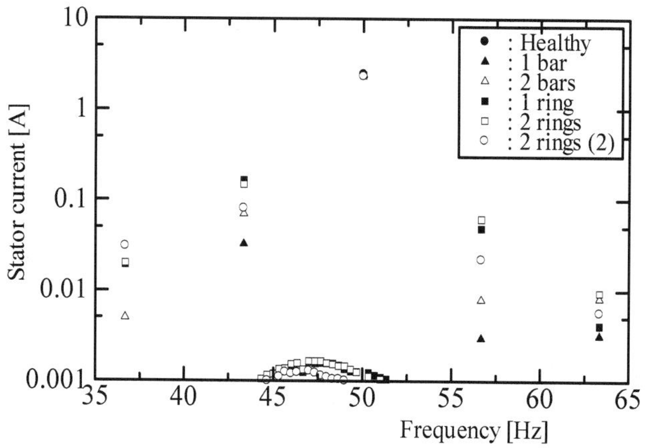

Figure 10. Fourier analysis of the calculated stator current.

	43.33 Hz	50 Hz	56.67 Hz
Healthy	0.0002 A	2.455 A	0.0003 A
1 bar	0.0326 A	2.434 A	0.0030 A
2 bars	0.0701 A	2.411 A	0.0079 A
1 ring	0.1637 A	2.390 A	0.0488 A
2 rings	0.1477 A	2.374 A	0.0618 A
2 rings (2)	0.0807 A	2.335 A	0.0222 A

Table 2. Fourier components of the measured stator current.

http://dx.doi.org/10.5772/60964

4.3. Electromagnetic field in the motor calculated by FEM

An example of the magnetic flux and eddy current distribution in the healthy motor under the rated speed with rated-load condition is shown in Fig. 11. Because this motor is a four-pole machine, the magnetic flux distribution is periodic in every one-fourth region, that is, in every nine stator slots. Here, we denote the number of stator slots in the same group of magnetic flux lines as *NS*. Then, the *NS* for each pole is 9, 9, 9, and 9, and there is no distortion in the magnetic flux distribution.

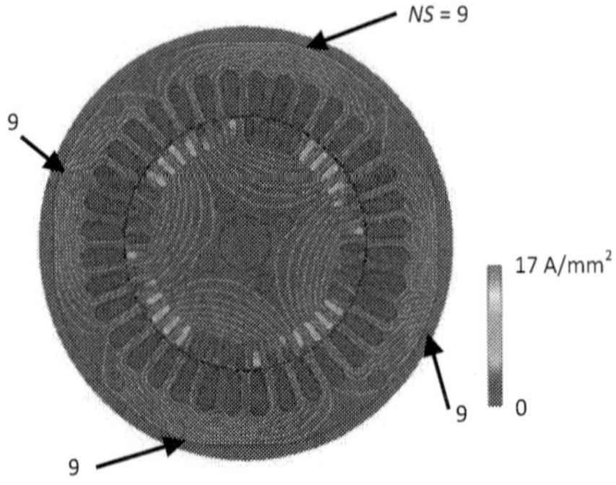

Figure 11. Distribution of the magnetic flux and eddy current in the healthy motor.

Figure 12 shows the distribution of the magnetic flux and eddy current in the motor with two broken bars under different rotor positions, namely, that where the magnetic flux does not pass through the broken bars [Fig. 12(a)] and that where it passes through the broken bars [Fig. 12(b)]. In Fig. 12(a), the eddy current distribution in the rotor bars is approximately the same as that in the healthy motor shown in Fig. 11, and the *NS* under each pole is approximately 8.5, 9.5, 8.5, and 9.5, indicating that magnetic flux distortion occurs. On the other hand, Fig. 12(b) shows a rotor bar where the eddy current density is very high, and the *NS* for each pole is approximately 9, 9, 10, and 8. These magnetic distortions, shown in Figs. 12 (a) and (b), are repeated.

Figure 13 shows the distribution of the magnetic flux and eddy current in the motor with a broken end ring segment under different rotor positions. No rotor bar exhibits a very high eddy current density, and the number of stator slots included in the flux lines for each pole is different. The *NS* for each pole is approximately 8.5, 9, 9, and 9.5 in Fig. 13(a) and approximately 8, 8.5, 10, and 9.5 in Fig. 12(b). These magnetic flux distortions are repeated. As explained in Chapter 3, these magnetic distortions produce a stator current frequency of $(1 \pm 2s)f$.

4.4. Effect of the number of broken bars and end ring segments

Next, we discuss the effect of the number of broken bars and broken end ring segments on the motor performance. Figure 14 shows the Fourier components of the stator current and torque

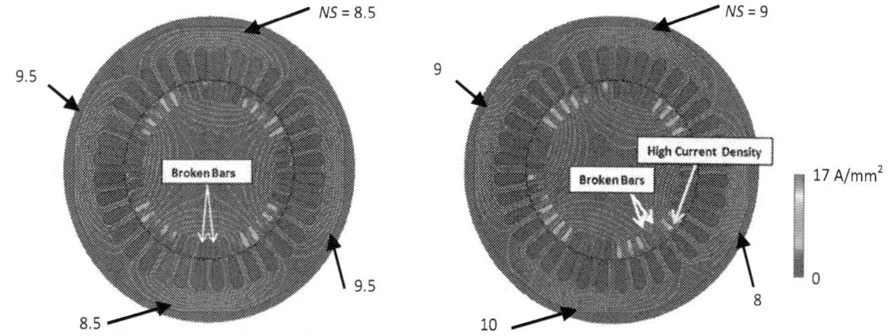

(a) Magnetic flux lines do not pass through the position of the broken bars

(b) Magnetic flux lines pass through the position of the broken bars

Figure 12. Distribution of the magnetic flux and eddy current in the motor with two broken bars under different rotor positions.

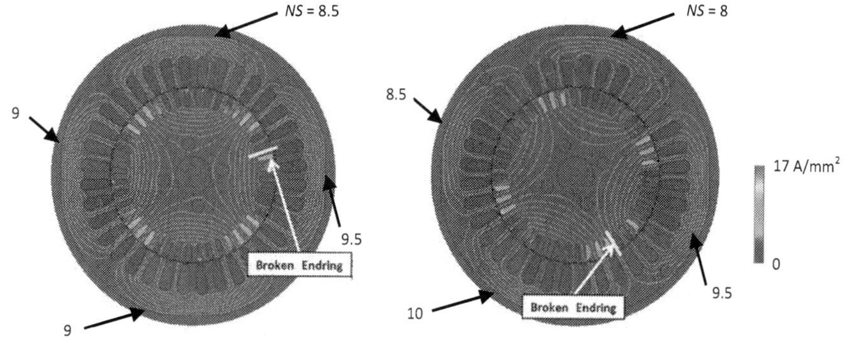

(a) Magnetic flux lines do not pass through the position of the broken end

(b) Magnetic flux lines pass through the position of the broken end ring

Figure 13. Distribution of the magnetic flux and eddy current in the motor with a broken end ring segment under different rotor positions.

for different numbers of broken bars. The fundamental components of the stator current and the average torque decrease, and the $(1 \pm 2s)f$ component of the stator current and the $2sf$ component of the torque increase when the number of broken bars increases. Figure 15 shows the Fourier components of the stator current and torque for different numbers of end ring segments. The fundamental components of the stator current and the average torque also decrease, and the $(1 \pm 2s)f$ component of the stator current and $2sf$ component of the torque approximately increase when the number of broken end ring segments increases. Tables 1 and

2 show that when the number of faults is small, the components for the different faults appear in the following order: [2 rings] > [1 ring] > [2 bars] > [1 bar] > [2 rings (2)] > [healthy]. In contrast, the comparison of Figs. 14 and 15 shows that the $(1 \pm 2s)f$ component of the stator current for the broken bars is larger than that for the broken end ring when the number of faults is greater than three.

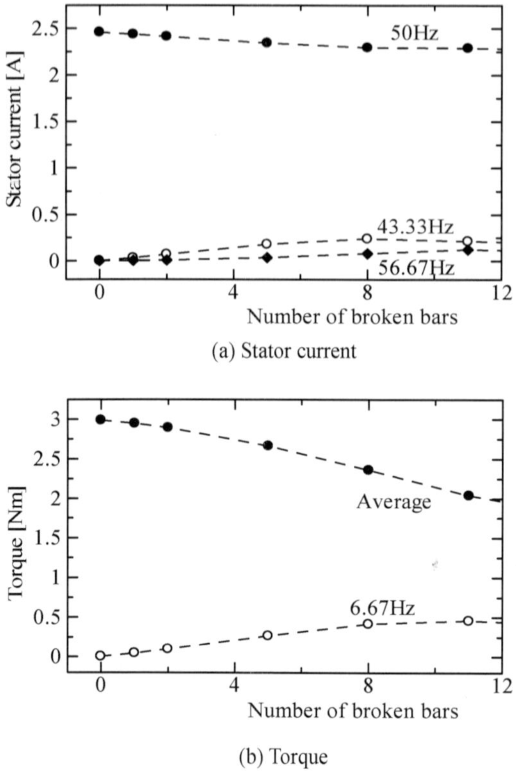

Figure 14. Stator current and torque for different numbers of broken bars.

5. Wavelet analysis

We have discussed about the failure diagnosis of broken end ring segments and broken bars in induction motor at the steady state using the Fourier analysis. In this section, the transient performance of an inverter-fed induction motor is discussed by using the wavelet analysis. There are two kinds of wavelet transform; continuous and discrete ones.

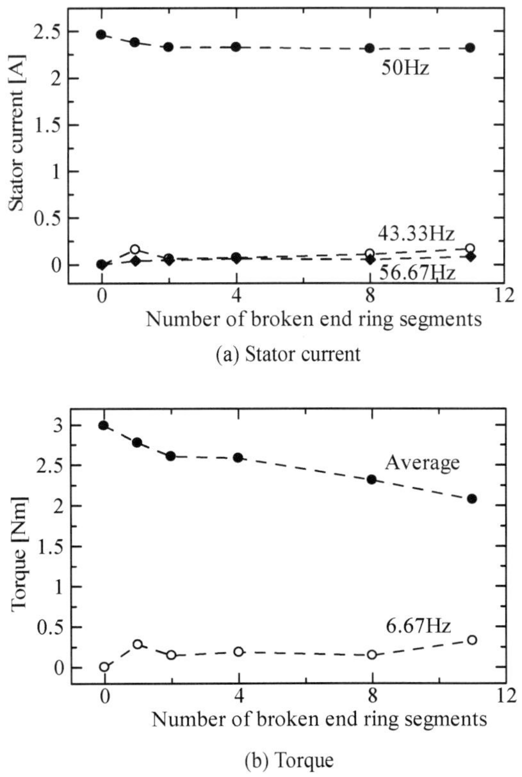

Figure 15. Stator current and torque for different numbers of broken end ring segments.

5.1. Continuous wavelet analysis

Let us make a brief introduction of continuous wavelet transform. Figure 16 shows the waveform of a signal and its wavelet analysis, which shows equipotential lines in the frequency and time plane. Although there are several kinds of Wavelet function – Morlet, Paul, and Derivative of Gaussian – Fig. 16 is the result of using the Morlet function, where the number of waves is 30. We can find high value region around 100 Hz and from 0.3 to 0.6 s and around 400 Hz and from 0.4 to 0.7 s.

We investigate transient response of an inverter-fed induction motor, where the control strategy is an open loop and the motor has no-load. The step responses of the stator voltage, stator current, and motor speed were measured when a start signal was input to the inverter. Figures 17 and 18 show the Wavelet analysis of the stator current i_u and stator voltage V_{uv} using the Morlet function, where the number of waves is two, for the healthy motor and for the motor with two broken bars. We can find a very slight difference in the Wavelet analysis of the stator

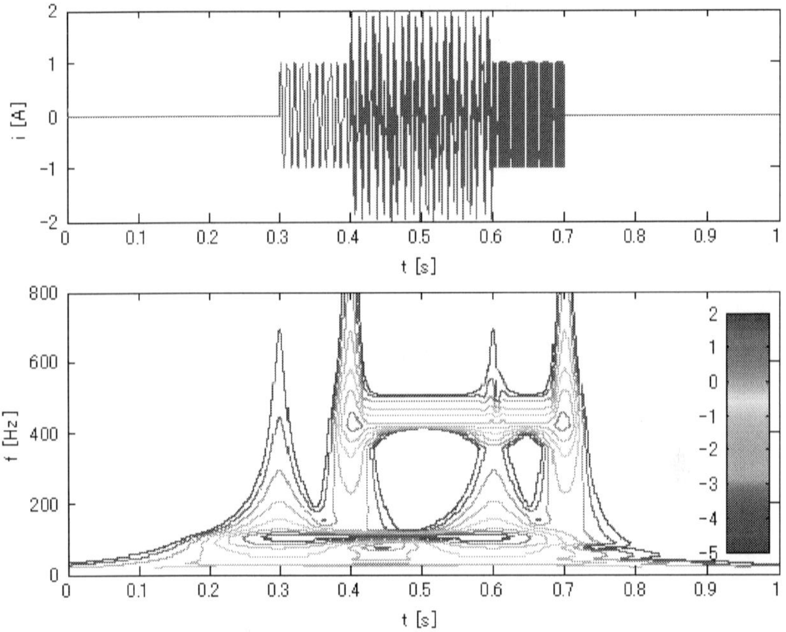

Figure 16. An example of the continuous wavelet analysis using the Morlet mother function.

current between the healthy rotor and two broken bars. There are large components at high frequency before the motor starts, namely, time is from 0 to 0.05 s. As discussed before, the rotor fault produces the components of $(1 \pm 2s)f$. The component becomes 150 Hz when the rotor is stopping, that is, $s = 1$. It is shown that there is no difference in the stator voltage. Therefore, it is difficult to distinguish the rotor defect by using the continuous wavelet transform of the stator current in the starting operation with no-load condition.

5.2. Discrete wavelet analysis

The discrete wavelet transform of a signal is calculated by passing it through a series of filters. As it is well known, the use of wavelet signals, that is, approximation and high-order details, resulting from discrete wavelet transform constitutes an interesting advantage because these signals act as filters. Moreover, the computational time of discrete wavelet transform is much shorter than that of continuous wavelet transform. Figure 19 shows the discrete wavelet analysis for the same signal as shown in Fig. 16. It is found that the component of 100 Hz appears in d6 and signal of 450 Hz component appears in d4 detail.

Table 3 shows frequency bands by decomposition in multi-levels. Figures 20 and 21 show discrete wavelet signals of stator current i_u of the motor with the healthy rotor and two broken bars. It is shown that there are large components of d7 at $t = 0.05$ s, and a large component of

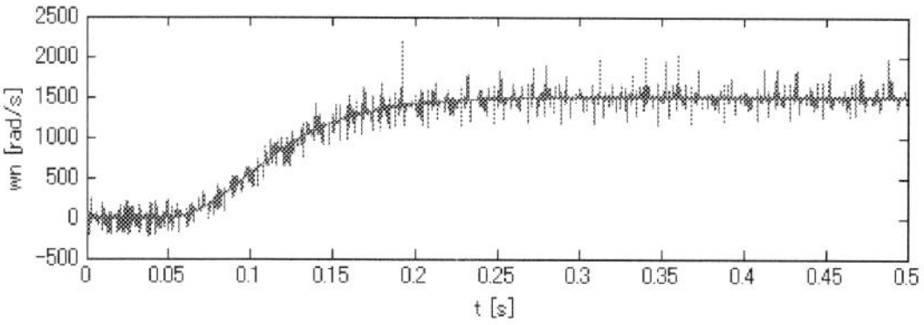

(a) Speed response.

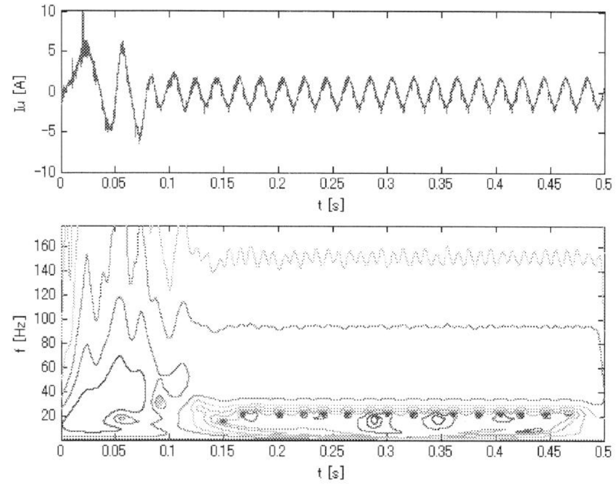

(b) Stator current.

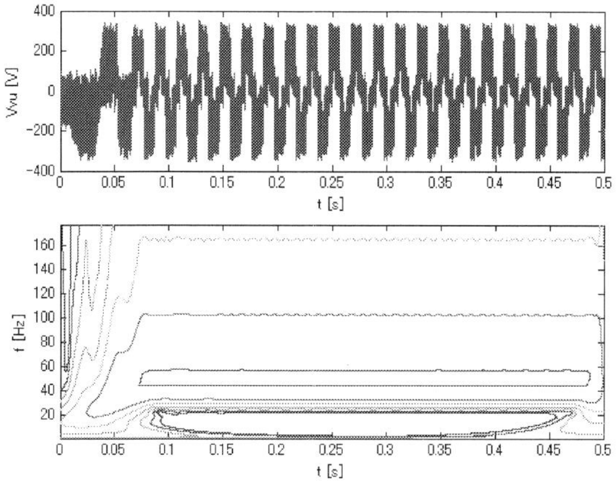

(c) Stator voltage.

Figure 17. Continuous wavelet analysis of healthy rotor.

http://dx.doi.org/10.5772/60964

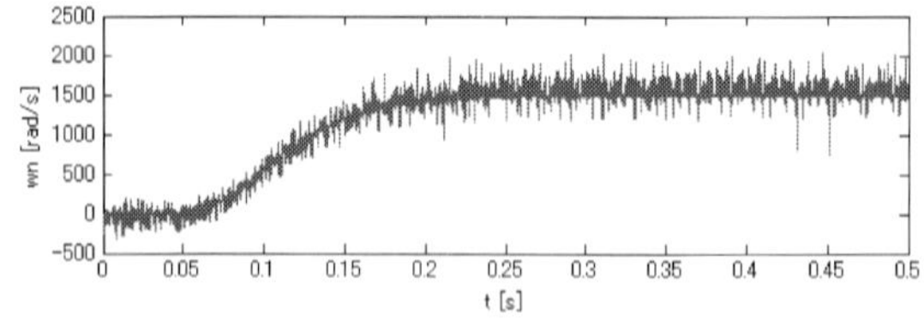

(a) Speed response.

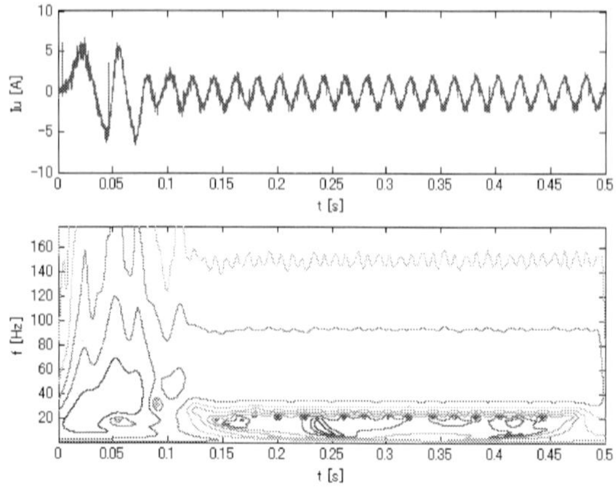

(b) Stator current.

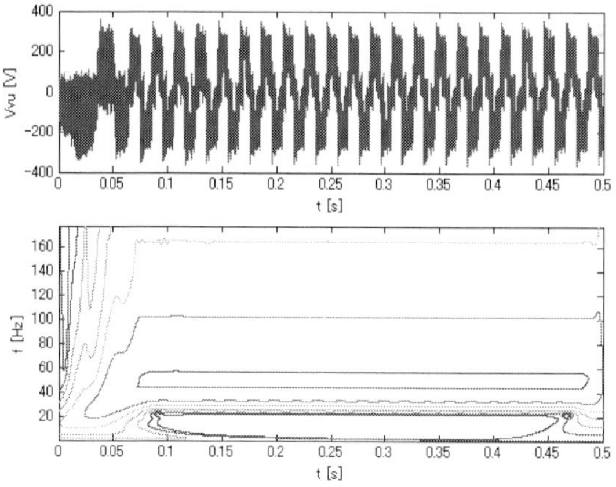

(c) Stator voltage.

Figure 18. Continuous wavelet analysis of the motor with two broken bars.

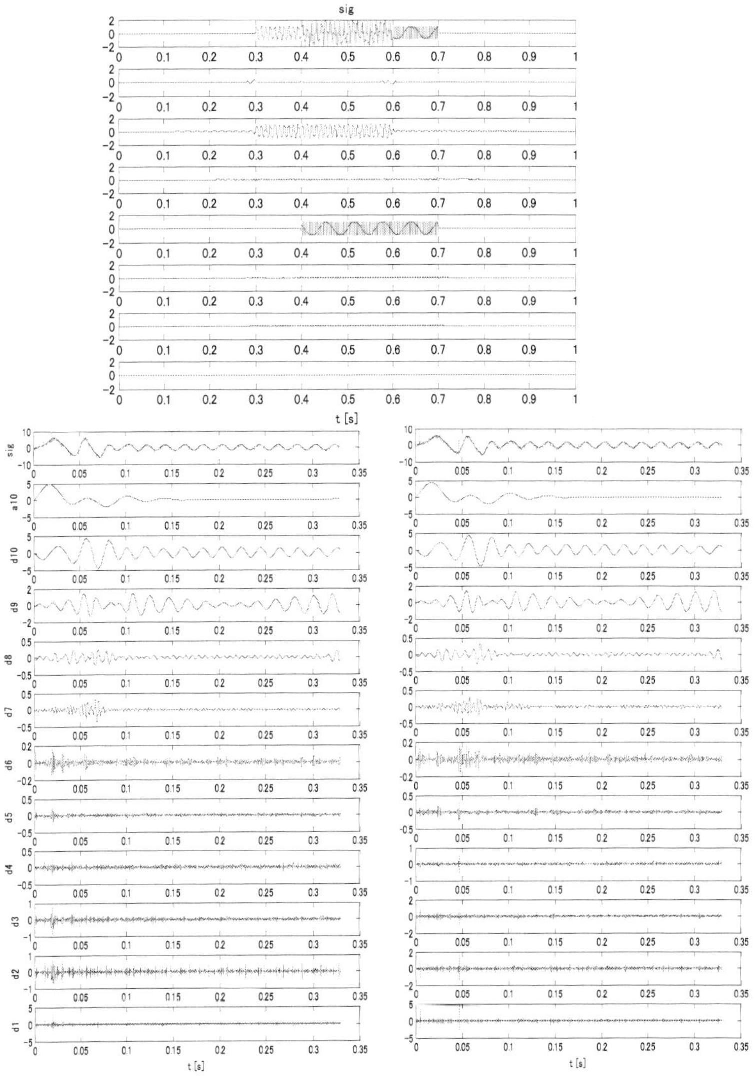

Figure 19. An example of the discrete wavelet analysis.

d8 at t = 0.07 s and a large component of d9 at t = 0.12 s. These components correspond to the frequency of $(1 + 2s)f$ Hz. We can find a slight difference of the components in these frequency bands between the healthy rotor and the rotor with two broken bars, especially in d2 through d6 details from 0 to 0.1 s.

http://dx.doi.org/10.5772/60964

Approximation and details	Frequency bands [Hz]
a10	0 – 24
d10	24 – 49
d9	49 – 98
d8	98 – 195
d7	195 – 391
d6	391 – 781
d5	781 – 1,563
d4	1,563 – 3,125
d3	3,125 – 6,250
d2	6,250 – 12,500
d1	12,500 – 25,000

Table 3. Frequency bands by decomposition in multi-levels.

6. Conclusions

This study has analyzed the Fourier components of broken end ring segments and compared them with those of the broken bars. We have verified, by both experiment and simulation, that the components of $(1 \pm 2s)f$ of the broken end ring segments are larger than those of the broken bars when the number of faults is one or two. The electromagnetic field in the motor with two broken bars and a broken end ring has been analyzed. Moreover, the effect of the number of broken bars and broken end ring segments on the motor performance has been clarified. The discrete wavelet analysis has shown that there are slight differences in the detail signals in high frequency bands between healthy rotor and the rotor with two broken bars.

Author details

Takeo Ishikawa

Address all correspondence to: ishi@el.gunma-u.ac.jp

Division of Electronics and Informatics, Faculty of Science and Technology, Gunma University, Japan

References

[1] A. H. Bonnett and G. C. Soukup, "Rotor Failures in Squirrel Cage Induction Motors," IEEE Transactions on Industry Applications, vol. IA-22, no. 6, pp. 1165–1173 (1986)

[2] W. T. Thomson and M. Fenger, "Current Signature Analysis to Detect Induction Motor Faults," IEEE Industry Applications Magazine, June/July, pp. 26–34 (2001)

[3] A. Daviu, R. Guasp, J. Folch, and M. Palomares, "Validation of a New Method for the Diagnosis of Rotor Bar Failures via Wavelet Transform in Industrial Induction Machines," IEEE Transactions on Industry Applications, vol. 42, no. 4, pp. 990–996, July–Aug (2006)

[4] O. Moreno, R. Troncoso, V. Frias, R. Gillen, and G. Perez, "Automatic Online Diagnosis Algorithm for Broken-Bar Detection on Induction Motors Based on Discrete Wavelet Transform for FPGA Implementation," IEEE Transactions on Industrial Electronics, vol. 55, no. 5, pp. 2193–2202, May (2008)

[5] R. Guasp, A. Daviu, P. Sanchez, P. Panadero, and P. Cruz, "A General Approach for the Transient Detection of Slip-Dependent Fault Components Based on the Discrete Wavelet Transform," IEEE Transactions on Industrial Electronics, vol. 55, no. 12, pp. 4167–4180, Dec. (2008)

[6] S. H. Kia, H. Henao, and G. A. Capolino, "Diagnosis of Broken-Bar Fault in Induction Machines Using Discrete Wavelet Transform without Slip Estimation," IEEE Transactions on Industry Applications, vol. 45, no. 4, pp. 1395–1404 (2009)

[7] Y. Gritli, C. Rossi, L. Zarri, F. Filippetti, A. Chatti, D. Casadei, and A. Stefani, "Advanced Diagnosis of Broken Bar Fault in Induction Machines by Using Discrete Wavelet Transform under Time-Varying Condition," 2011 IEEE International Electric Machines & Drives Conference (IEMDC), pp. 424–429, 15–18 May (2011)

[8] A. Bouzida, O. Touhami, R. Ibtiouen, A. Belouchrani, M. Fadel, and A. Rezzoug, "Fault Diagnosis in Industrial Induction Machines Through Discrete Wavelet Transform," IEEE Transactions on Industrial Electronics, vol. 58, no. 9, pp. 4385–4395 (2011)

[9] O. A. Mohammed, N. Y. Abed, and S. Ganu, "Modelling and Characterization of Induction Motor Internal Faults Using Finite-Element and Discrete," IEEE Transaction on Magnetics, vol. 42, no. 10, pp. 3434–3436 (2006)

[10] L. Weili, X. Ying, S. Jiafeng, and L. Yingli, "Finite-Element Analysis of Field Distribution and Characteristic Performance of Squirrel-Cage Induction Motor with Broken Bars," IEEE Transaction on Magnetics, vol. 43, no. 4, pp. 1537–1540 (2007)

[11] J. Faiz and B. M. Ebrahimi, "A New Pattern for Detecting Broken Rotor Bars in Induction Motors During Start-Up," IEEE Transaction on Magnetics, vol. 44, no. 12, pp. 4673–4683 (2008)

[12] T. Ishikawa, R. Ishigame, M. Matsunami, and N. Kurita, "Analysis and Failure Diagnosis of Squirrel-Cage Induction Motor with Broken Rotor," Proceedings of the 15th International Conference of Electrical Machines and Systems (ICEMS2012), DS2G2-9, Sapporo (2012)

http://dx.doi.org/10.5772/60964

[13] T. Ishikawa, S. Shinagawa, and N. Kurita, "Analysis and Failure Diagnosis of Squirrel-Cage Induction Motor with Broken Rotor Bars and End Rings," IEEE Journal of Industry Applications, vol. 2, no. 6, pp. 292–297 (2013)

[14] T. Ishikawa, Y. Seki, N. Kurita, and T. Matsuura, "Failure Diagnosis of Brushless DC Motor with Rotor Magnet Defect," IEEE 2011 International Electric Machines and Drives, pp. 1561–1565 (2011)

[15] T. Ishikawa, K. Kurihara, K. Shinohara, and Y. Sato, "Rotor Fault Diagnosis Techniques for Induction Machines," The 2007 Annual Meeting Record IEE Japan IA, 3-S4-3, pp. 9–14 (2007) (in Japanese)

Section 3

Control Strategies

Chapter 6

Fuzzy Direct Torque-controlled Induction Motor Drives for Traction with Neural Compensation of Stator Resistance

Mohammad Ali Sandidzadeh, Amir Ebrahimi and Amir Heydari

Additional information is available at the end of the chapter

http://dx.doi.org/10.5772/61545

Abstract

In this chapter, a new method for stator resistance compensation in direct torque control (DTC) drives, based on neural networks, is presented. The estimation of electromagnetic torque and stator flux linkages using the measured stator voltages and currents is crucial to the success of DTC drives. The estimation is dependent only on one machine parameter, which is the stator resistance. Changes of the stator resistances cause errors in the estimated magnitude and position of the flux linkage and therefore in the estimated electromagnetic torque. Parameter compensation by means of stator current phasor error has been proposed in literature. The proposed approach in this chapter is based on a principle that states the error between the measured current magnitude of the stator feedback and the stator's command, verified with neural network, is proportional to the variation of the stator resistance and is mainly caused by the motor temperature and the varying stator frequency. Then the correction value of stator resistance is achieved by means of a fuzzy controller. For the first time, a combination of neural control and fuzzy control approach in stator resistance variations based on the stator current is presented. The presented approach efficiently estimates the correct value of stator resistance.

Keywords: Fuzzy direct torque control, neural compensation, induction motor drives

1. Introduction

The direct torque control is one of the excellent control strategies available for torque control of induction machine. It is considered as an alternative to field oriented control (FOC) technique [1]. In fact, among all control methods for induction motor drives, direct torque control (DTC) seems to be particularly interesting being independent of machine rotor parameters and requiring no speed or position sensors [2].

A basic concept of direct torque control of induction motor drives is simultaneous control of the stator flux and electromagnetic torque of a machine. Compared to the conventional vector-controlled drives, the torque and flux of a DTC-based drive are controlled in a closed-loop system that does not use the current loops.

In principle, DTC-based drives require only the knowledge of stator resistance and thereby decrease the associated sensitivity to parameter variations [3, 4]. Moreover, compared to the conventional vector-controlled drives, DTC-based drives do not require fulfilling the coordinate transformation between stationary and synchronous frames. Depending on how the switching sectors are selected, two different DTC schemes become possible [5].

Since a DTC-based drive selects the inverter switching states using a switching table, neither the current controllers nor the pulse-width modulation (PWM) modulator is required. As a result, the DTC-based drive provides a fast torque response [6]. The conventional direct torque control (CDTC) suffers from some drawbacks such as high current, flux and torque ripple, difficulties in torque, and flux control at very low speeds [7]. However, the switching-table-based DTC approach has some disadvantages. If the switching frequency of the inverter is not high, the torque and flux pulsation could be high; moreover, there would be a sluggish response during the start-up or change of the reference flux or reference torque [8]. Hence, to improve the performance of the DTC drive during the start-up or changes in the reference flux and torque, a fuzzy-logic-based switching-vector process is developed in this chapter [9–15]. In DTC drives, the feedback of the electromagnetic torque and stator flux linkage is used as the input of controller. Using the measured stator currents and voltages, the electromagnetic torque and also stator flux linkages are estimated in stator reference frames [16, 17]. "The machine model is only dependent on stator resistance" [18]. There are different forms of direct torque control induction motor based on how currents and voltages are measured or estimated [19–21]. The stator current might be obtained using only the DC-link current sensor, and the motor line voltages could be reconstructed inexpensively using gate signals [22]. Nevertheless, all the measured values suffer from precision and low-speed operational problems caused by errors induced by the varying stator resistance in the flux and its angle calculator [23, 24]. The stator resistance change has a wide range, varying from 0.75 to 1.7 times the stator's nominal value. The variation is largely due to temperature variations, and to a small extent, due to the stator frequency variations [21]. The variation deteriorates the drive performance by introducing errors in the estimated magnitude and position of the flux linkage and therefore in the electromagnetic torque estimation, particularly at low speeds [25]. Note that at low speeds, the voltage drops of the stator resistance constitute a significant portion of the applied voltages. Only a few control schemes have been proposed so far for overcoming the mentioned parameter sensitivity (which restricts the speed control range of the drives). The stator resistance has problems such as convergence and slowness of response. A partial operating-frequency-dependent hybrid-flux estimator has been proposed for tuning the stator resistance [10]. Adjustment of the stator resistance, based on the difference between the flux current and its command, has problems in identifying the actual flux current [26, 27]. Finding the stator resistance based on the steady state voltage has the shortcoming of using direct axis flux linkages that are affected by the stator resistance variations.

In this chapter, a neural network estimator is developed to find the reference stator current values at each moment. Later, the error difference between the measured and the real stator current values is fed to a fuzzy logic controller, which then outputs the correct stator resistance value.

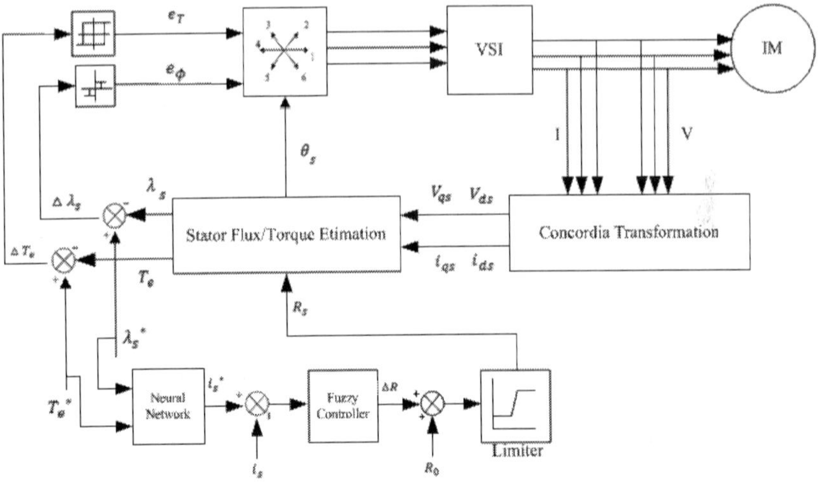

Figure 1. Block diagram of the fuzzy direct torque control of induction motor drives with a stator resistance estimator.

2. Fuzzy logic direct torque control

In this section, the concept and principle of direct torque control approach of an induction motor is briefly introduced. A schematic diagram of the proposed drive is shown in Fig. 1. The feedback control of torque and stator flux linkages, which are estimated from the measured voltages and currents of the motor, is used in the proposed drive scheme. In this approach, stator-reference frame model of the induction motor is used. To avoid the trigonometric operations faced in coordinate transformations of other reference frames, the same reference frame is used in the implementation [22]. This can be considered as one of the advantages of the control scheme. Through the integration of the difference between the phase voltage and the voltage drop in the stator resistance, Stator q and d and axis flux linkages λ_{qs}, λ_{ds} can be calculated as follows:

$$\lambda_{qs} = \int (V_{qs} - R_s i_{qs}) dt \tag{1}$$

$$\lambda_{ds} = \int (V_{ds} - R_s i_{ds}) dt \tag{2}$$

And the flux linkage phasor is as follows:

$$\lambda_s = \sqrt{\lambda_{qs}{}^2 + \lambda_{ds}{}^2)} \tag{3}$$

The stator flux linkage phasor position is:

$$\theta_s = \tan^{-1}(\frac{\lambda_{qs}}{\lambda_{ds}}) \tag{4}$$

And the electromagnetic torque is given by:

$$T_e = \frac{3}{2}\frac{P}{2}(i_{qs}\lambda_{ds} - i_{ds}\lambda_{qs}) \tag{5}$$

According to Fig. 2, the inverter switching states are selected based on the errors of the torque and the flux (as indicated by ΔT_e and $\Delta\lambda_s$, respectively). Provided that

$$\begin{aligned} \Delta T_e &= T_e^* - T_e \\ \Delta\lambda_s &= \lambda^*_s - \lambda_s \end{aligned} \tag{6}$$

The optimum switching vector is selected to decrease the errors [23, 25, 26]. Using a fuzzy-logic-based switching-vector selection process, it would be possible to improve the perform-ance of the DTC drive during start-up or changes in the reference flux and torque. For this, a Mamdani fuzzy-logic-based system is used. Using the flux and torque deviation from reference ones and the position of the stator flux linkage space vector, it is possible to select different voltages. Then a rule-base has to be formulated based on these states. Thus the aim of the approach is to use a fuzzy logic system to expand the system performance (i.e., gives faster torque and flux response), outputs the zero and non-zero voltage switching states (n), and uses three quantities as its inputs: e_Φ, the torque error (e_T), and the position of the stator flux space vector (θ_s). The stator flux linkage space vector can be located in any of the twelve sectors, each spanning a 60° wide region. These regions overlap each other as shown in Table 1.

http://dx.doi.org/10.5772/61545

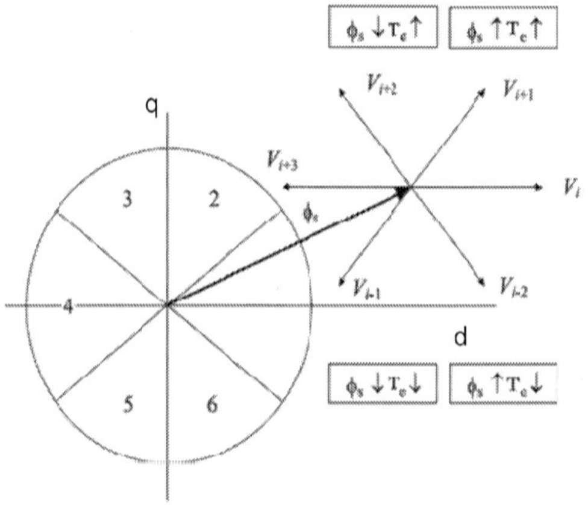

Figure 2. Changes of flux linkage space vectors due to the switching vectors.

S1	S2	S3	S4	S5	S6
315–75	345–45	15–75	45–105	75–135	105–165
S7	**S8**	**S9**	**S10**	**S11**	**S12**
135-195	165-225	195-255	225-285	255-315	285-345

Table 1. Overlaps between sectors

$e_\Phi e_t$	P	ZE	N
PL	1	2	2
PS	1	2	3
ZE	0	0	0
NS	6	0	4
NL	6	5	5

Table 2. Fuzzy vector selection in sector 1

Since it was assumed that there were three and five fuzzy sets for the flux error and the torque error, respectively, there will be 15 rules for every sector. Table 2 shows the various rules for sector 1. In particular, the stator flux error (e_Φ) can be positive (P), zero (ZE), or negative (N), corresponding to three overlapping fuzzy sets. The electromagnetic torque error can be positive large (PL), positive small (PS), zero (ZE), negative small (NS), or negative large (NL). This is because the intention is to make the torque variations smaller. Therefore, the universe of the torque is divided into five overlapping fuzzy sets. The various membership functions are shown in Fig. 3. Since there are 12 sectors, the total number of rules becomes 180. Each one of the rules can be described by the input variables and the control variable, which is the switching state (n). For example, Table 2 shows various rules for sector 1 as below:

Rule 1: If e_Φ is positive (P), e_T is positive large (PL) and θ_s is S1, then n is 1.

Rule 2: If e_Φ is positive (P), e_T is positive small (PS) and θ_s is S2, then n is 1.

Rule 3: If e_Φ is positive (P), e_T is ZE and θ_s is S3, then n is 0.

The goal of the fuzzy system is to obtain a crisp value (as the appropriate switching state) on its output. A general "*i*th rule" has the following form:

Rule *i*: If e_Φ is A_i, e_T is B_i and θ_s is C_i, then n is N_i

Thus, by using the minimum operation for the fuzzy *AND* operation and the firing strength of the *i*th rule, α_i can be obtained from

$$\alpha_i = \min[\mu_{Ai}(e_\Phi), \mu_{Bi}(e_T), \mu_{ci}(\theta_s)] \tag{7}$$

where $\mu_{Ai}(e_\Phi)$, $\mu_{Bi}(e_T)$, $\mu_{ci}(\theta_s)$ are membership functions of fuzzy sets A_i, B_i, and C_i of the variables flux error, the torque error, and the flux position, respectively. The output form of the *i*th rule is obtained from

$$\mu_{Ni}(n) = \min[\alpha_i, \mu_{Ni}(n)] \tag{8}$$

where $\mu_{Ni}(n)$ is the membership function of fuzzy set N_i of variable n. Therefore, the overall (or the combined) membership function of output n is gained by using the max operator as follows:

$$\mu_N(n) = \max[\mu_{Ni}(n)] \tag{9}$$

In this case, the outputs include crisp numbers, switching states, and for defuzzification, the maximum used criteria.

3. Stator resistance compensation

3.1. Scheme

A mismatch between the controller-set stator resistance and its actual value in the machine can create the instability shown in Fig. 6a. This figure shows the simulations for the changes of the

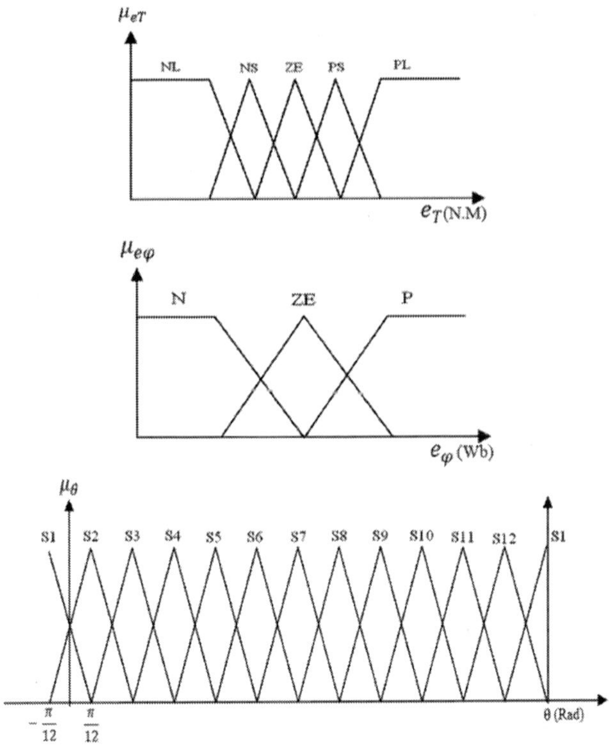

Figure 3. Membership functions.

step stator resistance from 100% to 50% of its nominal value at second 0.5. The drive system becomes unstable if the controller-instrumented stator resistance is higher than its actual value in the motor [11]. An explanation for this could be as follows: when motor resistance decrease in machine and the applied voltage is the same, the current increases, resulting in increased flux and electromagnetic torque [28]. The opposite effect occurs in controller. In fact, by current increments, which are inputs of the system, the stator resistance voltage drops will increase in the calculator. Therefore, lower flux linkages and electromagnetic torque estimations will present. Compared with their command values, they give large torque and flux linkages deviations, which result in commanding larger voltages and currents and leading to a run off condition as shown in Fig. 6a. "The parameter mismatch between the controller and machine will result in a nonlinear relation between the torque and the torque's reference, making it a non-ideal torque amplifier" [29]. This will have undesirable effect in a torque drive and speed-controlled drive systems. Therefore, it will be reasonable to design a motor resistance adaption law to overcome instability and to guarantee a linear torque amplifier in the DTC drive. A new approach is presented in the next section for stator resistance parameter adaption.

3.2. Stator current phasor command

A diagram of the applied stator resistance compensation is shown in Fig. 4. The presented technique is based on the principle that the error between the measured stator feedback current-phasor magnitude i_s and the stator's command i^*_s is proportional to the stator resistance variation, which is mainly caused by the motor temperature and the varying stator frequency. The correction value is obtained by means of a fuzzy controller. The final estimated value of $\hat{R}_s$ is obtained as the output of the limiter. The above algorithm requires the stator current phasor command, which is a function of the commanded torque and the commanded stator flux linkages.

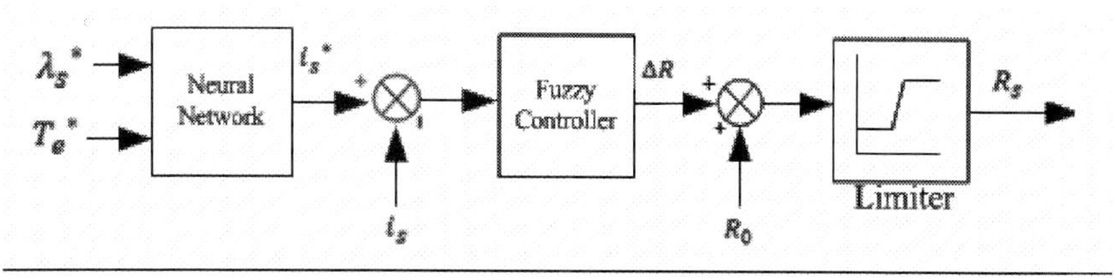

Figure 4. Block diagram of the adaptive stator resistance compensator.

A neural network estimator, presented in the following, is designed to evaluate the stator current command from the torque and stator flux linkage commands.

The stator feedback current phasor magnitude i_s is obtained from the q and d axis measured currents as

$$i_s = \sqrt{(i_{qs}^2 + i_{ds}^2)} \tag{10}$$

The stator command current phasor magnitude i^*_s is derived from the dynamic equations of the induction motor in the synchronous-rotating reference frame, using the torque command T^*_e and the stator flux linkage command λ^*_s and aligning the d axis with the stator flux linkage phasor as

$$\lambda^e_{qs} = 0, p\lambda^e_{qs} = 0, \lambda^e_{ds} = \lambda^*_s \tag{11}$$

where p is the number of poles. Substituting these equations in flux linkages and torque equations results in

$$T_e = \frac{3}{2}\frac{2}{p} i_{qs}^{e} \lambda_s \tag{12}$$

Then the q axis current command is directly obtained by using the torque command T^*_e and the stator flux linkage command λ^*_s as

$$i^{e*}_{qs} = \frac{2}{3}\frac{2}{p}\frac{T^*_e}{\lambda^*_s} \tag{13}$$

It can be shown that i^{e*}_{ds} is given by

$$L_s(i_{ds}^{e*})^2 - \lambda_s^*(1 - \frac{L_s L_r}{L^2_m - L_s L_r}) i_{ds}^{e*} + L_s(i_{qs}^{e*})^2 - \frac{(\lambda_s^*)^2 L_r}{L^2_m - L_s L_r} = 0 \tag{14}$$

Equation 14 gives two solutions for i^{e*}_{ds}, and the appropriate solution is the one that outputs a smaller value. Finally, the stator current command is calculated from

$$i^*_s = \sqrt{(i^{e*\,2}_{qs} + i^{e*\,2}_{ds})} \tag{15}$$

It is shown here that evaluation of the stator current command is a complicated and time-consuming process. Instead of using the numerical solution for the system, it is possible to perform the stator current command by using an artificial neural network (ANN) since it is known that ANN is a general nonlinear function estimator. As a result, a multilayer feed-forward back-propagation ANN, whose inputs are the torque and flux reference values, is trained to estimate the stator current command. A 2-8-8-1 structure, which has two hidden layers with 8 hidden nodes, is obtained by trial and error. The activation functions of the hidden layers are tan-sigmoid functions. Fig. 5 shows the structure of the ANN estimator. The neural estimator evaluates the reference stator current with less than 0.01% error. Furthermore, it is shown that more complicated ANN structures result in higher error rates.

4. Results

Dynamic simulations are performed to validate the performance of the proposed technique. The induction motor details, used in the simulation, are given in the appendix. Fig. 6a and 6b show the simulations for a step change in the stator resistance parameter-uncompensated and compensated torque drive system respectively. The system controller has the nominal value of the stator resistance, and after half a second, the stator resistance is changed to ½ of its nominal value. Then the corresponding effects are studied. In the compensated system, it is observed that the estimation of stator resistance has experienced an initial transient state, and after a short time, it converges gradually to its final actual value in a steady state. The similar transitions are observed in other variables. However, all variables reach to their steady state situation. A step variation in the stator resistance is rather an extreme test and not a significant case encountered in practice. In real operating conditions, the temperature change rate is very slow and so is the stator resistance.

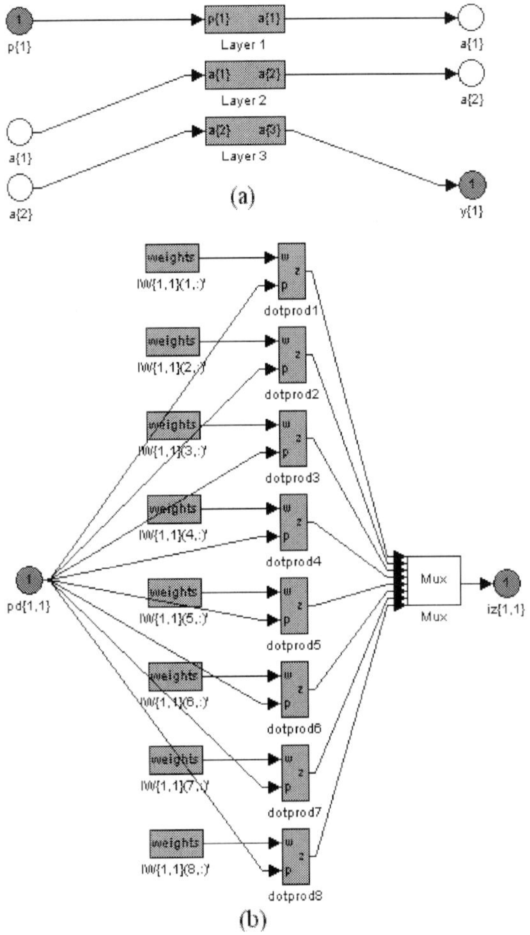

Figure 5. Neural network structure: (a) structure of layers and (b) structure of first hidden layer.

Stator flux linkages and the torque command are proportionally decreased and increased linearly from/to their original reference values. The tracking of motor variables and stator resistance is achieved, thus proving the effectiveness of the adaptive controller in the flux-weakening region. It also perfectly operates in stator resistance incremental case and in gradually stator resistance changes due to temperature changes. In these cases, there is not any oscillation even at the initial moments of resistance variations.

5. Conclusion

A fuzzy direct torque-controlled drive was introduced, and an adaptive stator resistance compensation scheme was applied to a typical three-phase induction motor. With this approach, the elimination of parameter sensitivity of the stator resistance by using only the existing stator current feedback occurred. The scheme was simple to implement, and its realization was indirectly dependent on stator inductances. Since the flux was controlled in the machine, the inductances used in the computation of stator phasor current command were constants. A procedure for finding the phasor command of the stator current from the torque and stator flux linkage commands was derived to realize the complication of this method. The ANN estimator was designed to effectively evaluate the reference stator current value. The scheme was verified via dynamic simulation for various operating conditions, including the flux-weakening mode. The scheme was successful despite rapid changes in the stator resistance, such as step changes. It was observed that the scheme adapted very well without transients even for simultaneous variations of the torque and flux linkages command while the stator resistance was varying. Finally, a simple fuzzy controller was used to generate the exact stator resistance value.

Variable	Symbol	Variable	Symbol
Stator q and d flux linkages	λ_{qs} , λ_{ds}	Stator command current phasor magnitude	i^*_s
Stator q and d and axis voltages	V_{qs} , V_{ds}	Torque command	T^*_e
Stator resistance	R_s	Stator flux linkage command	λ^*_s
Stator q and d and axis currents	i_{qs} , i_{ds}	Stator self-inductance	L_s
Stator flux linkage phasor position	θ_s	Mutual inductance	M
Motor torque	T_e	Rotor self-inductance	L_r
Number of poles	p		

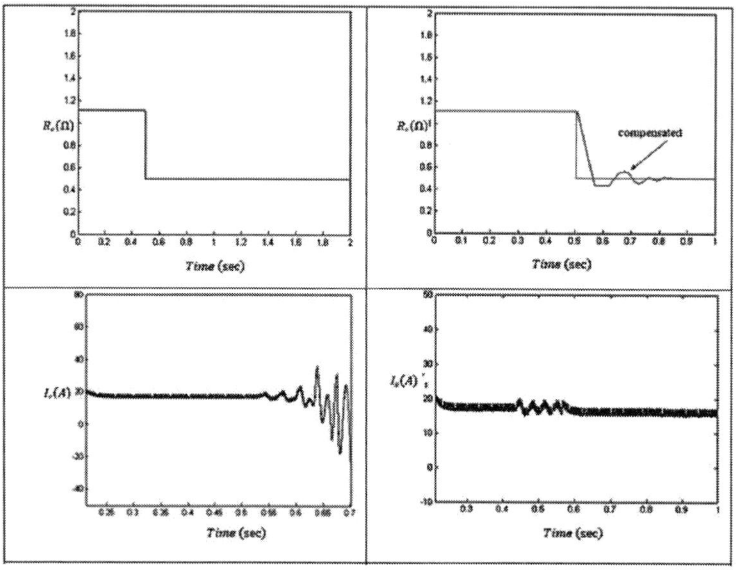

(a)

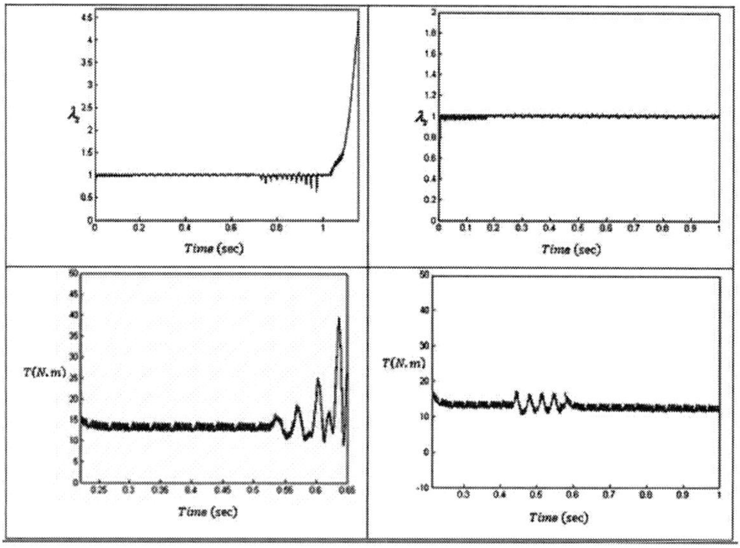

Figure 6. (a) The step response for a parameter uncompensated system. (b)The step response for a parameter compensated system.

http://dx.doi.org/10.5772/61545

6. Appendix

Induction motor parameters

Nominal operation condition		
Rated power	4	kW
Rated voltage	480	V
Rated speed	3000	rpm
Rated frequency	50	Hz

Parameters of simulated induction motor		
Stator resistance	1.115	Ω
Stator self-inductance	0.005974	H
Mutual inductance	0.2037	H
Friction factor	0.00051	N·m/s
Rotor resistance	1.083	Ω
Rotor self-inductance	0.005974	H
Rotor inertia	0.0333	kg/m^2
Number of poles	2	

Acknowledgements

The authors would like to thank the school of railway engineering in Iran University of Science and Technology for the great support of this project. The assistance of Mr. H. Zafari for revising this paper is appreciated as well.

Author details

Mohammad Ali Sandidzadeh[1*], Amir Ebrahimi[1] and Amir Heydari[2]

*Address all correspondence to: sandidzadeh@iust.ac.ir

1 School of Railway Engineering, Iran University of Science and Technology, Tehran, Iran

2 School of Railway Engineering, Iran university of Science and Technology, Tehran, Sepid Gatch Saveh, Saveh, Iran

References

[1] S. Allirani, V. Jagannathan, Direct torque control technique in induction motor drives —a review, Journal of Theoretical and Applied Information Technology, 28 February 2014, Vol. 60, No.3.

[2] D. Casadei, G. Serra, A. Tani, and L. Zarri, Assessment of direct torque control for induction motor drives, Bulletin of the Polish Academy of Sciences Technical Sciences, 2006, Vol. 54, No. 3.

[3] S. Zadeh and M. Khayamy, Efficiency-Optimizing Direct Torque Control of Interior Permanent Magnet Synchronous Machines with Fastest Start-up, 4th IET Conference on Power Electronics, Machines and Drives., Apr. 2008, pp. 218–224.

[4] G. Andreescu, C. Pitic, F. Blaabjerg and I. Boldea, Combined flux observer with signal injection enhancement for wide speed range sensor-less direct torque control of IPMSM drives, IEEE Transactions on Energy Conversion, Jun. 2008, Vol. 23, No. 2, 393–402.

[5] Nalin Kant Mohanty, Ranganath Muthu and M. Senthil Kumaran, A survey on controlled AC electrical drives, International Journal of Electrical and Power Engineering, 2009, Vol. 3, No. 3, 175–183.

[6] A. Jidin, N. Idris and A. Yatim, Study on Stability and Performances of DTC Due to Stator Resistance Variation, 5th Student Conference on Research and Development, Dec. 2007, pp. 1–6.

[7] Soufien Gdaim, Abdellatif Mtibaa and Mohamed Faouzi Mimouni, Direct torque control of induction machine based on intelligent techniques, International Journal of Computer Applications, November 2010, Vol. 10, No. 8, 0975–8887.

[8] C. Yongjun, H. Shenghua, W. Shanming and W. Fang, A Stator Resistance Compensator for a Direct Torque Controlled Low Speed and High Torque Permanent Magnet Synchronous Motor, 42nd International Universities Power Engineering Conference, Sept. 2007, pp. 174–177.

[9] L. Wang and Y. Gao, A Novel Strategy of Direct Torque Control for PMSM Drive Reducing Ripple in Torque and Flux, IEEE International Conference on Electric Machines & Drives, May 2007, Vo. 1, pp. 403–406.

[10] D. Chen, C. Liao and K. Yao, Direct Torque Control for a Matrix Converter Based on Induction Motor Drive Systems, Second International Conference on Innovative Computing, Information and Control, Sept. 2007, pp. 101–102.

[11] H. Soliman and M. Elbuluk, Direct Torque Control of a Three Phase Induction Motor Using a Hybrid PI/Fuzzy Controller, Industry Applications Conference, 42nd IAS Annual Meeting Conference, Sept. 2007, pp. 1681–1685.

[12] Y. S. Lai and J. H. Chen, A new approach to direct torque control of induction motor drives for constant inverter switching frequency and torque ripple reduction, IEEE Transactions on Energy Conversion, Sept. 2001, Vol. 16, No. 3, 220–227.

[13] Hongkui Li, Fuzzy DTC for induction motor with optimized command stator flux, Intelligent Control and Automation (WCICA), 2010 8th World Congress, 7–9 July 2010, pp. 4958–4961.

[14] R.Toufouti, S.Meziane and H. Benalla, Direct torque control for induction motor using fuzzy logic, ACSE Journal, June 2006, Vol. 6, No. 2, 19–26.

[15] J. Viola, J. Restrepo and M. Diaz, Fuzzy-DTC applied to dynamic load emulation Industrial Electronics, 2006 IEEE International Symposium, 9–13 July 2006, pp. 2326–2331.

[16] B. S. Lee and R. Krishnan, Adaptive Stator Resistance for High Performance Direct Torque Controlled Induction Motor Drives, 33rd IAS Annual Meeting, IEEE Industry Applications Conference, Oct. 1998, Vol. 1, pp. 423–430.

[17] Y. S. Lai and J. C. Lin, New hybrid fuzzy controller for direct torque control induction motor drives, IEEE Transactions on Power Electronics, Sept. 2003, Vol. 18, No. 5, 1211–1219.

[18] M. Zalman and I. Kuric, Direct torque and flux control of induction machine and fuzzy controller, Journal of Electrical Engineering, 2005, Vol. 56, No. 9–10, 278–280.

[19] P. Z. Grabowski and F. Blaabjerg, Direct Torque Neuro-fuzzy Control of Induction Motor Drive, DSP Implementation, Proceedings of the 24th Annual Conference of the IEEE Industrial Electronics Society, 1998, Vol. 2, pp. 657–661.

[20] Y. S. Lai and J. H. Chen, Novel switching techniques for reducing the speed ripple of AC drives with direct torque control, IEEE Transactions on Industrial Electronics, 2004, Vol. 51, No. 4, 744–757.

[21] N. M. Silva, A. P. Martins and A. S. Carvalho, Torque and Speed Modes Simulation of a DTC Controlled Induction Motor, Proceedings of the 10th Mediterranean conference on control and automation, Portugal, 2002.

[22] M. P. Kazmierkowski and G. Buja, Review of Direct Torque Control Methods for Voltage Source Inverter-Fed Induction Motors, 29th Annual Conference of the IEEE Industrial Electronics Society, Nov. 2003, Vol. 1, pp. 981–991.

[23] Z. Tan, Y. Li and Y. Zeng, A Three-Level Speed Sensor-Less DTC Drive of Induction Motor Based on a Full-Order Flux Observer, International Conference on Power System Technology, Oct. 2002, Vol. 2,pp. 1054–1058.

[24] Norazila Jaalam, Ahmed M. A. Haidar, Nor Laili Ismail ans Rohana Abdul Karim, A Fuzzy Logic Approach for Stator Resistance Estimation of an Induction Motor, Inter-

national Conference on Instrumentation, Control & Automation ICA 2009, Bandung, Indonesia, October 2009, pp. 109–114.

[25] J. Kiang, Torque Ripple Minimization Strategy for Direct Torque Control of Induction Motor, Conference Record of IEEE-IAS Annual Meeting, 1998, pp. 438–443.

[26] F. Zidani and R. N. Said, Direct torque control of induction motor with fuzzy minimization torque ripple, Journal of Electrical Engineering, 2005, Vol. 56, No. 7–8, 183–188.

[27] L. Romeral, A. Arias, E. Aldabas and M. Jayne, Novel direct torque control (DTC) scheme with fuzzy adaptive torque-ripple reduction, IEEE Transactions on Industrial Electronics, June 2003, Vol. 50, No. 3, 487–492.

[28] B.H. Kenny and R. D. Lorenz, Stator and rotor flux based deadbeat direct torque control of induction machines, IEEE Transactions on Industry Applications, 2003, Vol. 39, No. 4, 1093–1101.

[29] G. Poddar and V. T. Ranganthan, Direct torque and frequency control of double inverter fed slip ring induction motor drive, IEEE Transactions on Industrial Electronics, Dec. 2004, Vol. 51, No. 6, 1329–1337.

Chapter 7

Development of Fuzzy Applications for High Performance Induction Motor Drive

Ali Saghafinia and Atefeh Amindoust

Additional information is available at the end of the chapter

http://dx.doi.org/10.5772/61071

Abstract

This chapter develops a sliding mode and fuzzy logic-based speed controller, which is named adaptive fuzzy sliding-mode controller (AFSMC) for an indirect field-oriented control (IFOC) of an induction motor (IM) drive. Essentially, the boundary layer approach is the most popular method to reduce the chattering phenomena, which leads to trade-off between control performances, and chattering elimination for uncertain nonlinear systems. For the proposed AFSMC, a fuzzy system is assigned as the reaching control part of the fuzzy sliding-mode controller so that it improves the control performances and eliminates the chattering completely despite large and small uncertainties in the system. A nonlinear adaptive law is also implemented to adjust the control gain with uncertainties of the system. The adaptive law is developed in the sense of Lyapunov stability theorem to minimize the control effort. The applied adaptive fuzzy controller acts like a saturation function in the thin boundary layer near the sliding surface to guarantee the stability of the system. The proposed AFSMC-based IM drive is implemented in real-time using digital signal processor (DSP) board TI TMS320F28335. The experimental and simulation results show the effectiveness of the proposed AFSMC-based IM drive at different operating conditions such as load disturbance, parameter variations, etc.

Keywords: Boundary layer approach, sliding mode control, adaptive fuzzy sliding mode controller, speed controller, induction motor

1. Introduction

The electrical motors convert more than 50% of electrical energy into mechanical energy. Over the years, induction motors (IMs) have been widely used in industry for variable speed drives due to some of the advantages such as robust construction, low cost, low maintenance, and reliability [1]. Electric motor drives are applied widely in industrial applications such as pumps, fan, paper and textile mills, elevators, electric vehicle and subway transportation, home applications, servos and robotics, ship propulsion, etc. Nowadays, high performance electric motor drives are an essential requirement for new industrial applications. The high performance electric motor drives have some characteristics such as high reliability, high-energy transformation efficiency, and quick response of torque, position, and speed, robust to parameter variations and external load disturbance, low weight, and less expensive.

The field-oriented control (FOC) technique decouples the flux and torque controls so that the central task becomes easier in both steady and transient states. Thus, the IM behaves like a separately excited DC motor while maintaining its general advantages over DC motor. The indirect FOC as an alternative method to measure the flux position without using any flux sensor, and utilizes the rotor speed and the angular slip frequency to make the unit vector signals for achieving the flux orientation. In fact, the motor current components are decoupled by estimation of the slip speed, which requires a suitable knowledge of the rotor time constant. The accuracy of this method depends on the precise estimation of the motor parameters and rotor time constant [2, 3]. However, changes in these parameters often cause field orientation detuning and degrade the electrical drive performance. Thus, the torque capability and efficiency of the drive are reduced in steady state. Also, the torque/ampere capability is significantly decreased due to the inverter current limits and causes unsatisfactory drive performance especially to fast dynamic speed command [4, 5]. Moreover, disturbances such as external load torque and unmodeled dynamics have effect on the drive performance [6, 7]. In order to achieve indirect vector control of high performance from IM drive sophisticated speed control method is required [8].

Generally, in field of drives control, the methods of control can be classified into three main categories such as fixed gain or linear methods, adaptive methods, artificial intelligence methods, and a combination of them may also be used depending on their applications. Linear controllers, which include the proportional-integral controller (PI), proportional-integral-derivative (PID), and pseudo-derivative-feedback (PDF), are used as the most common approaches in industrial applications. However, the IM drives are nonlinear, time variant, complex, and uncertain systems and system conditions may be changed while the PI controller as a linear controller is valid to operate within certain specific range and consequently it is unable to deal with uncertainties [9, 10].

Several nonlinear adaptive methods are able to adapt or upgrade the PI controllers such as model reference adaptive control (MRAC) [11], variable structure control (VSC) [12], self-tuning PI controllers [13], etc. To implement the mentioned speed controller methods, it is required to understand the exact mathematical model. To overcome these problems, over the past four decades, the field of fuzzy controller applications has spread to include many industrial control applications, and significant research work has supported the development of fuzzy controllers [14, 15]. Fuzzy logic controllers (FLCs) have been developed and can be divided into two groups [16]. The first group focuses on improving the design and performance of the standard FLC [16, 17]. The second group of approaches combines the advantages of FLC and those of conventional nonlinear adaptive techniques. Thus, they are adopted in the fuzzy control area, such as fuzzy sliding control [18-20], fuzzy gain scheduling, various forms of self-tuning and self-organizing FLCs [21, 22], and adaptive fuzzy control [23], in order to alleviate difficulties in constructing the fuzzy rule base and improve the performance of the drive under severe perturbations of model parameters and operating conditions. Hence, in this chapter, the intelligent speed controller based on the second aforementioned group is developed to high performance IM drive.

Due to some important characteristics of VSC, or in particular sliding-mode control (SMC) system such as robustness to system parameters and fast dynamic response, it is applied in IM drives [24]. However, the SMC-based system suffers from a chattering problem in steady state. The chattering makes various undesirable effects such as current harmonics and torque pulsation. To reduce or eliminate the chattering some methods have already been proposed by the researchers [25, 26]. Generally, introducing a thin boundary layer around the sliding surface can solve the chattering problem by interpolating a continuous function inside the boundary layer of switching surface [27]. However, the slope of the continuous function is a compromise between control performance and chattering elimination [28]. Also, asymptotic stability is not guaranteed and may cause a steady-state error [29]. To tackle these problems and due to the advantages of the fuzzy controllers based on the SMC system, the fuzzy controllers were used to improve the chattering and the dynamic performance of the SMC-based speed controller drives [30-32].

In this chapter, the fuzzy approach is applied to cope with the saturation function in reaching the control part of the control effort in the SMC system to guarantee the stability of the system so that the switching control law acts like a saturation function technique with a nonlinear slope inside the thin boundary layer near the sliding surface. The applied fuzzy controller improves the tracking performance despite the system uncertainties while the chattering is reduced significantly. The salient advantages of the designed fuzzy controller on the basis of the SMC system are decreasing the number of fuzzy rules and relaxation of the uncertainty bound requirement [33]. Moreover, an adaptive law is developed to estimate the unknown bound of uncertainty, which is obtained in the sense of Lyapunov stability theorem to minimize the control effort. Therefore, in this chapter a fuzzy sliding mode control (FSMC) technique is developed for IM drive to handle the large uncertainties. The performance of the proposed adaptive FSMC (AFSMC)-based IM drive is tested in both the simulation and experiment. The performance of the proposed AFSMC is also compared to the tuned PI controller and the conventional SMC-based IM drive.

2. Mathematical model of IM for IFOC

By using the vector control, the IM can be represented as a two-phase motor in a stationary reference frame (d^s, q^s) and then transform in synchronously dynamic reference frame (d^e, q^e) by applying Park's transformation.

Figure 1 shows the block diagram of the IFOC induction motor drive. The concept of *d-q* representations can be utilized to develop the basic machine equations for vector control application in a dynamic model [34]. Equation (1) shows the *d-q* axes model of an induction motor using reference axis rotating at synchronous speed.

$$\begin{bmatrix} v_{ds}^e \\ v_{qs}^e \\ 0 \\ 0 \end{bmatrix} = \begin{bmatrix} R_s + \sigma L_s p & -\sigma L_S \omega_e & \frac{L_m}{L_r} p & -\frac{L_m}{L_r} \omega_e \\ \sigma L_S \omega_e & R_s + \sigma L_s p & \frac{L_m}{L_r} \omega_e & \frac{L_m}{L_r} p \\ -L_m \frac{R_r}{L_r} & 0 & \frac{R_r}{L_r} + p & -\omega_{sl} \\ 0 & -L_m \frac{R_r}{L_r} & \omega_{sl} & \frac{R_r}{L_r} + p \end{bmatrix} \begin{bmatrix} i_{ds}^e \\ i_{qs}^e \\ \phi_{dr}^e \\ \phi_{qr}^e \end{bmatrix} \tag{1}$$

where

$$\sigma = 1 - \frac{L_m^2}{L_s L_r}, p = \frac{d}{dt}, \omega_{sl} = (\omega_e - \omega_r) \tag{2}$$

The electromagnetic torque of 3-phase and P-pole induction motor is obtained as,

$$T_e = \frac{3}{2}\frac{P}{2}\frac{L_m^{\ 2}}{L_r}\left(i_{qs}^e \phi_{dr}^e - i_{ds}^e \phi_{qr}^e\right) \tag{3}$$

The FOC mentions that the stator current components are oriented in flux and torque component in reference axis rotating at synchronous speed. For vector control, ϕ_{qr} is set to zero so that the flux is oriented in the *d*-axis:

$$\phi_r^e = \phi_{dr}^e \tag{4}$$

The slip frequency is obtained as,

$$\omega_{sl} = \frac{L_m}{\phi_r^e}\left(\frac{R_r}{L_r}\right) i_{qs}^e \tag{5}$$

The electromagnetic developed torque equation is given by:

http://dx.doi.org/10.5772/61071

$$T_e = \frac{3}{2}\frac{P}{2}\frac{L_m^2}{L_r} i_{qs}^{*e} i_{ds}^{*e} = k_t i_{qs}^{*e} \tag{6}$$

Where

$$k_t = \frac{3}{2}\frac{P}{2}\frac{L_m^2}{L_r} i_{ds}^{e*} \tag{7}$$

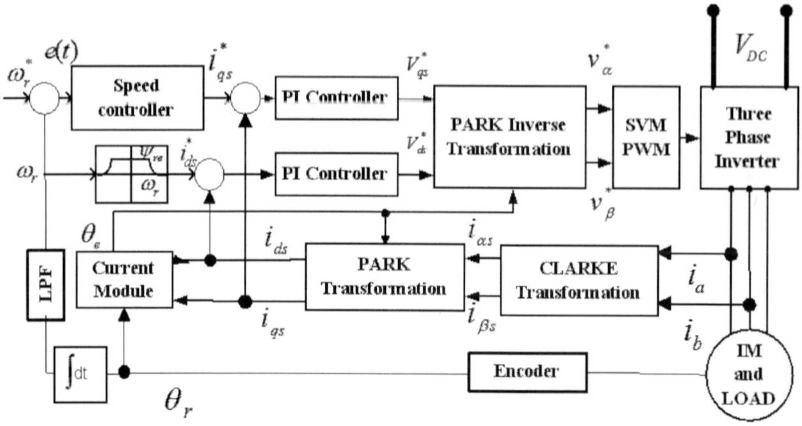

Figure 1. Control structure of an IFOC induction motor.

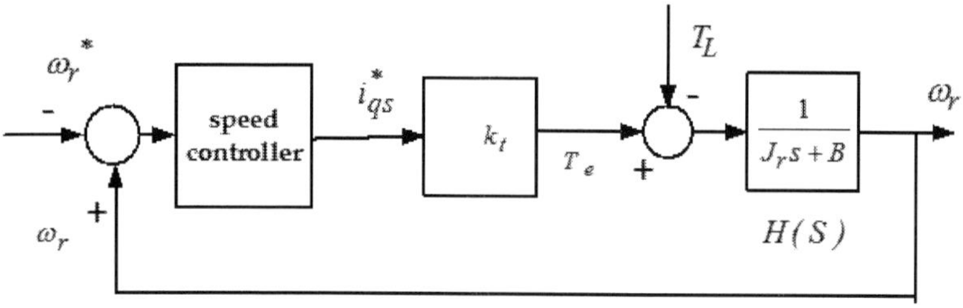

Figure 2. Simplified control block diagram of IFOC induction motor.

Considering the implementation of IFOC, the induction motor drive can be simplified as shown in Figure and the mechanical equation of an induction motor can be presented as follows [34]:

$$J_r \dot{\omega}_r(t) + B\omega_r(t) = T_e - T_L \tag{8}$$

where J_r, B, and T_L are represented as rotor inertia, friction factor, and the external load disturbance, respectively. Substituting Eq. (6) and Eq. (7) in Eq. (8) yields:

$$\begin{aligned}\dot{\omega}_r(t) &= -\frac{B}{J_r}\omega_r(t) + \frac{k_t}{J_r}i_{qs}^{*e} - \frac{T_L}{J_r} \\ &= B_p\omega_r + A_p i_{qs}^{*e} + D_p T_L\end{aligned} \tag{9}$$

where $A_p = k_t / J_r > 0$, $B_p = -B / J_r < 0$, and $D_p = -1 / J_r > 0$.

To achieve the nominal model of IM drive, the nominal value of the parameters must be considered without any disturbances. Thus, the nominal model of the IM drive given by Eq. (9) can be written as,

$$\dot{\omega}_r(t) = \bar{B}\omega_r + \bar{A}i_{qs}^{*e} \tag{10}$$

where $\bar{A} = \bar{K}_t / \bar{J}_r$ and $\bar{B} = -\bar{B} / \bar{J}_r$ are the nominal values of A_p and B_p, respectively. To handle the uncertainties, they must be considered and added to the nominal model for real-time induction motor (IM) drive [35]. So, the dynamic Eq. (10) considering structured and unstructured uncertainties and the unmodeled dynamics for the actual induction motor drive can be written as follows:

$$\dot{\omega}_r(t) = (\bar{B} + \Delta B)\omega_r(t) + (\bar{A} + \Delta A)i_{qs}^{*e} + D_p T_L + \delta = \bar{B}\omega_r(t) + \bar{A}_{qs}^{*e} + L(t) \tag{11}$$

where $L\ (t) = \Delta B\omega_r(t) + \Delta A i_{qs}^{*e} + D_p T_L\ + \delta$

In Eq. (11), $L\ (t)$is called lumped uncertainty and the uncertainties are shown by ΔA & ΔB. In addition, unstructured uncertainty due to detuning field-orientation in the transient state and the unmodeled dynamics in practical applications are shown as δ.

3. The conventional SMC system

Considering the equation $e(t) = \omega_r(t) - \omega_r^*(t)$ as speed tracking error, time-varying surface of sliding mode in the state of space $\Re^2$ is introduced as shown in Eq. (12):

http://dx.doi.org/10.5772/61071

$$S(t) = h\left(Ce(t) + \dot{e}(t)\right) \tag{12}$$

where h in the preceding scalar equation is a positive constant and $h < 1$. Substituting Eq. (11) in Eq. (12) the first derivative of $S(t)$ is obtained as,

$$\dot{S}(t) = h\left(C\dot{e}(t) + \bar{B}\dot{\omega}_r(t) + \bar{A}_u(t) + \dot{L}(t) - \ddot{\omega}_r^*(t)\right) \tag{13}$$

where $u(t) = \dot{i}_{qs}^{*e}(t)$.

In the preceding equation, $u(t)$is called the control effort or control law. In Eq. (13) it is assumed to be the bound of $\dot{L}(t)$, which is unknown but is limited as $|\dot{L}(t)| < M$.

By solving $\dot{S}(t) = 0$ without consideration of lumped uncertainty $\left(\dot{L}(t) = 0\right)$, the desired performance under nominal system model can be achieved. In this situation $u(t)$ equals $u_{eq}(t)$ and can be obtained using Eq. (12):

$$u_{eq}(t) = -\bar{A}^{-1}\left[\left(C + \bar{B}\right)\dot{e}(t) + \bar{B}\,\dot{\omega}_r^* - \ddot{\omega}_r^*\right] \tag{14}$$

where $u_{eq}(t)$ is called the equivalent control part of the control effort. In order to have suitable performance despite uncertainties on the dynamic of the system (lumped uncertainty), a discontinuous term must be added to the equivalent control across the sliding surface $S(t)$. The term discontinuous is called hitting control part or reaching control part of the control effort [36]. It is given as,

$$u_r(t) = -\left(\bar{A}h\right)^{-1} k(t)\,\mathrm{sgn}\left(S(t)\right) \tag{15}$$

where $k(t)$ is called the control gain, and Sgn is the sign function as follows:

$$\mathrm{sgn}(S(t)) = \begin{cases} 1 & if \quad S(t) > 0 \\ -1 & if \quad S(t) < 0 \end{cases}$$

Therefore, the control law is obtained as shown in Eq. (16):

$$u(t) = u_{eq}(t) + u_r(t) = u_{eq}(t) - (\bar{A}h)^{-1} k(t)\operatorname{sgn}(S(t)) \tag{16}$$

$$i_{qs}^{*} = \frac{1}{\tau}\int_0^t u(t)dt \tag{17}$$

where τ is the integral constant.

By defining the Lyapunov function as Eq. (18), stability of the system is guaranteed by Eq. (19):

$$V = \frac{1}{2}S^2(t) \tag{18}$$

$$\dot{V} \le -\eta|S(t)| \tag{19}$$

The stability condition Eq. (19) guarantees stability of the system considering lumped uncertainties by keeping the scalar $S(t)$ at zero. Then, substituting Eq. (11) in Eq. (12) and considering Eq. (13), stability condition is obtained as,

$$\begin{aligned}\dot{V} &= S(t)\dot{S}(t) = -S(t)h\bar{A}u_r(t) + hS(t)\dot{L}(t)\\ &\le -k|S(t)| + h|S(t)||\dot{L}(t)|\\ &\le -|S(t)|(k(t) - hM)\end{aligned} \tag{20}$$

A comparison between Eq. (19) and Eq. (20) yields,

$$k(t) \ge hM + \eta \tag{21}$$

Then by choosing Eq. (21), stability of the system Eq. (16) is guaranteed.

4. The proposed AFSMC controller

In case of the designed AFSMC, the "sgn" function in Eq. (15) is replaced by the fuzzy system so that the control law for the AFSMC is obtained as shown in Eq. (22):

$$u(t) = u_{eq}(t) - (\bar{A}h)^{-1} k(t) u_{fsmc} \tag{22}$$

where u_{fsmc}=$FSMC(S(t),\ \Delta S(t))$ and $u_{eq}(t)$ is presented in Eq. (14).

Triangular type inputs membership function (MF) with fuzzy sets negative (N), zero (Z), and positive (P) and triangular and trapezoidal type output MF with fuzzy sets negative big (NB), negative medium (NM), negative small (NS), zero (ZE), positive small (PS), positive medium (PM), and positive big (PB) on the common interval [-1 1] have been defined for the AFSMC as shown in Figure 3 and Figure 4, respectively. According to these figures, the thickness of the boundary layer can be changed by varying the range of the fuzzy sets "Z" and "ZE" on the interval [0 1] in the input and output membership functions, respectively. Since the proposed fuzzy system structure is based on the saturation function technique, the best thickness of the boundary layer can be derived from a fixed boundary layer sliding-mode controller that selects saturation function as reaching the control part of its effort control. Thus, boundary layer thickness can be adjusted in two steps as follows.

Step 1: Varying the slope of a saturation function so that the best performance is achieved for the fixed boundary layer sliding mode-based speed controller.

Step 2: Varying the range of the fuzzy sets "Z" and "ZE" to settle the selected value of boundary layer in the first step for the proposed FSMC-based speed controller.

According to these steps, considering Figure 4 and Figure 5 4, the obtained value of the boundary layer thickness ψ is 0.75 in this chapter. In fact, the structure of the designed fuzzy controller is shown in Figure 5.

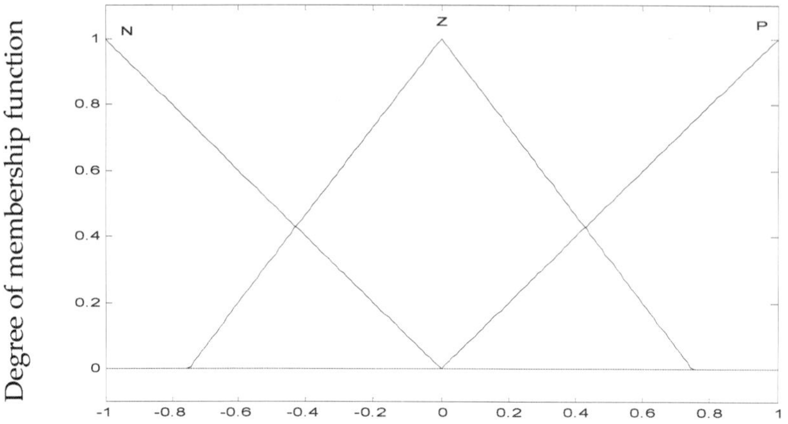

Figure 3. Membership functions of the inputs $(S(t),\ \ \Delta S(t))$, for the proposed AFSMC.

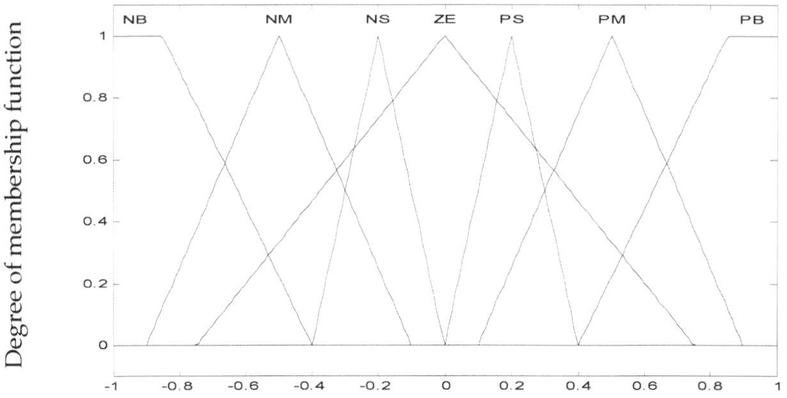

Figure 4. Output membership functions for the proposed AFSMC.

The Mamdani type fuzzy inferring method with nine rules is applied to have an appropriate tracking response, have a fast dynamic response, eliminate chattering phenomenon, and satisfy the requirement of stability condition Eq. (19) as shown in Table 1. The center of the area (COA) method has used defuzzification method.

In case of the designed FSMC speed controller, the upper band of lumped uncertainty ($L(t)$) is not available and the control effort cannot be minimized without the knowledge of $L(t)$. It is necessary to estimate $k(t)$ by using the adoption law to assure existing sliding-mode conditions. Then, $k(t)$can be estimated with $\hat{k}$ to minimize the control effort Eq. (16) so that the stability condition through Lyapunov stability theorem is guaranteed Eq. (19). In this chapter, to achieve the mentioned targets $\dot{k}(t)$ is chosen as follows:

$$\dot{k}(t) = \lambda_k |S(t)| \tag{23}$$

where λ_k is a strictly positive constant. In fact, $k(t)$acts like an adaptive filter to minimize the control effort.

Considering the following Lyapunov candidate function Eq. (24), $\hat{k}$can be an estimation value of $k(t)$.

$$V(t) = \frac{1}{2} S(t)^2 + \frac{1}{2\lambda_k}\left(k(t) - \hat{k}\right)^2 \tag{24}$$

Substituting Eq. (12) and Eq. (24) in Eq. (19) for $|S(t)| > \psi(t)$ yields,

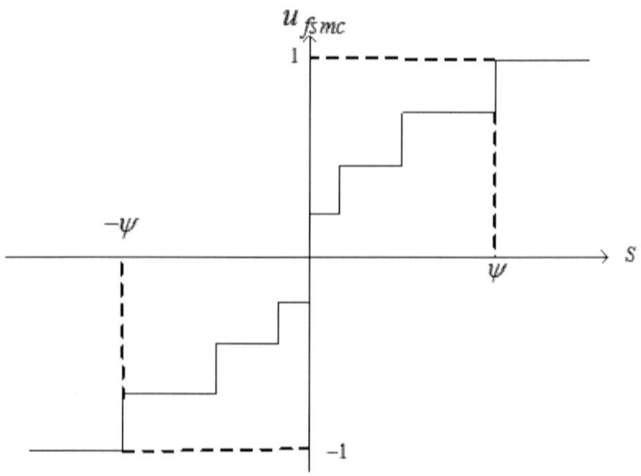

Figure 5. The performance of the proposed AFSMC system.

		Switching variable s(t)		
		N	Z	P
Alteration of switching variable Δs(t)	N	NB	NS	PM
	Z	NB	ZE	PB
	P	NM	PS	PB

Table 1. Fuzzy rule-based matrix for AFSMC

$$\begin{aligned}\dot{V}(t) &= S(t)h\left(\bar{A}u_r(t)+\dot{L}(t)\right)+\frac{1}{\lambda_k}\left(k(t)-\hat{k}\right)\dot{k}(t)\\ &= S(t)h\left(-\bar{A}k(t)\left(h\bar{A}\right)^{-1}\operatorname{sgn}(S)+\dot{L}(t)\right)+\frac{1}{\lambda_k}\left(k(t)-\hat{k}\right)\dot{k}(t)\\ &= -S(t)k(t)\operatorname{sgn}(S)+hS(t)\dot{L}(t)+\frac{1}{\lambda_k}\left(k(t)-\hat{k}\right)\dot{k}(t)\end{aligned} \tag{25}$$

Substituting Eq. (25) in Eq. (23) and considering Eq. (19) yields:

$$\begin{aligned}\dot{V}(t) &\le -\left|k(t)+\hat{k}-\hat{k}\right|\left|S(t)\right|+h\left|\dot{L}(t)\right|\left|S(t)\right|+\left|k(t)-\hat{k}\right|\left|S(t)\right|\\ &< -\left|k(t)-\hat{k}\right|\left|S(t)\right|-\hat{k}\left|S(t)\right|+hm\left|S(t)\right|+\left|k(t)-\hat{k}\right|\left|S(t)\right|\\ &< (-\hat{k}+hm)\left|S(t)\right|\end{aligned} \tag{26}$$

A comparison between Eq. (19) and Eq. (26) yields:

$$\dot{V}(t) < (-\hat{k} + mh)|S(t)| \leq -\eta|S(t)| \tag{27}$$

Consequently, $\hat{k}$ can be chosen so that the value of $-\hat{k} + mh + \eta$ remains negative. In other words, the stability of IFOC induction motor drive through the proposed AFSMC speed controller is guaranteed by selecting $\hat{k} \geq mh + \eta$. In this stage, by applying the proposed AFSMC system along with the designed fuzzy rules and the mentioned conditions, the stability condition Eq. (19) is satisfied and consequently the stability of system is guaranteed.

The overall control block diagram of the proposed AFSMC model for IFOC of IM drive is shown in Figure 6.

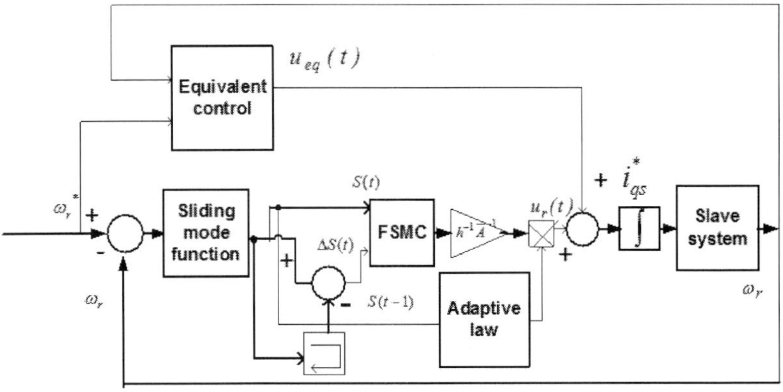

Figure 6. The control block diagram of the proposed AFSMC for IFOC of IM drive.

5. Experimental implementation of the proposed AFSMC

The block diagram of hardware schematic for space vector modulated-voltage source inverter (SVM-VSI) fed IM drive is shown in Figure 7. Three phase power inverters with 380V DC bus voltage, 20 kHz and 10 kHz SVM-PWM switching frequency for non-fuzzy controllers (the tuned PI, the conventional SMC), and the proposed controllers, respectively, are used for the drive system. A 2 μs dead time is also used for short circuit protection of the inverter. As shown in Figure, the ezdspF28335 board is used to implement the IM drive. An optical rotary encoder E60H20 with resolution of 5000 pulses per turn is used to sense a position of rotor, which is mounted on the rotor shaft and is provided as feedback to ezdspF28335 through the I/O expansion. The motor speed is calculated from the rotor position using backward difference equation. According to Figure 7, two HX 10-P/SP2 current sensors are employed to sense the phase currents. The current signals are fed back to ezdspF28335 board through A/D channels. The control algorithms are made by Simulink models based on Figure 1 using the digital motor control (DMC) and IQMath libraries from TI and Mathworks to optimize the Simulink blocks. Then, a fully automatic process is provided to compile, assemble, and download of the real-time algorithms to the DSP board through Code Composer Studio (CCStudio) TI software and MATLAB Simulink. The outputs of the board are six logic signals, which are fed to the inverter through get drive circuit. The sampling frequencies of experimental implementation are used as 10 kHz and 4 kHz for non-fuzzy (the tuned PI, the conventional SMC) and the proposed controllers, respectively. The necessary data is saved on DSP's memory with 400 Hz sampling frequency. A DC generator is coupled to the IM, which is used as a load. The experimental setup for the proposed AFSM-based prototype 1 kW IM drive system is shown in Figure 8.

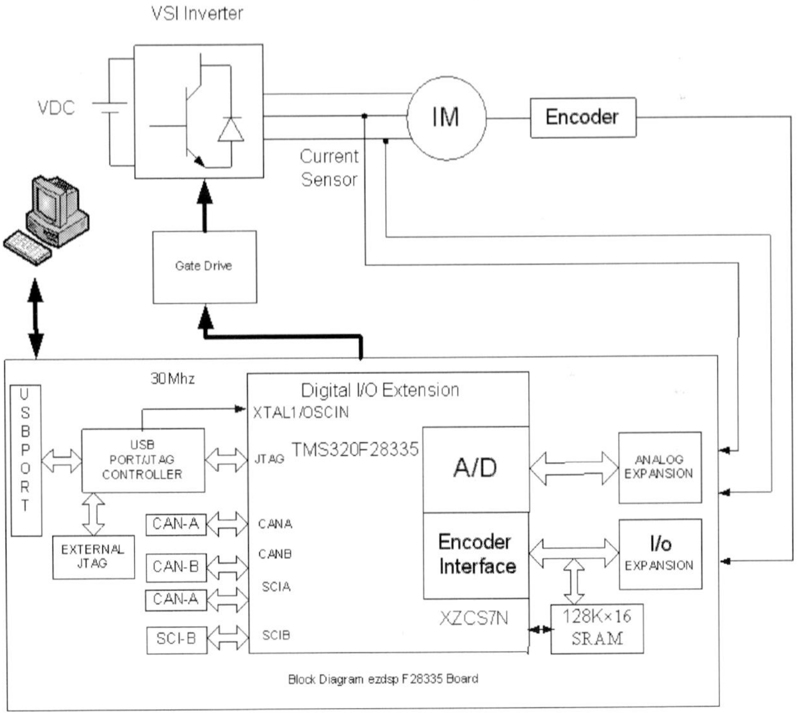

Figure 7. Block diagram of the hardware schematic for real-time implementation of VSI fed IM drive.

The performance of the proposed AFSMC controller-based IM drives have been investigated extensively both simulation and experiment. In order to show the superiority, the performance of the proposed AFSMC is also compared with the tuned PI and conventional SMC controllers.

Parameters of the 1kw 2-pole IM are given in Table 2. The PI controllers are initially tuned by the Ziegler–Nichols method based on stability boundary. The saturation of the controller is avoided by adding a correction of the integral term (K_C) as shown in Figure 9 [26]. This method is suggested by TI and Math Works. The gains, K_P(proportional gain), K_I(integral gain), and K_C (integral correction) are tuned based on the method suggested by TI [37]. The PI parameters are found as K_P=0.3, K_I=0.0001, and K_C=0.0001.

Parameters	Value	Parameters	value
Rated power	1000W	Rated torque	3.37 Nm
R_s	6	J_r	0.0055 kg.m^2
R_r	5.72	B	0.001 kg.m^2/s
L_s	428.7e-3H	n	2
L_r	428.7e-3H	Rated speed	2830rpm
L_m	416.6e-3H		

Table 2. IM parameters (referred to stator side)

The PI coefficients of the d-axis current and the q-axis current controllers including the proportional term (K_{Pd}, K_{Pq}), the integral term (K_{Id}, K_{Iq}), and the correction of the integral term (K_{Cd}, K_{Cq}) are obtained as (0.3, 0.05), (0.00625, 0.00625), and (0.00625, 0.00625), respectively. The PI coefficients of the speed controller including the proportional term (K_{Pw}), the integral term (K_{Iw}), and the correction of the integral term (K_{Cw}) are also obtained as 0.3, 0.0001, and 0.0001, respectively. The parameters of control are adjusted so that the restriction of the control effort, the requirement of system stability, and the best transient control performance are provided. So, to achieve these goals, the parameters for the c proposed AFSMC controller are chosen as C=1500, $h=\bar{A}^{-1}$, τ=1, and λ_k=400.

6. Simulation study of the proposed AFSMC

According to the block diagram of closed-loop vector control of IM drive shown in Figure 1, the SVM-VSI type inverter is modeled based on fast switching Insulated Gate Bipolar Transistors (IGBTs) by the Simulink toolbox along with the existing libraries in MATLAB (for more details, reader are refereed to [38]). Simulation results are provided at different operating conditions such as load variation, inertia, and friction factor variations of the induction motor. Their advantages are indicated in comparison with the conventional SMC and tuned PI controller.

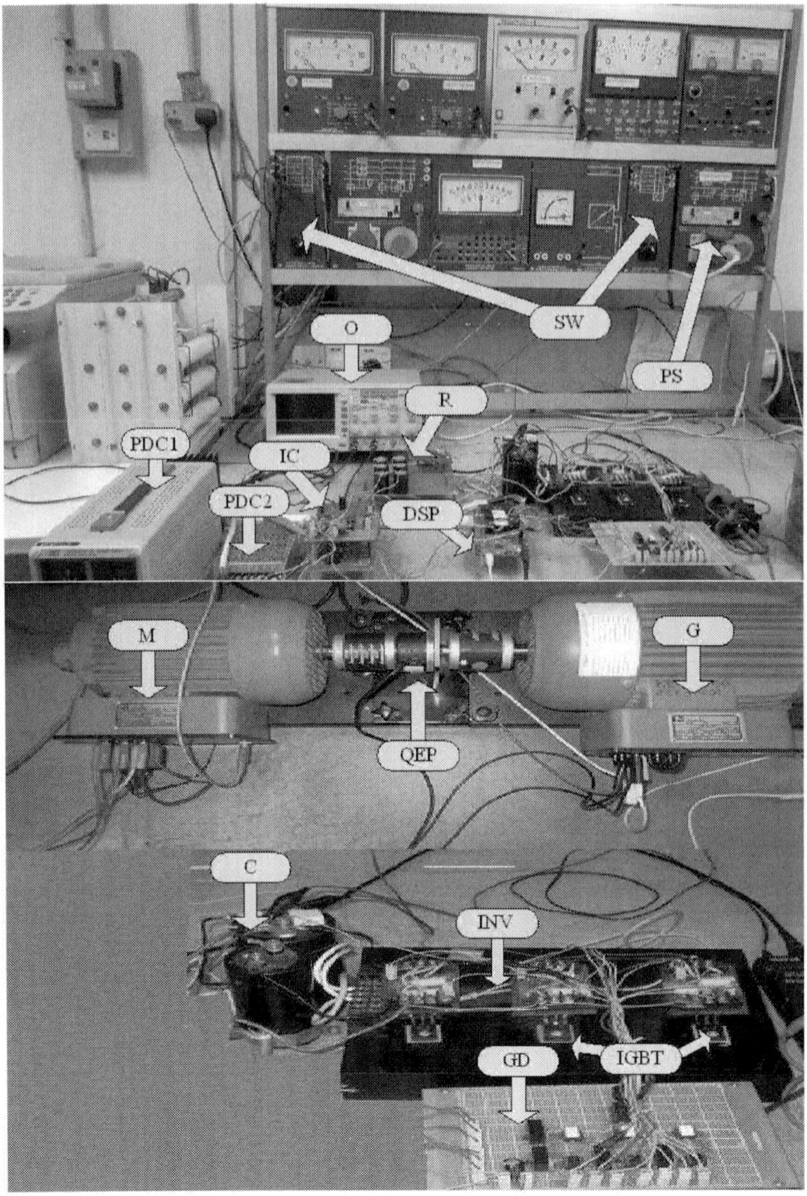

Figure 8. Experimental setups of the proposed speed controller of the IM drive.

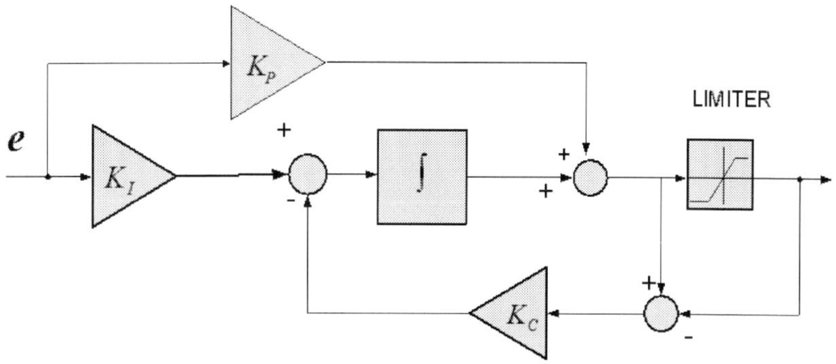

Figure 9. PI controller structure with anti-windup correction term.

For simulation tests, the following cases including parameter variations and external load disturbance are considered. If not mentioned, all other parameters are considered to be nominal in all the cases.

Case-1: Using nominal parameters for simulation at no load and full load.

Case-2: Inertia and friction factor is chosen two times of nominal value, which is applied at t=7 sec while the motor starts with half rated load.

Case-1 (no load and full load condition with nominal parameters):

Simulation results are illustrated in Figures 10–12(a)–(b) at no load and full load. It can be seen that dynamic and steady state performance of the conventional SMC and the proposed AFSMC controller are better than those of the tuned PI controller. In terms of overshoot and settling time, the proposed AFSMC controller exhibits the best performance among all controllers. It can also be seen from Figures 10–12(b) that the suitable tracking response has been obtained for the conventional SMC at the expense of undesirable chattering in both currents and speed. In addition, from these figures, the favorable tracking response has been obtained for the proposed AFSMC controller without any chattering in currents and speed. From Figure 11 (a)–(b), it can be seen that motor currents and the components (Iq) are affected by chattering, which appear in motor speed.

Case-2 (sinusoidal command speed with inertia and friction coefficient variation):

The sinusoidal command is selected as speed command to show the tracking error properly. The other parameter variations are also tested in this case. Simulation results are shown in Figures 13–14(a)–(b)for the conventional SMC and the proposed AFSMC, respectively. From these figures, the proposed AFSMC has favorable tracking performance without any chattering while the conventional SMC suffers from chattering for both speed and current. In addition, the tracking error is obtained close to zero for the proposed AFSMC controller despite large uncertainty in system, which shows the high accuracy tracking of this proposed controller.

http://dx.doi.org/10.5772/61071

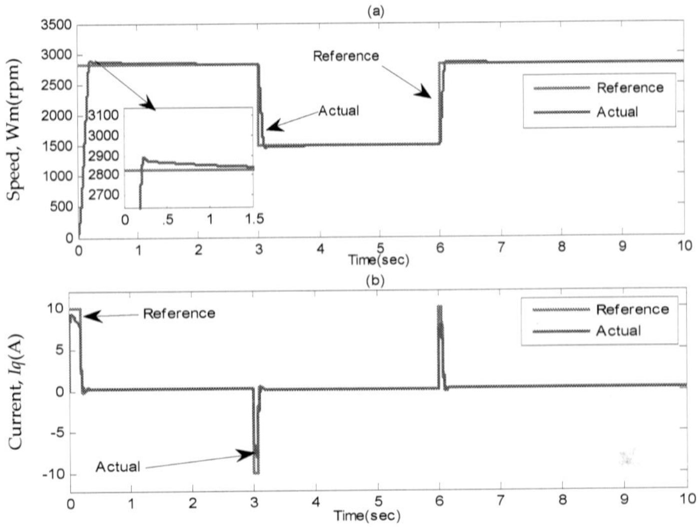

Figure 10. Simulated response of the tuned PI controller-based IM drive at no load in Case-1: (a) speed and (b) q-axis current.

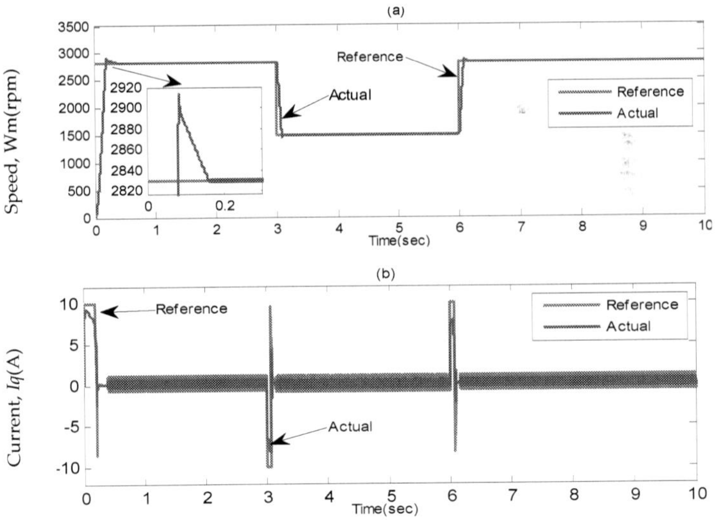

Figure 11. Simulated response of the conventional SMC controller-based IM drive at full rated load in Case-1: (a) speed and (b) q-axis current.

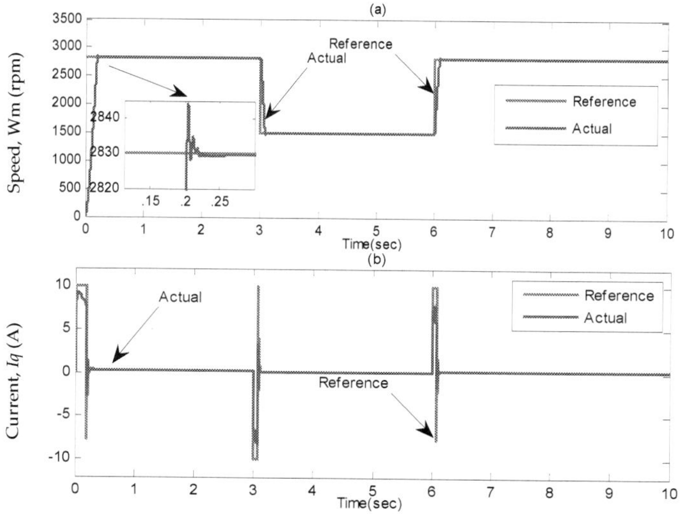

Figure 12. Simulated response of the proposed AFSMC controller-based IM drive at full rated load in Case-1: (a) speed and (b) q-axis current.

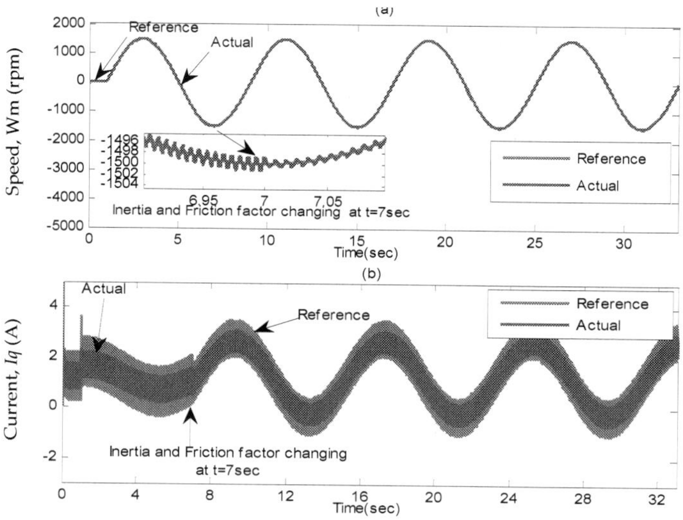

Figure 13. Simulated response of the conventional SMC controller-based IM drive with Jr and Br variations at t=7 in Case-2: (a) speed and (b) q-axis current.

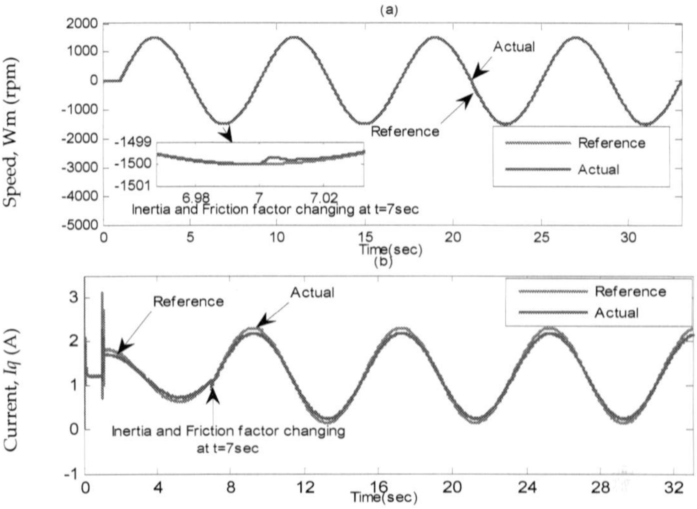

Figure 14. Simulated response of the proposed AFSMC controller-based IM drive with *Jr* and *Br* variations at *t*=7 in Case-3: (a) speed, (b) *q*-axis current, and (d) tracking error.

Simulation results in these cases indicate that the tracking capability, dynamic, and steady-state performances of the proposed AFSMC controller are the best among the conventional SMC and tuned PI controllers. In addition, the proposed AFSMC is found robust to parameter variations and external load disturbances. Moreover, the chattering is absent in the control effort of the proposed sliding-mode controller despite parameter variations and external load disturbances.

7. Experimental study of the proposed AFSMC

For experimental tests, the following cases are considered and if not mentioned, all other parameters are considered to be nominal in all the cases.

Case-1: Step changes in command speed with no load

Case-2: Step increase in load from "0" to full load.

Case-3: Inertia coefficient is increased three times of nominal value while full rated load is applied from the beginning.

For case-1, experimental results are illustrated in Figures 15–17(a)–(b)at no load. From Figures 15–17(a), it can be seen that dynamic and steady-state performance of the proposed AFSMC controller are better than those of the conventional SMC and the tuned PI controllers. It can also be seen from Figures 15–17(c)that the suitable tracking response has been obtained for the conventional SMC at the expense of undesirable chattering in both currents and speed while the proposed intelligent speed controller shows the favorable tracking response without any chattering in currents and speed. In terms of rising time, the proposed AFSMC controller exhibits the best performance as compared to the conventional SMC and the tuned PI controllers.

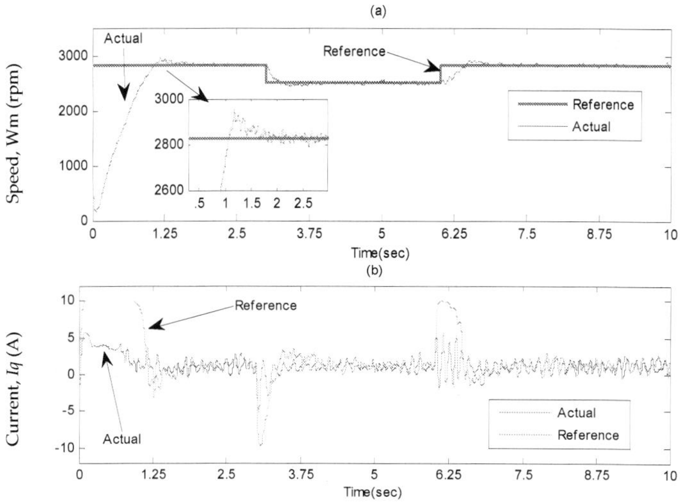

Figure 15. The experimental response of the tuned PI controller-based IM drive at no load in Case-1: (a) speed and (b) q-axis current.

For case-2, experimental results are illustrated in Figures 18–19(a)–(b)for a step increase full load. The motor was running at no load condition, and then suddenly full load disturbance is applied to the motor. From these figures, it can be seen that the PI controller suffers from a significant dip in speed ($\approx$ 120 rpm) when the step increase in load is applied. On the other hand, the proposed AFSMC is found almost insensitive when a step increase in full load is applied. The proposed AFSMC is also free from chattering in steady-state. Moreover, the settling time for the proposed AFSMC-based IM drive is faster than the PI-based IM drive.

For case-3, experimental results are illustrated in Figure 20(a) –(b).It is found from Figure 20(a) and 20(b) that the proposed AFSMC controller provides nearly the same responses with both nominal inertia and three times of nominal inertia. Thus, the proposed AFSMC controller is found insensitive to parameter variation in real-time.

Experimental results in this case indicate that the proposed AFSMC-based IM drive ensures smooth operation of the motor and results in less harmonic losses in the motor and shows superior performance in terms of tracking and transient responses without any chattering in currents and speed. Moreover, the experimental results validate the obtained simulation results in similar cases.

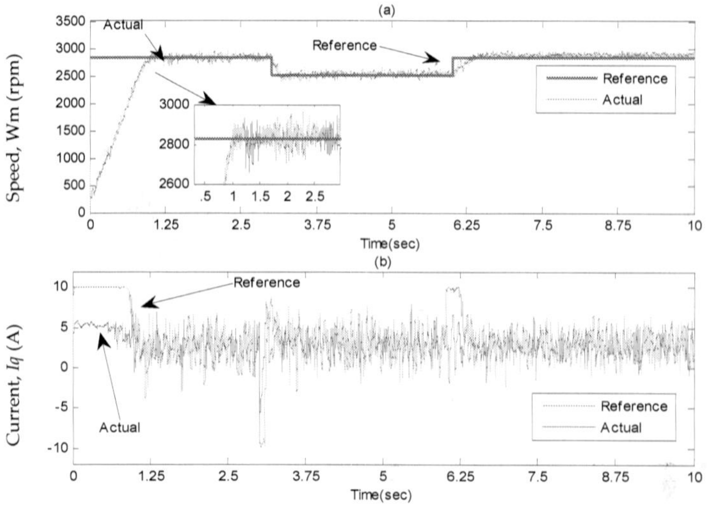

Figure 16. The experimental response of the conventional SMC controller-based IM drive at no load in Case-1: (a) speed and (b) q-axis current.

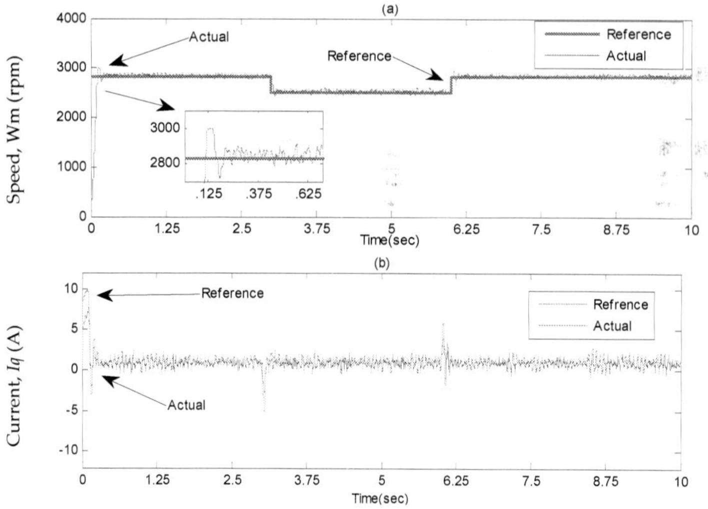

Figure 17. The experimental response of the proposed AFSMC controller-based IM drive at no load in Case-1: (a) speed and (b) q-axis current.

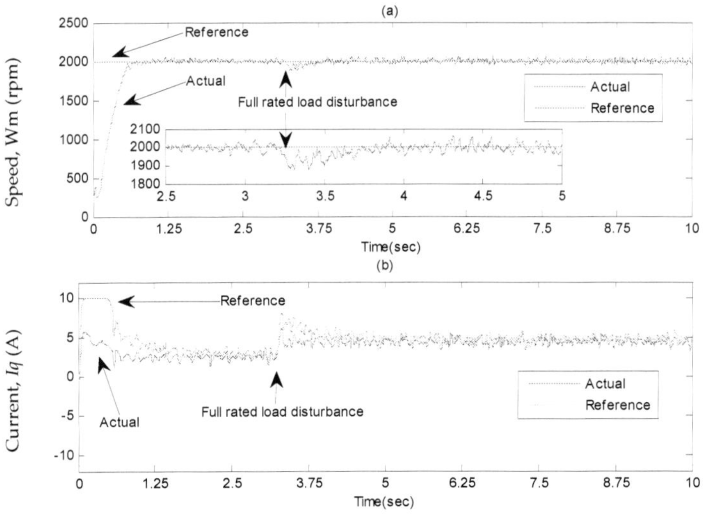

Figure 18. The experimental responses of the tuned PI controller-based IM drive in Case-2: (a) speed and (b) q-axis current.

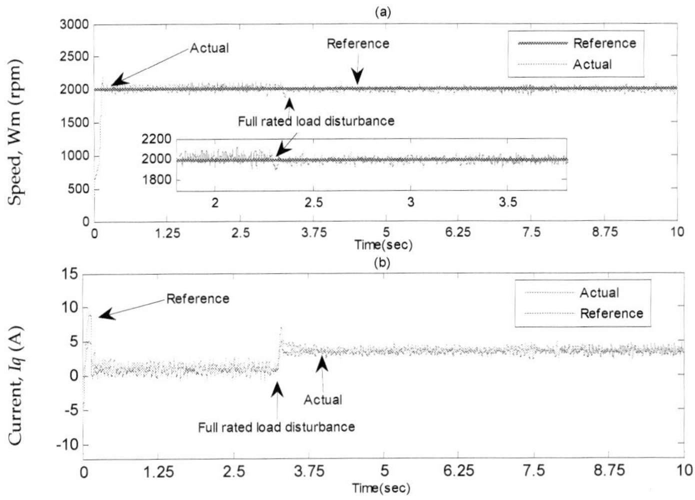

Figure 19. The experimental responses of the proposed AFSMC controller-based IM drive in Case-2: (a) speed and (b) q-axis current.

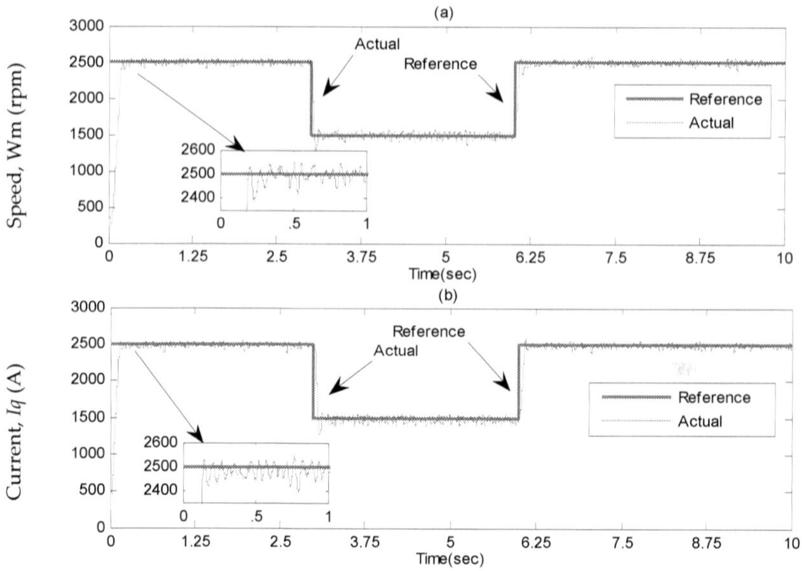

Figure 20. Experimental speed responses of the proposed AFSMC-based IM drive: (a) nominal inertia with full load, (b) three times of nominal inertia with full load (Case-3).

8. Conclusion

A sliding-mode and fuzzy logic controller-based IFOC of IM drive has been presented in this chapter. The structure of the proposed controller is based on smoothing out the control discontinuity in a thin boundary layer near the sliding surface. The proposed fuzzy system acts like the saturation function technique with a nonlinear slope inside the thin boundary layer. The proposed AFSMC-based IM drive has been successfully implemented in real time using DSP board ezdspF28335 for a prototype 1.5 hp motor. The performance of the proposed AFSMC has been tested in both simulation and experiment. The performance of the proposed AFSMC controller was found superior to the tuned PI and conventional SMC controllers at different operating conditions such as step change in command speed, load disturbance, and parameter variations. Furthermore, the proposed AFSMC reduced the steady state chattering in current. Thus, the proposed AFSMC ensures less harmonic loss and associated heat dissipation in the motor.

Author details

Ali Saghafinia[1*] and Atefeh Amindoust[2]

*Address all correspondence to: Saghafi_Ali@yahoo.com

1 Electrical Engineering Department, Majlesi Branch, Islamic Azad University, Esfahan, Iran

2 Industrial Engineering Department, Najafabad Branch, Islamic Azad University, Esfahan, Iran

References

[1] Bose B. K., *Modern power electronics and AC drives*. Upper Saddle River: Prentice Hall, 2002.

[2] Moallem M., Mirzaeian B., Mohammed O. A., and Lucas C., "Multi-objective genetic-fuzzy optimal design of PI controller in the indirect field oriented control of an induction motor," *IEEE Transactions on Magnetics,* vol. 37, pp. 3608–3612, 2001.

[3] Saghafinia A. and Ping H. W., "High performance induction motor drive using fuzzy self-tuning hybrid fuzzy controller," in *2010 IEEE International Conference on Power and Energy (PECon),* 2010, pp. 468–473.

[4] Nordin K. B., Novotny D. W., and Zinger D. S., "The influence of motor parameter deviations in feedforward field orientation drive systems," *IEEE Transactions on Industry Applications,* vol. 4, pp. 1009–1015, 1985.

[5] Zhen L. and Xu L., "On-line fuzzy tuning of indirect field-oriented induction machine drives," *IEEE Transactions on Power Electronics,* vol. 13, pp. 134–141, 1998.

[6] Masiala M., Vafakhah B., Salmon J., and Knight A. M., "Fuzzy self-tuning speed control of an indirect field-oriented control induction motor drive," in *41st Annual Meeting of the IEEE-Industry-Applications-Society*, Tampa, 2006, pp. 1732–1740.

[7] Chao K. H. and Liaw C. M., "Fuzzy robust speed controller for detuned field-oriented induction motor drive," *IEE Proceedings on Electric Power Applications,* vol. 147, pp. 27–36, 2000.

[8] Saghafinia A., Ping H. W., and Uddin M. N., "Sensored field oriented control of a robust induction motor drive using a novel boundary layer fuzzy controller," *Sensors,* vol. 13, pp. 17025–17056, 2013.

[9] Wang Y. and Shao H., "Optimal tuning for PI controller," *Automatica-Oxford,* vol. 36, pp. 147–152, 2000.

[10] Saghafinia A., Wooi Ping H., and Rahman M., "High performance induction motor drive using hybrid fuzzy-PI and PI controllers: a review," *International Review of Electrical Engineering-Iree,* vol. 5, pp. 2000–2012, 2010.

[11] Maiti S., Chakraborty C., Hori Y., and Ta M. C., "Model reference adaptive controller-based rotor resistance and speed estimation techniques for vector controlled induction motor drive utilizing reactive power," *IEEE Transactions on Industrial Electronics,* vol. 55, pp. 594–601, 2008.

[12] Abad G., Rodríguez M. A., and Poza J., "Two-level VSC based predictive direct torque control of the doubly fed induction machine with reduced torque and flux ripples at low constant switching frequency," *IEEE Transactions on Power Electronics,* vol. 23, pp. 1050–1061, 2008.

[13] Kong X. H., Zhang B. J., Mao X. H., Chen Y. F., and Song C. Y., "Design and application of self-tuning PI controller," *Applied Mechanics and Materials,* vol. 43, pp. 160–164, 2011.

[14] Wai R. J. and Su K. H., "Adaptive enhanced fuzzy sliding-mode control for electrical servo drive," *IEEE Transactions on Industrial Electronics,* vol. 53, pp. 569–580, 2006.

[15] Cheng M., Sun Q., and Zhou E., "New self-tuning fuzzy PI control of a novel doubly salient permanent-magnet motor drive," *IEEE Transactions on Industrial Electronics,* vol. 53, pp. 814–821, 2006.

[16] Masiala M., Vafakhah B., Salmon J., and Knight A. M., "Fuzzy self-tuning speed control of an indirect field-oriented control induction motor drive," *IEEE Transactions on Industry Applications,* vol. 44, pp. 1732–1740, 2008.

[17] Saghafinia A., Pinga H. W., and Uddin M. N., "Designing self-tuning mechanism on hybrid fuzzy controller for high performance and robust induction motor drive," *The International Journal of Advanced Technology and Engineering Research(IJATER),* vol. 3, pp. 63–72, 2013.

[18] Kar B. N., Choudhury S., Mohanty K. B., and Singh M., "Indirect vector control of induction motor using sliding-mode controller," *International Conference on Sustainable Energy and Intelligent Systems,* pp. 507–511, 2011.

[19] Gadoue S. M., Giaouris D., and Finch J. W., MRAS sensorless vector control of an induction motor using new sliding-mode and fuzzy-logic adaptation mechanisms," *IEEE Transactions on Energy Conversion,* vol. 25, pp. 394–402, 2010.

[20] Saghafinia A., Ping H. W., and Uddin M. N., "Fuzzy sliding mode control based on boundary layer theory for chattering-free and robust induction motor drive," *The International Journal of Advanced Manufacturing Technology,* vol. 71, pp. 57–68, 2014.

[21] Palm R., "Scaling of fuzzy controllers using the cross-correlation," *IEEE Transactions on Fuzzy Systems,* vol. 3, pp. 116–123, 1995.

[22] Procyk T. J. and Mamdani E. H., "A linguistic self-organizing process controller," *Automatica,* vol. 15, pp. 15–30, 1979.

[23] Wang S. Y., Tseng C. L., and Chiu C. J., "Design of adaptive TSK-fuzzy observer for vector control induction motor drives," pp. 5220–5223, 2011.

[24] Ahmed A. H. O., Ajangnay M. O., Mohamed S. A., and Dunnigan M. W., Speed control of induction motor using new sliding mode control technique," 2010, pp. 111–115.

[25] Shahnazi R., Shanechi H. M., and Pariz N., "Position control of induction and DC servomotors: a novel adaptive fuzzy PI sliding mode control," *IEEE Transactions on Energy Conversion,* vol. 23, pp. 138–147, 2008.

[26] Jinhui Z., Peng S., and Yuanqing X., "Robust adaptive sliding-mode control for fuzzy systems with mismatched uncertainties," *IEEE Transactions on Fuzzy Systems,* vol. 18, pp. 700–711, 2010.

[27] Lorenz R., "A simplified approach to continuous on-line tuning of field-oriented induction machine drives," *IEEE Transactions on Industry Applications,* vol. 26, pp. 420–424, 2002.

[28] Cupertino F., Naso D., Mininno E., and Turchiano B., "Sliding-mode control with double boundary layer for robust compensation of payload mass and friction in linear motors," *IEEE Transactions on Industry Applications,* vol. 45, pp. 1688–1696, 2009.

[29] Kim Y. K. and Jeon G. J., "Error reduction of sliding mode control using sigmoid-type nonlinear interpolation in the boundary layer," *International Journal of Control, and Systems,* vol. 2, pp. 523–529, 2004.

[30] Kuo-Yang T., Tsu-Tian L., and Chi-Hsu W., "Design of a new fuzzy suction controller using fuzzy modeling for nonlinear boundary layer," *IEEE Transactions on Fuzzy Systems,* vol. 13, pp. 605–616, 2005.

[31] Orowska-Kowalska T., Kaminski M., and Szabat K., "Implementation of a sliding-mode controller with an integral function and fuzzy gain value for the electrical drive with an elastic joint," *IEEE Transactions on Industrial Electronics,* vol. 57, pp. 1309–1317, 2010.

[32] Viet Quoc L., Han Ho C., and Jin-Woo J., "Fuzzy sliding mode speed controller for PM synchronous motors with a load torque observer," *IEEE Transactions on Power Electronics,* vol. 27, pp. 1530–1539, 2012.

[33] Yagiz N., Hacioglu Y., and Taskin Y., "Fuzzy sliding-mode control of active suspensions," *IEEE Transactions on Industrial Electronics,* vol. 55, pp. 3883–3890, 2008.

[34] Leonhard W., *Control of electrical drives*. Berlin: Springer Verlag, 2001.

http://dx.doi.org/10.5772/61071

[35] Kung C. C. and Su K. H., "Adaptive fuzzy position control for electrical servodrive via total-sliding-mode technique," *IEE Proceedings on Electric Power Applications*, vol. 152, pp. 1489–1502, 2005.

[36] Wang W. J. and Chen J. Y., "A new sliding mode position controller with adaptive load torque estimator for an induction motor," *IEEE Transactions on Energy Conversion*, vol. 14, pp. 413–418, 1999.

[37] Akin B. and Bhardwaj M., "Sensored field oriented control of 3-phase induction motors," *Texas Instrument Guide*, 2013.

[38] Saghafinia A., Ping H. W., Uddin M. N., and Amindoust A., "Teaching of simulation an adjustable speed drive of induction motor using MATLAB/Simulink in advanced electrical machine laboratory," *Pro*cedia-Social and Behavioral Sciences, vol. 103, pp. 912–921, 2013.

Chapter 8

A Robust Induction Motor Control using Sliding Mode Rotor Flux and Load Torque Observers

Oscar Barambones, Patxi Alkorta,
Jose M. Gonzalez de Duran and Jose A. Cortajarena

Additional information is available at the end of the chapter

Abstract

A sliding mode position control for high-performance real-time applications of induction motors is developed in this work. The design also incorporates a sliding mode based flux and load torque observers in order to avoid this sensors, that increases the cost and reduces the reliability. Additionally, the proposed control scheme presents a low computational cost and therefore can be implemented easily in a real-time applications using a low cost DSP-processor.

The stability analysis of the controller under parameter uncertainties and load disturbances is provided using the Lyapunov stability theory. Finally simulated and experimental results show that the proposed controller with the proposed observer provides a good trajectory tracking and that this scheme is robust with respect to plant parameter variations and external load disturbances.

Keywords: Position Control, Sliding Mode Control, Robust Control, Induction Machines, Lyapunov Stability, Nonlinear Control

1. Introduction

AC induction motors have been widely used in industrial applications such machine tools, steel mills and paper machines owing to their good performance provided by their solid architecture, low moment of inertia, low ripple of torque and high initiated torque. Some control techniques have been developed to regulate these induction motor servo drives in high-performance applications. One of the most popular technique is the indirect field oriented control method [1, 2].

The field-oriented technique guarantees the decoupling of torque and flux control commands of the induction motor, so that the induction motor can be controlled linearly as a separated excited D.C. motor. However, the control performance of the resulting linear system is still influenced by uncertainties, which usually are composed of unpredictable parameter variations, external load disturbances, and unmodelled and nonlinear dynamics.

In the last decades the proportional integral derivative (PID) controller has been widely used in the vector control of induction motors due to its good performance and its simple structure. However in some applications the PID controller may not meet the concerned robustness under parameter variations and external load disturbances. Therefore, many studies have been made on the motor drives in order to preserve the performance under these parameter variations and external load disturbance, such as nonlinear control, multivariable control, optimal control, H-∞ control and adaptive control [3]-[7], etc. However usually these controllers present a high computational cost and cannot be implemented over a low cost DSP processor to perform a real time control.

The sliding-mode control can offer many good properties, such as good performance against unmodelled dynamics, insensitivity to parameter variations, external disturbance rejection and fast dynamic response. These advantages of the sliding-mode control may be employed in the position and speed control of an AC servo system [8]. In [9] an integral sliding mode speed control for induction motor based on field oriented control theory is proposed. In the work of [10], an integrated sliding mode controller (SMC) based on space vector pulse width modulation method is proposed to achieve high-performance speed control of an induction motor. In this work using a field-oriented control principle, a flux SMC is first established to achieve fast direct flux control and then a speed SMC is presented to enhance speed control by the direct torque method. However in this work the performance of the proposed controller is not validated over a real induction motor.

Position control is often used in some applications of electrical drives like robotic systems, conveyor belts, etc. In these applications uncertainty and external disturbances are present and therefore a robust control system that maintain the desired control performance under this situations are frequently required [11]-[15].

The variable structure control strategy using the sliding-mode has also been focussed on many studies and research for the position control of the induction motors [16, 17].

The induction motor position control problem has been studied in [18] using a discrete time sliding mode control. The field oriented control theory is also used in order to decouple the flux and the electromagnetic torque. In this paper the authors calculates the rotor flux vector angular position using the slip estimates which is very sensitive to the rotor resistance variation. In contrast, a rotor flux sliding mode observer is proposed in our control scheme, in order to calculate an accurate value for the angular position of the rotor flux vector in the presence of system uncertainties.

On the other hand in the last decade remarkable efforts have been made to reduce the number of sensors in the control systems [19]-[22]. The sensors increases the cost and also reduces the reliability of the control system because this elements are generally expensive, delicate and difficult to instal.

In this chapter a robust approach for induction motor position control is presented. The proposed sliding mode control may overcome the system uncertainties and load disturbances that usually are present in the real systems. In the controller design, the field oriented control theory is used to simplify the system dynamical equations. Moreover, the proposed controller does not present a high computational cost and therefore can be implemented easily in a real-time applications using a low cost DSP-processor.

In this work a sliding mode flux observer is proposed in order to avoid the flux sensors. The estimated rotor flux is used to calculate the rotor flux vector angular position, whose value is essential in order to apply the field oriented control principle. A load torque estimation algorithm, based on a sliding mode observer, is also presented in order to improve the adaptive robust position control performance. Additionally, the overall control scheme does not involve a high computational cost and therefore can be implemented easily in a real time applications.

Moreover, the control scheme presented in this chapter is validated in a real test using a commercial induction motor of 7.5 kW in order to demonstrate the real performance of this controller. The experimental validation has been implemented using a control platform based on a DS1103 PPC Controller Board that has been designed and constructed in order to carry out the experimental validation of the proposed controller.

This manuscript is organized as follows. The sliding mode flux observer is developed in Section 2 and the sliding mode load torque observer is designed in Section 3. Then, the proposed sliding mode position control is presented in Section 4. In the Section 5, the experimental control platform is presented and some simulation and experimental results are carried out. Finally, concluding remarks are stated in Section 6.

2. Sliding mode observer for rotor flux estimator

Many schemes based on simplified motor models have been devised to estimate some internal variables of the induction motor from measured terminal quantities. This procedure is frequently used in order to avoid the presence of some sensors in the control scheme. In order to obtain an accurate dynamic representation of the motor, it is necessary to base the calculation on the coupled circuit equations of the motor.

Since the motor voltages and currents are measured in a stationary frame of reference, it is also convenient to express the induction motor dynamical equations in this stationary frame.

The system state space equations in the stationary reference frame can be written in the form [23]:

$$\begin{aligned}
\dot{i}_{ds} &= \frac{-1}{\sigma L_s}\left(R_s + \frac{L_m^2}{L_r^2}R_r\right)i_{ds} + \frac{L_m}{\sigma L_s L_r}\frac{1}{T_r}\psi_{dr} \\
&\quad + \frac{L_m}{\sigma L_s L_r}w_r\psi_{qr} + \frac{1}{\sigma L_s}V_{ds} \\
\dot{i}_{qs} &= \frac{-1}{\sigma L_s}\left(R_s + \frac{L_m^2}{L_r^2}R_r\right)i_{qs} - \frac{L_m}{\sigma L_s L_r}w_r\psi_{dr} \\
&\quad + \frac{L_m}{\sigma L_s L_r}\frac{1}{T_r}\psi_{qr} + \frac{1}{\sigma L_s}V_{qs} \\
\dot{\psi}_{dr} &= \frac{L_m}{T_r}i_{ds} - \frac{1}{T_r}\psi_{dr} - w_r\psi_{qr} \\
\dot{\psi}_{qr} &= \frac{L_m}{T_r}i_{qs} + w_r\psi_{dr} - \frac{1}{T_r}\psi_{qr}
\end{aligned} \tag{1}$$

where V_{ds}, V_{qs} are stator voltages; i_{ds}, i_{qs} are stator currents; ψ_{dr}, ψ_{qr} are rotor fluxes; w_r is motor speed; R_s, R_r are stator and rotor resistances; L_s, L_r are stator and rotor inductances; L_m, is mutual inductance; $\sigma = 1 - \frac{L_m^2}{L_s L_r}$ is leakage coefficient; $T_r = \frac{L_r}{R_r}$ is rotor-time constant.

From singular perturbation theory [24], and based on the well-known induction motor model dynamics [23], the slow variables of the system are ψ_{dr}, ψ_{qr} and the fast variables are i_{ds}, i_{qs}. Therefore, the corresponding singularly perturbed model of eqn.(1) is:

$$\begin{aligned}
\varepsilon \dot{i}_{ds} &= -L_m \alpha_r i_{ds} + \alpha_r \psi_{dr} + w_r \psi_{qr} + \frac{L_r}{L_m}\left(V_{ds} - R_s i_{ds}\right) \\
\varepsilon \dot{i}_{qs} &= -L_m \alpha_r i_{qs} - w_r \psi_{dr} + \alpha_r \psi_{qr} + \frac{L_r}{L_m}\left(V_{qs} - R_s i_{qs}\right) \\
\dot{\psi}_{dr} &= L_m \alpha_r i_{ds} - \alpha_r \psi_{dr} - w_r \psi_{qr} \\
\dot{\psi}_{qr} &= L_m \alpha_r i_{qs} + w_r \psi_{dr} - \alpha_r \psi_{qr}
\end{aligned} \tag{2}$$

where $\varepsilon = \frac{\sigma L_s L_r}{L_m}$ and $\alpha_r = \frac{1}{T_r}$.

The proposed sliding mode observer is a copy of the original system model, which has corrector terms with switching functions based on the system outputs. Therefore, considering the measured stator currents as the system outputs, the corresponding sliding-mode-observer can be constructed as follows:

$$\begin{aligned}
\varepsilon \dot{\hat{i}}_{ds} &= -L_m \alpha_r i_{ds} + \alpha_r \hat{\psi}_{dr} + w_r \hat{\psi}_{qr} + \frac{L_r}{L_m}\left(V_{ds} - R_s i_{ds}\right) \\
&\quad -k_1 e_{i_d} + g_{i_d}\,\mathrm{sgn}(e_{id}) \\
\varepsilon \dot{\hat{i}}_{qs} &= -L_m \alpha_r i_{qs} - w_r \hat{\psi}_{dr} + \alpha_r \hat{\psi}_{qr} + \frac{L_r}{L_m}\left(V_{qs} - R_s i_{qs}\right) \\
&\quad -k_2 e_{iq} + g_{i_q}\,\mathrm{sgn}(e_{iq}) \\
\dot{\hat{\psi}}_{dr} &= L_m \alpha_r i_{ds} - \alpha_r \hat{\psi}_{dr} - w_r \hat{\psi}_{qr} + g_{\psi_d}\,\mathrm{sgn}(e_{id}) \\
\dot{\hat{\psi}}_{qr} &= L_m \alpha_r i_{qs} + w_r \hat{\psi}_{dr} - \alpha_r \hat{\psi}_{qr} + g_{\psi_q}\,\mathrm{sgn}(e_{iq})
\end{aligned} \tag{3}$$

where $\hat{i}$ and $\hat{\psi}$ are the estimations of i and ψ; k_1 and k_2 are positive constant gains; g_{i_d}, g_{i_q}, g_{ψ_d} and g_{ψ_q} are the observer gain matrix; $e_{i_d} = \hat{i}_{ds} - i_{ds}$ and $e_{i_q} = \hat{i}_{qs} - i_{qs}$ are de current errors, and sgn() is the sign function.

Subtracting eqn.(2) from eqn.(3), the estimation error dynamics are:

$$\begin{aligned}
\varepsilon \dot{e}_{i_d} &= \alpha_r e_{\psi_d} + w_r e_{\psi_q} - k_1 e_{i_d} + g_{i_d}\,\mathrm{sgn}(e_{id}) \\
\varepsilon \dot{e}_{i_q} &= -w_r e_{\psi_d} + \alpha_r e_{\psi_q} - k_2 e_{iq} + g_{i_q}\,\mathrm{sgn}(e_{iq}) \\
\dot{e}_{\psi_d} &= -\alpha_r e_{\psi_d} - w_r e_{\psi_q} + g_{\psi_d}\,\mathrm{sgn}(e_{id}) \\
\dot{e}_{\psi_q} &= w_r e_{\psi_d} - \alpha_r e_{\psi_q} + g_{\psi_q}\,\mathrm{sgn}(e_{iq})
\end{aligned} \tag{4}$$

where $e_{\psi_d} = \hat{\psi}_{dr} - \psi_{dr}, e_{\psi_q} = \hat{\psi}_{qr} - \psi_{qr}$

The previous equations can be expressed in matrix form as:

$$\begin{aligned} \varepsilon \dot{e}_i &= +Ae_\psi + K_i e_i + G_i Y_e \\ \dot{e}_\psi &= -Ae_\psi + G_\psi Y_e \end{aligned} \tag{5}$$

where $A = \alpha_r I_2 - w_r J_2$, $e_i = [e_{i_d}\ e_{i_q}]^T$, $e_\psi = [e_{\psi_d}\ e_{\psi_q}]^T$, $Y_e = [\mathrm{sgn}(e_{id})\ \mathrm{sgn}(e_{iq})]^T$,

$$G_i = \begin{bmatrix} g_{i_d} & 0 \\ 0 & g_{i_q} \end{bmatrix}, G_\psi = \begin{bmatrix} g_{\psi_d} & 0 \\ 0 & g_{\psi_q} \end{bmatrix}$$

$$I_2 = \begin{bmatrix} 1 & 0 \\ 0 & 1 \end{bmatrix}, J_2 = \begin{bmatrix} 0 & -1 \\ 1 & 0 \end{bmatrix}, K_i = \begin{bmatrix} -k_1 & 0 \\ 0 & -k_2 \end{bmatrix}$$

Following the two-time-scale approach, the stability analysis of the above system can be considered determining the observer gains G_i and K_i of the fast subsystem or measured state variables (i_{ds}, i_{qs}), to ensure the attractiveness of the sliding surface $e_i = 0$ in the fast time scale. Thereafter, the observer gain G_ψ of the slow subsystem or inaccessible state variables (ψ_{dr}, ψ_{qr}), are determined, such that the reduced-order system obtained when $e_i \cong \dot{e}_i \cong 0$ is locally stable [24].

From singular perturbation theory, the fast reduced-order system of the observation errors can be obtained by introducing the new time variable $\tau = (t - t_0)/\varepsilon$ and thereafter setting $\varepsilon \to 0$ [24]. In the new time scale τ, taking into account that $d\tau = dt/\varepsilon$, eqn.(5) becomes:

$$\begin{aligned} \frac{d}{d\tau} e_i &= Ae_\psi + K_i e_i + G_i Y_e \\ \frac{d}{d\tau} e_\psi &= 0 \end{aligned} \tag{6}$$

Therefore, if the observer gains G_i and K_i are adequately chosen, the sliding mode occurs in eqn.(6) along the manifold $e_i = [e_{i_d}\ e_{i_q}]^T = 0$.

The attractivity condition of the sliding surface $e_i = 0$ given by:

$$e_i^T \frac{de_i}{d\tau} < 0 \tag{7}$$

is fulfilled with the following inequalities,

$$g_{i_d} < -\left|\alpha_r e_{\psi_d} + w_r e_{\psi_q}\right| - k_1 \left|e_{id}\right| \tag{8}$$

$$g_{i_q} < -\left|-w_r e_{\psi_d} + \alpha_r e_{\psi_q}\right| - k_2 \left|e_{iq}\right| \tag{9}$$

Proof:

Let us define the following Lyapunov function candidate,

$$V = \frac{1}{2}e_i^T e_i$$

whose time derivative is,

$$\begin{aligned} \frac{dV}{d\tau} &= e_i^T \frac{de_i}{d\tau} \\ &= e_i^T \left[Ae_\psi + K_i e_i + G_i Y_e \right] \\ &= \begin{bmatrix} e_{id} \left\{ g_{i_d} \operatorname{sgn}(e_{id}) + \alpha_r e_{\psi_d} + w e_{\psi_q} - k_1 e_{id} \right\} \\ e_{iq} \left\{ g_{i_q} \operatorname{sgn}(e_{iq}) - w e_{\psi_d} + \alpha_r e_{\psi_q} - k_2 e_{iq} \right\} \end{bmatrix} \end{aligned} \tag{10}$$

Taking into account that all states and parameters of induction motor are bounded, then there exist sufficiently large negative numbers g_{i_d}, g_{i_q}, and positive numbers k_1 and k_2 so that the inequalities defined in eqn.(9) are verified and then the attractivity condition defined in eqn.(7) is fulfilled.

Then, once the currents trajectory reaches the sliding surface $e_i = 0$, the observer error dynamics given by eqn.(6) behaves, in the sliding mode, as a reduced-order subsystem governed only by the rotor-flux error e_ψ, assuming that $e_i = \dot{e}_i = 0$.

The slow error dynamics (when $e_i = 0$ and $\dot{e}_i = 0$), can be obtained setting $\varepsilon = 0$ in the system equation presented in eqn.(5):

$$\begin{aligned} 0 &= +Ae_\psi + G_i Y_e \\ \dot{e}_\psi &= -Ae_\psi + G_\psi Y_e \end{aligned} \tag{11}$$

In order to demonstrate de stability of the previous system, the following Lyapunov function candidate is proposed:

$$V = \frac{1}{2}e_\psi^T e_\psi \tag{12}$$

The time derivative of the Lyapunov function candidate is:

$$\frac{dV}{dt} = \dot{e}_\psi^T e_\psi \tag{13}$$

From eqn.(11) it is deduced that:

$$e_\psi = -A^{-1}G_iY_e \tag{14}$$
$$\dot{e}_\psi = (G_i + G_\psi)Y_e \tag{15}$$

Then from eqn.(13), (14) and (15)

$$\begin{aligned}
\frac{dV}{dt} &= -Y_e^T(G_i + G_\psi)^T A^{-1}G_iY_e \\
&= -Y_e^T(G_i + G_\psi)^T A^{-1}G_iY_e \\
&= -Y_e^T\left[(I_2 + G_\psi G_i^{-1})G_i\right]^T A^{-1}G_iY_e \\
&= -Y_e^T G_i^T(I_2 + G_\psi G_i^{-1})^T A^{-1}G_iY_e \\
&= -Y_e^T G_i^T(A^{-1})^T A^T(I_2 + G_\psi G_i^{-1})^T A^{-1}G_iY_e \\
&= -(A^{-1}G_iY_e)^T A^T(I_2 + G_\psi G_i^{-1})^T A^{-1}G_iY_e \\
&= -e_\psi^T A^T(I_2 + G_\psi G_i^{-1})^T e_\psi \\
&= -e_\psi^T(I_2 + G_\psi G_i^{-1})A\, e_\psi
\end{aligned} \tag{16}$$

To ensure that $\dot{V}$ is negative definite the following sufficient condition can be requested:

$$(I_2 + G_\psi G_i^{-1})A \geq \varrho I_2 \tag{17}$$

where ϱ is a positive constant

Solving the gain matrix G_ψ in eqn.(17) yields:

$$(I_2 + G_\psi G_i^{-1}) \geq \varrho I_2 A^{-1} \tag{18}$$
$$(I_2 + G_\psi G_i^{-1}) \geq \varrho A^{-1} \tag{19}$$
$$G_\psi G_i^{-1} \geq \varrho A^{-1} - I_2 \tag{20}$$
$$G_\psi \leq (\varrho A^{-1} - I_2)G_i \tag{21}$$

Therefore, the time derivative of the Lyapunov function will be negative definite if the observer gain G_ψ is chosen taking into account eqn.(21). As a result from eqn.(16) it is concluded that the equilibrium point ($e_\psi = 0$) of the flux observer error dynamic given by eqn.(11) is exponentially stable; that is, the flux observer error converges to zero with exponential rate of convergence.

3. Load Torque Observer

In the traditional sliding mode control schemes, the load torque should be known or should be measured using a torque sensors in order to compensate this load torque. On the other hand, the load torque could be also considered as a system uncertainty, but in this case the control system should be robust under all load torque values that would appear over the time and therefore the sliding gain should be adequately high in order to compensate the these load torque values. Obviously these high values for the sliding gain will increase the control activity and are undesirable in the real applications.

Therefore, when the load torque is unknown or is very variable over time, and the system has no torque sensors, a good solution could be the use of a load torque estimator. In this chapter a sliding mode observer is proposed in order to obtain the load torque applied to the induction motor without requiring the use of the load torque sensor.

The mechanical equation of an induction motor can be written as:

$$J\ddot{\theta}_m + B\dot{\theta}_m + T_L = T_e \tag{22}$$

where J and B are the inertia constant and the viscous friction coefficient of the induction motor respectively; T_L is the external load; θ_m is the rotor mechanical position, which is related to the rotor electrical position, θ_r, by $\theta_m = 2\theta_r/p$ where p is the pole numbers and T_e denotes the generated torque of an induction motor, defined as [23]:

$$T_e = \frac{3p}{4}\frac{L_m}{L_r}(\psi^e_{dr} i^e_{qs} - \psi^e_{qr} i^e_{ds}) \tag{23}$$

where ψ^e_{dr} and ψ^e_{qr} are the rotor-flux linkages, with the subscript 'e' denoting that the quantity is referred to the synchronously rotating reference frame; i^e_{ds} and i^e_{qs} are the d-q stator current components, and p is the pole numbers.

The relation between the synchronously rotating reference frame and the stationary reference frame is performed by the so-called reverse Park's transformation:

$$\begin{bmatrix} x_a \\ x_b \\ x_c \end{bmatrix} = \begin{bmatrix} \cos(\theta_e) & -\sin(\theta_e) \\ \cos(\theta_e - 2\pi/3) & -\sin(\theta_e - 2\pi/3) \\ \cos(\theta_e + 2\pi/3) & -\sin(\theta_e + 2\pi/3) \end{bmatrix} \begin{bmatrix} x^e_d \\ x^e_q \end{bmatrix} \tag{24}$$

where θ_e is the angular position between the d-axis of the synchronously rotating reference frame and the a-axis of the stationary reference frame, and it is assumed that the quantities are balanced.

Using the field-orientation control principle, the current component i^e_{ds} is aligned in the direction of the rotor flux vector $\bar{\psi}_r$, and the current component i^e_{qs} is aligned in the perpendicular direction to it. At this condition, it is satisfied that:

$$\psi^e_{qr} = 0, \qquad \psi^e_{dr} = |\bar{\psi}_r| \tag{25}$$

Taking into account the results presented in eqn.(25), the induction motor torque of eqn.(23) is simplified to:

$$T_e = \frac{3p}{4}\frac{L_m}{L_r}\psi^e_{dr} i^e_{qs} = K_T\, i^e_{qs} \tag{26}$$

where K_T is the torque constant, defined as follows:

$$K_T = \frac{3p}{4}\frac{L_m}{L_r}\psi^{e^*}_{dr} \tag{27}$$

where $\psi^{e^*}_{dr}$ denotes the command rotor flux.

With the above mentioned proper field orientation, the rotor flux dynamics is given by [23]:

$$\frac{d\psi^e_{dr}}{dt} + \frac{\psi^e_{dr}}{T_r} = \frac{L_m}{T_r}\, i^e_{ds} \tag{28}$$

From the system mechanical equation eqn.(22) and the induction motor torque equation eqn.(26), the following dynamic equation is obtained:

$$\dot{w}_m = -\frac{B}{J} w_m + \frac{K_T}{J}\, i^e_{qs} - \frac{1}{J} T_L \tag{29}$$

where $w_m = \dot{\theta}_m$

It is assumed that the load torque only changes at certain moments, and therefore the load torque can be considered as a quasi-constant signal:

$$\dot{T}_L = 0 \tag{30}$$

Therefore, the system state space equations are:

$$\begin{aligned} \dot{w}_m &= -\frac{B}{J} w_m + \frac{K_T}{J}\, i^e_{qs} - \frac{1}{J} T_L \\ \dot{T}_L &= 0 \end{aligned} \tag{31}$$

Taking into account that the load torque T_L is taken as a quasi-constant signal, the load torque can be considered the slow component of this system. Therefore, from singular perturbation theory [24], the stability of the above system can be demonstrated assuring the asymptotic stability of the fast component of this system (the rotor speed), and thereafter the convergence of the slow component (the load torque) for the reduced system, when the rotor speed estimation error is zero.

Then, from eqn.(31) the sliding-mode-observer can be constructed as:

$$\begin{aligned}\dot{\hat{w}}_m &= \frac{-B}{J}w_m + \frac{K_T}{J}i_{qs}^e - \frac{1}{J}\hat{T}_L + k_{w_1}e_w + h_1\,\text{sgn}(e_w)\\ \dot{\hat{T}}_L &= -k_{w_2}e_w - h_2\,\text{sgn}(e_w)\end{aligned} \tag{32}$$

where $e_w = w_m - \hat{w}_m$, and k_{w_1}, k_{w_2}, h_1 and h_2 are a positive constants.

Subtracting eqn. (32) from (31), the estimation error dynamic is obtained:

$$\begin{aligned}\dot{e}_w &= -\frac{1}{J}e_T - k_{w_1}e_w - h_1\,\text{sgn}(e_w)\\ \dot{e}_T &= k_{w_2}e_w + h_2\,\text{sgn}(e_w)\end{aligned} \tag{33}$$

where $e_T = T_L - \hat{T}_L$

In order to demonstrate the stability of the fast component of the system the following Lyapunov function candidate is proposed:

$$V = \frac{1}{2}e_w^2 \tag{34}$$

The time derivative of this Lyapunov function candidate is:

$$\dot{V} = e_w\dot{e}_w \tag{35}$$

$$= e_w\left(-\frac{1}{J}e_T - k_{w_1}e_w - h_1\,\text{sgn}(e_w)\right) \tag{36}$$

$$= -h_1|e_w| - k_{w_1}e_w^2 - \frac{1}{J}e_we_T \tag{37}$$

To ensure that $\dot{V}$ is negative definite the following sufficient condition can be requested:

$$h_1 \geq \left|\frac{1}{J}e_T\right| - k_{w_1}|e_w| + \eta_w \tag{38}$$

where $\eta_w > 0$

Therefore,

$$\dot{V} \leq -\eta_w|e_w| \tag{39}$$

From eqn.(39) it is deduced that the equilibrium point $e_w = 0$ is asymptotically stable, and from this equation it can be also deduced that the maximum time in order to reach the equilibrium point $e_w = 0$ is:

$$t_{reach} \leq \frac{e_w(t=0)}{\eta_w} \tag{40}$$

When the speed observation error reaches the equilibrium point, $e_w = 0$ and $\dot{e}_w = 0$, and then from eqn.(33) it is obtained that the observer error dynamics behaves as the reduced-order subsystem presented below:

$$0 = -\frac{1}{J}e_L - h_1 \operatorname{sgn}(e_w) \tag{41}$$

$$\dot{e}_T = h_2 \operatorname{sgn}(e_w) \tag{42}$$

From the previous equations it is obtained:

$$\dot{e}_T = -\frac{1}{J}\frac{h_2}{h_1}e_T \tag{43}$$

The solution of the previous differential equation is:

$$e_T(t) = C \exp\left(-\frac{1}{J}\frac{h_2}{h_1}t\right) \tag{44}$$

Consequently, the load torque estimation error tends exponentially to zero.

Therefore, if the observer gains h_1, h_2 k_{w_1} and k_{w_2} are adequately chosen, then the estimation error converges to zero. Consequently the estimated states $\hat{w}_m$, $\hat{T}_L$ converges to the real states w_m, T_L as t tends to infinity. Hence, the load torque may be obtained from the states observer given by eqn.(32), that uses the rotor speed and the stator current in order to obtain the load torque applied to the induction motor.

4. Variable structure robust position control

The mechanical equation of an induction motor presented in equation (29) can be rewritten as:

$$\ddot{\theta}_m + a\,\dot{\theta}_m + \bar{f} = b\,i_{qs}^e \tag{45}$$

where the parameters are defined as:

$$a = \frac{B}{J}, \quad b = \frac{K_T}{J}, \quad \bar{f} = \frac{T_L}{J}; \tag{46}$$

Now, the previous mechanical equation (45) is considered with uncertainties as follows:

$$\ddot{\theta}_m = -(a + \Delta a)\dot{\theta}_m - (f + \Delta f) + (b + \Delta b)i^e_{qs} \tag{47}$$

where $f = \frac{\hat{T}_L}{J}$, and the terms Δa, Δb and Δf represents the uncertainties of the terms a, b and f respectively.

It should be noted that the load torque T_L has been replaced by the estimated load torque $\hat{T}_L$ and the difference between the real and the estimated value is taken as an uncertainty.

Let us define the position tracking error as follows:

$$e(t) = \theta_m(t) - \theta^*_m(t) \tag{48}$$

where θ^*_m is the rotor position command.

Taking the second derivative of the previous equation with respect to time yields:

$$\ddot{e}(t) = \ddot{\theta}_m - \ddot{\theta}^*_m = u(t) + d(t) \tag{49}$$

where the following terms have been collected in the signal $u(t)$,

$$u(t) = b\, i^e_{qs}(t) - a\,\dot{\theta}_m(t) - f(t) - \ddot{\theta}^*_m(t) \tag{50}$$

and the uncertainty terms have been collected in the signal $d(t)$,

$$d(t) = -\Delta a\, w_m(t) - \Delta f(t) + \Delta b\, i^e_{qs}(t) \tag{51}$$

Now, the sliding variable $S(t)$ is defined as:

$$S(t) = \dot{e}(t) + k\,e(t) + k_i \int e(t)\,dt \tag{52}$$

where k and k_i are a positive constant gains.

Then, the sliding surface is defined as:

$$S(t) = \dot{e}(t) + k\,e(t) + k_i \int e(t)\,dt = 0 \tag{53}$$

The sliding mode controller is designed as:

$$u(t) = -k\dot{e} - k_i e - \beta\,\mathrm{sgn}(S) \tag{54}$$

where k and k_i are the previously defined positive constant gains, β is the switching gain, S is the sliding variable defined in eqn. (52) and $\mathrm{sgn}(\cdot)$ is the sign function.

Assumption. In order to obtain the position trajectory tracking, the gain β must be chosen so that $\beta \geq \bar{d}$ where $\bar{d} \geq \sup_{t \in R^{0+}} |d(t)|$. Note that this condition only implies that the system uncertainties are bounded magnitudes.

Theorem. Consider the induction motor given by equation (47), the control law (54) leads the rotor mechanical position $\theta_m(t)$ so that the position tracking error $e(t) = \theta_m(t) - \theta_m^*(t)$ tends to zero as the time tends to infinity.

Proof: Define the Lyapunov function candidate:

$$V(t) = \frac{1}{2}S(t)S(t) \tag{55}$$

Its time derivative is calculated as:

$$\begin{aligned}
\dot{V}(t) &= S(t)\dot{S}(t) \\
&= S \cdot [\ddot{e} + k\dot{e} + k_i e] \\
&= S \cdot [u + d + k\dot{e} + k_i e] \\
&= S \cdot [-k\dot{e} - k_i e - \beta\,\mathrm{sgn}(S) + d + k\dot{e} + k_i e] \\
&= S \cdot [d - \beta\,\mathrm{sgn}(S)] \\
&\leq -(\beta - |d|)|S| \\
&\leq 0
\end{aligned} \tag{56}$$

It should be noted that the eqns. (52), (49) and (54) have been used in the proof.

Using the Lyapunov's direct method, since $V(t)$ is clearly positive-definite, $\dot{V}(t)$ is negative definite and $V(t)$ tends to infinity as $S(t)$ tends to infinity, then the equilibrium at the origin $S(t) = 0$ is globally asymptotically stable. Therefore $S(t)$ tends to zero as the time t tends to infinity. Moreover, all trajectories starting off the sliding surface $S = 0$ must reach it in finite time and then they will remain on this surface. This system's behavior, once on the sliding surface is usually called *sliding mode.*

When the sliding mode occurs on the sliding surface (53), then $S(t) = \dot{S}(t) = 0$, and therefore the dynamic behavior of the tracking problem (49) is equivalently governed by the following equation:

$$\dot{S}(t) = 0 \quad \Rightarrow \quad \ddot{e}(t) + k\,\dot{e}(t) + k_i e(t) = 0 \tag{57}$$

Then, like k and k_i are a positive constants, the tracking error $e(t)$ and its derivatives $\dot{e}(t)$ and $\ddot{e}(t)$ converges to zero exponentially.

It should be noted that, a typical motion under sliding mode control consists of a *reaching phase* during which trajectories starting off the sliding surface $S = 0$ move toward it and reach it in finite time, followed by *sliding phase* during which the motion will be confined to this surface and the system tracking error will be represented by the reduced-order model (57), where the tracking error tends to zero.

Finally, the torque current command, $i_{qs}^{e*}(t)$, can be obtained directly substituting eqn. (54) in eqn. (50):

$$i_{qs}^{e*}(t) = \frac{1}{b}\left[-k\,\dot{e} - k_i\,e - \beta\,\mathrm{sgn}(S) + a\,\dot{\theta}_m + \ddot{\theta}_m^* + f(t)\right] \tag{58}$$

It should be noted that the current command is a bounded signal because all its components are bounded.

Therefore, the proposed variable structure position control resolves the position tracking problem for the induction motor in presence of some uncertainties in mechanical parameters and load torque.

It should be pointed out that, as it is well known, the variable structure control signals may produce the so-called chattering phenomenon, caused by the discontinuity that appear in eqn.(58) across the sliding surface. Chattering is undesirable in practice, since it involves high control activity and further may excite high-frequency dynamics. However, in the induction motor system, this high frequency changes in the electromagnetic torque will be filtered by the mechanical system inertia. Nevertheless, in order to reduce the chattering effect, the control law can also be smoothed out. In this case a simple and easy solution (proposed in [25]) could be to replace the sign function by a tansigmoid function in order to avoid the discontinuity in the control signal.

5. Simulation and Experimental Results

In this section the position regulation performance of the proposed sliding-mode field oriented control versus reference and load torque variations is analyzed by means of different simulation examples and real test using a commercial induction motor.

The block diagram of the proposed robust position control scheme is presented in Figure 1, and the function of the blocks that appear in this figure are explained below:

The block 'SMC Controller' represents the proposed sliding-mode controller, and it is implemented by equations (52) and (58). The block 'limiter' limits the current applied to the motor windings so that it remains within the limit value, being implemented by a saturation function. The block '$dq^e \rightarrow abc$' makes the conversion between the synchronously rotating

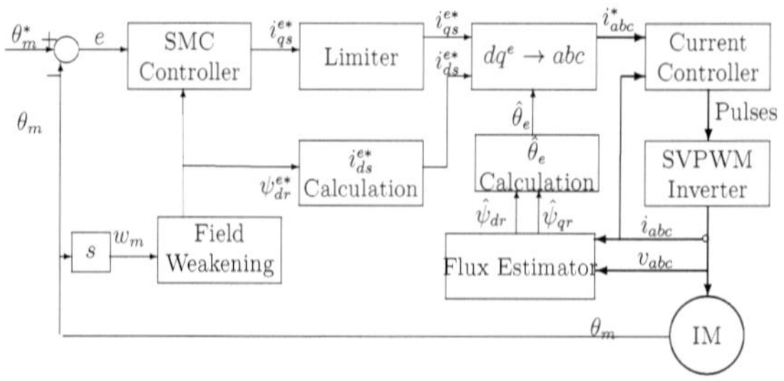

Figure 1. Block diagram of the proposed sliding-mode field oriented control

and stationary reference frames (Park's Transformation). The block 'Current Controller' consists of a SVPWM current control. The block 'SVPWM Inverter' is a six IGBT-diode bridge inverter with 540 V DC voltage source. The block 'Field Weakening' gives the flux command based on rotor speed, so that the PWM controller does not saturate. The block 'i_{ds}^{e*} Calculation' provides the current reference i_{ds}^{e*} from the rotor flux reference through the equation (28). The block 'Flux Estimator' represents the proposed sliding mode flux estimator, and it is implemented by the eqn.(3). The block '$\hat{\theta}_e$ Calculation' provides the angular position of the rotor flux vector. Finally, the block 'IM' represents the induction motor.

In order to carry out the real experimental validation of the proposed control scheme, the control platform show in figure 2 is used . The block diagram of this experimental platform is shown in figure 3.

This control platform allows to verify the real time performance of the induction motor controls in a real induction motor. The platform is formed by a PC with Windows XP in which it is installed MatLab7/Simulink R14 and ControlDesk 2.7 and the DS1103 Controller Board real time interface of dSpace. The power block is formed of a three-phase rectifier connected to 380 V/50 Hz AC electrical net and a capacitor bank of 27.200 μF in order to get a DC bus of 540 V. The platform also includes a three-phase IGBT/Diode bridge of 50A, and the M2AA 132M4 ABB induction motor of 7.5kW of die-cast aluminium squirrel-cage type and 1440 rpm, with the following parameters given by the manufacturer:

- w_N, nominal speed, 1440 rpm
- T_N, nominal torque, 49.3 Nm
- R_s, stator resistance, 0.81 Ω
- R_r, rotor resistance, 0.57 Ω
- L_m, magnetizing inductance, 0.118 mH
- L_s, stator inductance, 0.120 mH
- L_r, rotor inductance, 0.122 mH
- p, pair of poles, 2
- J, moment of inertia, 0.057 $kg\, m^2$
- B, viscous friction coefficient, 0.015 $N\,m/(rad/s)$

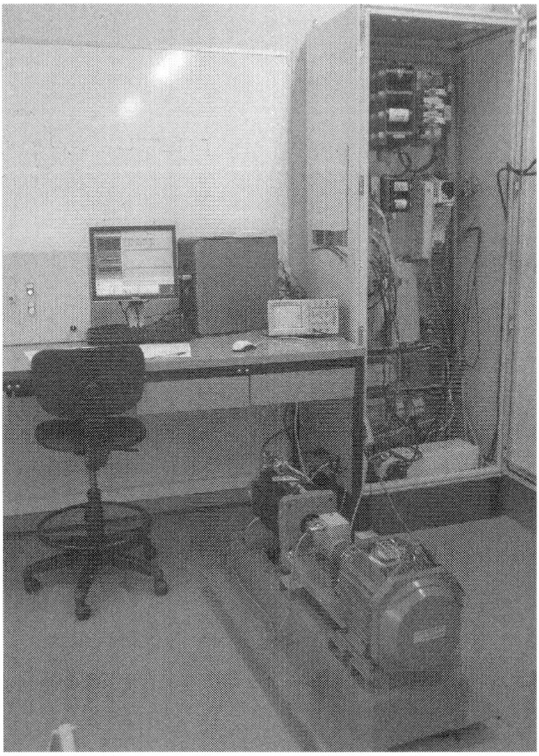

Figure 2. Induction motor experimental platform

- α_{Al}, temperature coefficient of Aluminium, $0.0039K^{-1}$

The rotor position of this motor is measured using the G1BWGLDBI LTN incremental rotary encoder of 4096 square impulses per revolution. This pulses are quadruplicated in a decoder, giving a resolution of 16384 ppr which gives an angle resolution of 0.000385 rad (0.022 deg).

The platform also includes a 190U2 Unimotor synchronous AC servo motor of 10.6 kW connected to the induction motor to generate the load torque (controlled in torque). This servo motor is controlled by its VSI Unidrive inverter module.

The sample time used to realize the real implementation of the the position control is $100\mu s$, and the processor used for the real tests is a floating point PowerPC at 1MHz, located in the real time DS1103 hardware of dSpace. This target incorporates the TMS320F240 DSP working as slave to generate the SVPWM pulses for the inverter. Finally, the position and currents control algorithms, the θ_e angle and flux estimator, the SVPWM calculations, and the Park's transformations have been realized in C programming language in a unique S-Builder module of Simulink, in order to obtain a compact and portable code.

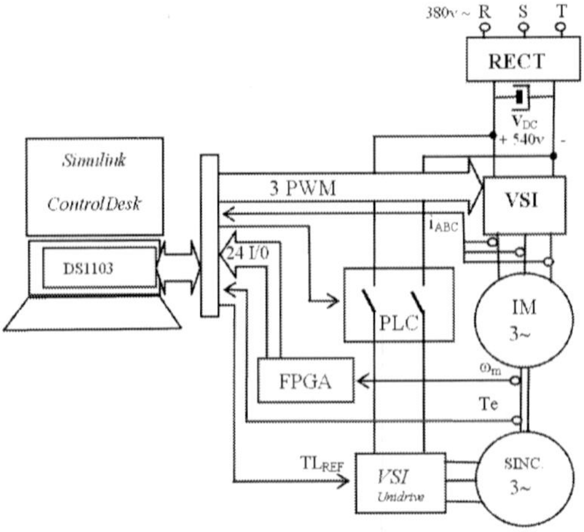

Figure 3. Block diagram of the induction motor experimental platform

In the experimental validation it is assumed that there is an uncertainty around 50% in the system mechanical parameters, that will be overcome by the proposed sliding mode control. The nominal value of the rotor flux is 1.01 Wb and it is obtained for a flux current command value of $i_{sd}^{*} = 8.61A$. However, in some cases, for a very high rotor speed, the flux command should be reduced so that the PWM controller does not saturate.

On the other hand, the electromagnetic torque current command, i_{sq}^{*}, has been limited to 30 A, in order to provide a protection against overcurrents in the induction motor's stator feed. Finally, the frequency of commutation of VSI module of the platform is limited to 8 kHz.

In this example the motor starts from a standstill state and it is required that the rotor position follows a position command, whose amplitude varies between 0 and 2π *rad*.

The system starts with an initial load torque $T_L = 0\,N.m$, and at time $t = 0.1\,s$, the load torque steps from $T_L = 0\,N.m$ to $T_L = 20\,N.m$, then at time $t = 1\,s$, the load torque steps from $T_L = 20\,N.m$ to $T_L = 40\,N.m$ and finally at time $t = 2\,s$, the load torque steps from $T_L = 40\,N.m$ to $T_L = 60\,N.m$, which is a 20% above the nominal torque value (49 Nm).

In these examples the values for the controller parameters are: $k = 46$, $k_i = 160$ and $\beta = 20$, the values for the flux observer parameters are: $g_{i_d} = -44.5$, $g_{i_q} = -44.5$, $g_{\psi_d} = -50$, $g_{\psi_q} = -50$, $k_1 = 100$ and $k_2 = 100$, and the values for the load torque observer parameters are: $k_{w_1} = 25$, $k_{w_2} = 250$ $h_1 = 100$ and $h_2 = 100$.

Figure 4 shows the simulation test of the proposed adaptive variable structure position control. The first graph shows the reference and the real rotor position, and the second graph shows the rotor position error. As it can be observed, after a transitory time in which the sliding gain is adapted, the rotor position tracks the desired position in spite of system uncertainties. Nevertheless, at time $t = 1s$ and $t = 2s$ a little position error can be observed. This error appears because there is a torque increment at this time, and then the controlled system lost the so called Śsliding modeŠ because the actual sliding gain is too small for the new uncertainty introduced in the system due to the load torque increment. However, after a short time, the new load torque value is adapted and then the sliding gain value can compensate the system uncertainties, and hence the rotor position error is eliminated. The third graph shows the real and the estimated rotor flux. In this figure it can be observed that the proposed sliding mode observer provides an accurate and fast rotor flux estimation. The fourth graph shows the motor torque, the load torque and the estimated load torque. As it can be seen in this graph, after a transitory time, the load torque observer estimates the load torque value with a small estimation error. This figure also shows that the so-called chattering phenomenon appears in the motor torque. Although this high frequency changes in the torque will be reduced by the mechanical system inertia, they could cause undesirable vibrations in the real rotor, which may be a problem for certain systems. However, for the systems that do not support this chattering, it may be eliminated substituting the sign function by the saturation function in the control signal. The fifth graph shows the stator current i_A. This graph shows that the current signal increases when the load torque increases in order to increment the motor torque. The sixth graph shows the time evolution of the sliding variable. In this figure it can be seen that the system reaches the sliding condition $(S(t) = 0)$ at time $t = 0.25s$, but then the system lost this condition at time $t = 1s$ and $t = 2s$ due to the load torque increment which produces an increment in the system uncertainties.

Figure 5 shows the real test of the variable structure position control using the experimental platform. In this figure, a small noise can be observed in the signals due to the sensors used to make the real measurements in the system. The first graph shows the reference and the real rotor position. Like in the previous case (simulation test), the rotor position tracks the reference position in spite of system uncertainties. The second graph shows the rotor position error. In this experimental validation a small position error is obtained in the presence of a high load torque. It should be noted that this performance is not an easy task to achieve for an induction motor. The third graph shows the estimated rotor flux and the fourth graph shows the motor torque, the load torque and the estimated load torque. It can be noted that the proposed sliding mode observers also perform very well in a practice. The fifth graph shows the stator current i_A, and finally the sixth graph shows the sliding variable S.

6. Conclusion

In this chapter an induction motor position regulation using a sliding mode control for a real-time applications has been presented. In the design a field oriented vector control theory is employed in order to simplify the system dynamic equations.

Additionally, in order to avoid the flux sensors, because the flux sensors increase the cost and reduces the reliability, a rotor flux estimator is proposed. This flux estimator is a sliding mode observer and employs the measured stator voltages and currents in the stationary reference frame. The design incorporates also a load torque observer, based on sliding mode theory, in order to improve the controller performance.

In order to demonstrate the performance of the proposed design over a commercial induction motor of 7.5 kW, a new experimental platform has been designed and constructed in order to text the proposed robust controller in a real time application over a high power commercial induction motor

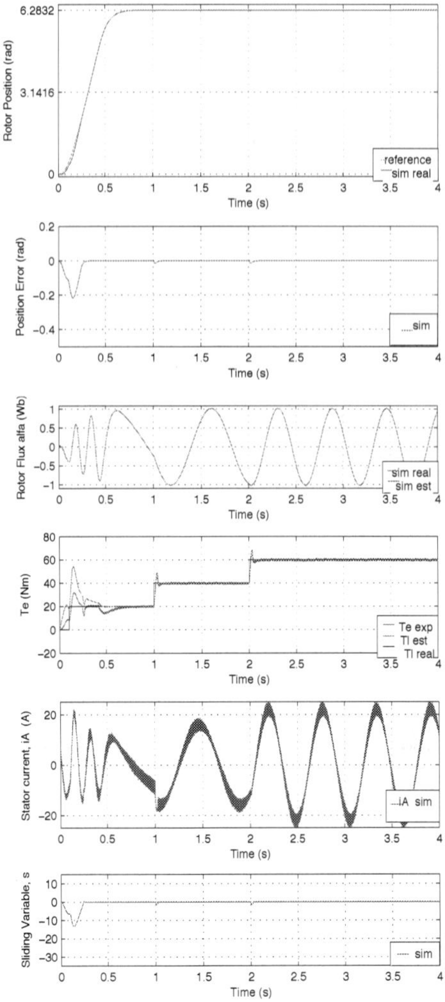

Figure 4. Position tracking simulation results

Due to the nature of the sliding mode control this control scheme is robust under system uncertainties and changes in the load torque applied to the induction motor. The closed loop stability of the presented design has been proved through Lyapunov stability theory.

Finally, by means of simulation and real examples, it has been confirmed that the proposed position control scheme presents a good performance in practice, and that the position tracking objective is achieved under parameter uncertainties and under load torque variations.

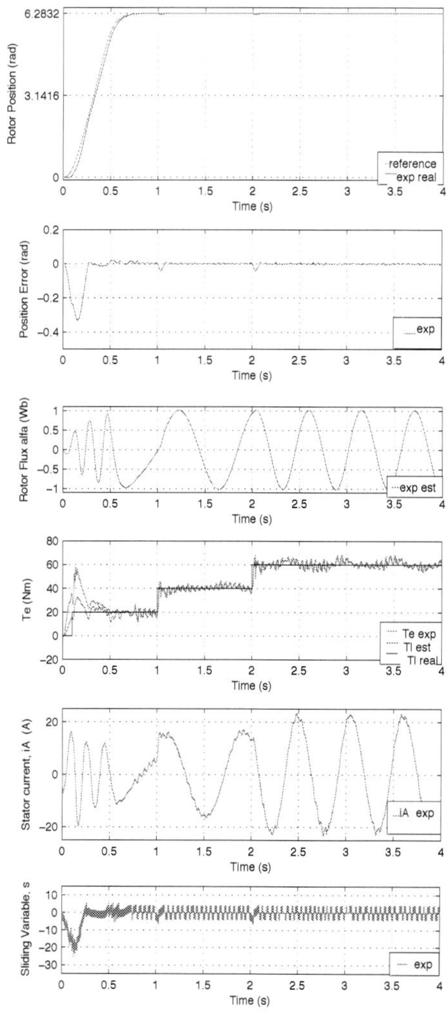

Figure 5. Position tracking experimental results

Acknowledgements

The authors are very grateful to the Basque Government by the support of this work through the project S-PE12UN015 and S-PE13UN039 and to the UPV/EHU by its support through the projects GIU13/41 and UFI11/07.

Author details

Oscar Barambones[1*], Patxi Alkorta[2], Jose M. Gonzalez de Durana[1] and Jose A. Cortajarena[2]

*Address all correspondence to: oscar.barambones@ehu.es

1 Vitoria Engineering School, Vitoria, Spain

2 Eibar Engineering School, Eibar, Spain

References

[1] P. Vas. *Vector Control of AC Machines*. Oxford Science Publications, Oxford. 1994

[2] W. Lehonhard. *Control of Electrical Drives*. Springer, Berlin. 1996

[3] R. Marino, P. Tomei, C.M. Verrelli. (2005) A nonlinear tracking control for sensorless induction motors, *Automatica*, vol. 41, pp. 1071-1077.

[4] R. Yazdanpanah, J. Soltani, G.R. Arab Markadeh. Nonlinear torque and stator flux controller for induction motor drive based on adaptive input-output feedback linearization and sliding mode control *Energy Conversion and Management*, vol. 49, pp. 541-550. 2008

[5] H. Ouadi, F.Giri, A.Elfadili, L.Dugard. (2010) Induction machine speed control with flux optimization, *Control Engineering Practice*, vol. 18, pp. 55-66.

[6] P-H Choua, C-S Chen, F-J Lin. (2012) DSP-based synchronous control of dual linear motors via Sugeno type fuzzy neural network compensator, *Journal of the Franklin Institute*, Vol. 349, pp.792-812.

[7] M.A. Hamidaa, A. Glumineau,n, J. de Leon. (2012) Robust integral backstepping control for sensorless IPM synchronous motor controller, *Journal of the Franklin Institute*, Vol. 349, pp.1734-1757.

[8] V.I. Utkin. Sliding mode control design principles and applications to electric drives, IEEE Trans. Indus. Electro., vol. 40, pp. 26-36. 1993

[9] O. Barambones and P. Alkorta, A robust vector control for induction motor drives with an adaptive sliding-mode control law, Journal of the Franklin Institute, vol. 348, pp. 300-314. 2011

[10] C.-Y Chen, Sliding mode controller design of induction motor based on space-vector pulse width modulation method, Int. J. of Innovative Computing, Information and Control 5, pp. 3603-3614. 2009.

[11] G.R. Lii, C.L. Chiang, C.T. Su and H.R. Hwung. An induction motor position controller optimally designed with fuzzy phase-plane control and genetic algorithms, Electric Power Systems Research, vol. 68, pp. 103-112. 2004.

[12] D. Naso, F. Cupertino and B. Turchiano. Precise position control of tubular linear motors with neural networks and composite learning, Control Engineering Practice, vol. 18, pp. 515-522. 2010.

[13] P. Alkorta, O. Barambones, J.A. Cortajarena, and A. Zubizarrreta, "Efficient Multivariable Generalized Predictive Control for Sensorless Induction Motor Drives", *IEEE Trans. Ind. Electron.*, vol. 61. pp.5126-5134. 2014.

[14] O. Barambones and P. Alkorta,, "Position Control of the Induction Motor using an adaptive sliding mode controller and observers", *IEEE Trans. Ind. Electron.*, vol. 61. pp.6556-6565. 2014.

[15] O. Barambones, P. Alkorta and J.M. Gonzalez de Durana, A real-time estimation and control scheme for induction motors based on sliding mode theory, Journal of the Franklin Institute, vol. 351, pp. 4251-4270. 2014

[16] A. Benchaib and C. Edwards. Nonlinear sliding mode control of an induction motor, *Int. J. of Adaptive Control and Signal Procesing*, vol. 14, pp. 201-221. 2000.

[17] W.J. Wang and J.Y. Chen . Passivity-based sliding mode position control for induction motor drives *IEEE Trans. on Energy conversion*, vol. 20, pp. 316-321. 2005.

[18] B. Veselić, B. Peruničit'ć-Draženović, and Č. Milosavljević, High-Performance Position Control of Induction Motor Using Discrete-Time Sliding-Mode Control, *IEEE Trans. Ind. Electron.*, vol. 55, no. 11, pp. 3809-3817, Nov. 2008.

[19] M. Menaa, O. Touhami, R. Ibtiouen and M. Fadel. Sensorless direct vector control of an induction motor, *Control Engineering Practice*, vol. 16, pp. 67-77. 2008

[20] O. Barambones, A.J. Garrido and I. Garrido. (2008) Robust speed estimation and control of an induction motor drive based on artificial neural networks, *Int. J. Adapt. Control Signal Process.*, Vol. 22, pp.440-464.

[21] H. Yang, Y. Xiaa, P. Shi. (2010) Observer-based sliding mode control for a class of discrete systems via delta operator approach, *Journal of the Franklin Institute*, Vol. 347, pp.1199-1213.

[22] A.Y.Alanisa, E.N.Sanchez, A.G.Loukianov, E.A. Hernandez. (2010) Discrete-time recurrent high order neural networks for nonlinear identification, *Journal of the Franklin Institute*, Vol. 347, pp.1253-1265.

[23] B.K. Bose. *Modern Power Electronics and AC Drives.*, Prentice Hall, New Jersey. 2001.

[24] P.V. Kokotovic, H. Khalil, J. OŚReilly. (1996) Singular Perturbation Methods in Control: Analysis and Design *Academic Press*, New York.

[25] O. Barambones, V. Etxebarria, Robust neural control for robotic manipulators, Automatica 38 pp.235-242. 2002

Chapter 9

An Optimized Hybrid Fuzzy-Fuzzy Controller for PWM-driven Variable Speed Drives

Nordin Saad, Muawia A. Magzoub, Rosdiazli Ibrahim and Muhammad Irfan

Additional information is available at the end of the chapter

http://dx.doi.org/10.5772/61086

Abstract

This paper discusses the performance and the impact of disturbances onto a proposed hybrid fuzzy-fuzzy controller (HFFC) system to attain speed control of a variable speed induction motor (IM) drive. Notably, to design a scalar controller, the two features of field-oriented control (FOC), i.e., the frequency and current, are employed. Specifically, the features of fuzzy frequency and fuzzy current amplitude controls are exploited for the control of an induction motor in a closed-loop current amplitude input model; hence, with the combination of both controllers to form a hybrid controller. With respect to finding the rule base of a fuzzy controller, a genetic algorithm is employed to resolve the problem of an optimization that diminishes an objective function, i.e., the Integrated Absolute Error (IAE) criterion. Furthermore, the principle of HFFC, for the purpose of overcoming the shortcoming of the FOC technique is established during the acceleration-deceleration stages to regulate the speed of the rotor using the fuzzy frequency controller. On the other hand, during the steady-state stage, the fuzzy stator current magnitude controller is engaged. A simulation is conducted via MATLAB/Simulink to observe the performance of the controller. Thus, from a series of simulations and experimental tests, the controller shows to perform consistently well and possesses insensitive behavior towards the parameter deviations in the system, as well as robust to load and noise disturbances.

Keywords: Indirect field-oriented control (IFOC), hybrid fuzzy-fuzzy control (HFFC), hybrid fuzzy-PI control (HFPIC), disturbances, genetic algorithm (GA)

1. Introduction

During the last forty years, induction motors have been largely utilized in applications that use variable speeds. In the industry, the term workhorse is used to refer to an induction motor.

With the development in the field of silicon-rectifier devices, the variable speed induction motor drives techniques began to emerge in the late 1960s. At that time, the principal of speed control was only based on the steady state aspects of an induction machine. The rigorous research in this field has made possible the emergence of more techniques in industrial drives. One of the techniques that previously used for drives control is the V/f ratio that has been applied in an open-loop speed control of drives that normally need low dynamics. In addition, the slip frequency control technique is also effective for producing improved dynamics. Until the emergence of field-oriented controls (FOCs), this technique, based on all the high-performance induction motor (IM) drives, was considered as an industry standard for AC drives whose dynamics resemble DC motors [1–2, 19]. Hence, the invention of vector control and FOC are considered as the most significant factors in AC motor drives that boosted programs related to the development and research for improving control performance. Several benefits of process control can be obtained by adjusting the speed of the drive motors, such as varying speeds for its operation process, better control of speed variation, work efficiency, precise control for positioning, tension or torque, energy saving, and compensation of process varying variables [2].

Zadah, in 1965, described Fuzzy Logic as a novel kind of mathematical set approach consisting of a fuzzy set theory that is considered as a basic theory of the fuzzy logic. A fuzzy control system is established by applying a principle of fuzzy logic that consists of three phases: fuzzification, inference engine, and defuzzification. The inputs are converted into fuzzy sets in an initial phase. An inference engine defines the fuzzy rules in the next phase that links the outputs by using the sets of inputs via explicit rules. Finally, the conclusions are inferred by combining the outcomes of the fuzzy rules, which is then transformed into a sharp value from the fuzzy sets [3, 13].

In order to provide an effective means for a variable speed drive (VSD) control, several research studies exist that are based on control techniques and commercially available tools yielding a high degree of performance and reliability. Hence, a PLC-based hybrid-fuzzy control for pulse-width modulation PWM driven VSD is examined that depends upon the s-domain transfer function in a scientifically presented model of an original plant by keeping the V/f ratio at a constant value as in [5]. Notably in [5], the optimizations of the controller's performance against the parameter variations and external disturbances are not fully considered. Hence, the disadvantages of the FOC method and the results gained from the simulation are overcame by implementing two stage controllers as explain in [6–8], though all the practical implementations are quite satisfactory. Moreover, the satisfactory results are achieved by applying some controller algorithms for controlling the speed of an IM as in [9, 10].

Generally, by using simple solutions, fuzzy systems are capable of managing non-linear, complicated, and at times mathematically intangible dynamic systems [11]. Though, it is not an easy task to get an optimal set of fuzzy membership rules and functions, the designer needs to invest some times, skills, and experience for the tiresome fuzzy tuning exercise. Even though an iterative and heuristic process for transforming the membership functions to enhance performance has been recommended, in principle, there is no common method or rule for the fuzzy logic setup [11]. A lot of researchers have recently considered numerous intelligent schemes for the purpose of tuning the fuzzy system [12–17]. For instance, the genetic algorithm (GA) and neural network (NN) approach to optimize either the membership rules or functions have turned into a trend in the development of the fuzzy logic system [11]. The benefits of the GA approach, apart from involving less cost, it is easy to implement the procedures and requires a single objective to be assessed [22]. Figure 1 presents the basic configuration of GA for a fuzzy control system with fuzzifier, defuzzifier, and an inference engine.

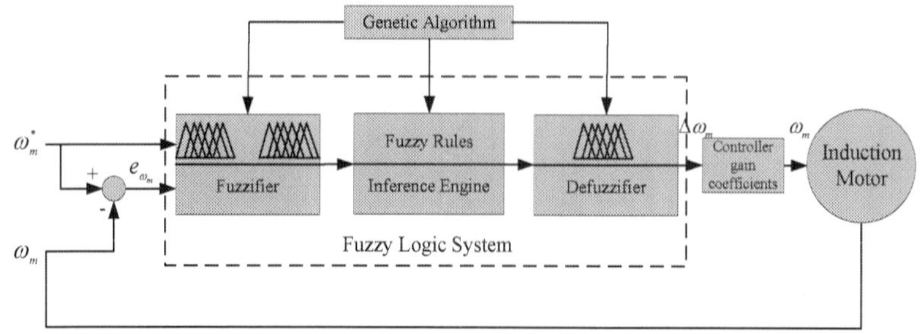

Figure 1. Genetic algorithm for a fuzzy control system.

A GA is employed in this research to attain the rules of the fuzzy inference system. However, the key aim of this study is to compare the performance of a fuzzy controller built on heuristics with a controller developed via the optimization technique.

Therefore, the best combination between the fuzzy input-output variables is needed to be discovered to enhance the inference rules of a fuzzy controller for a particular range of the fuzzy logic controller operation.

The FOC has two features [4, 6] that have been implemented in this work. Firstly, it is not able to do frequency control directly, due to the fact that the supply frequency changes during the period of acceleration-deceleration of the FOC while the slip frequency remains the same. Furthermore, in the presence of a torque command, the magnitude of a supply current magnitude remains stable.

The application of the FOC method has commonly been effective in achieving elevated performances in adjustable speed induction motor drives, however, it still suffers the following disadvantages [6]:

i. When evaluating certain integrals (i.e., the error accumulation), the degradation will be in the steady-state if the control time is extended and the outcome will be a transient responses owing to too much drift and accumulation of error in the values of the parameters;

ii. Susceptibility of the parameter deviations;

iii. There should be a steadiness in control and the calculation must be set off right from an initial state;

iv. Complex calculation is included in the final step.

The simulation and modeling of an induction motor controller constructed using MATLAB/ Simulink and the examination of the performance of the controllers (i.e., hybrid fuzzy-fuzzy controller (HFFC), hybrid fuzzy-PI controller (HFPIC), and indirect field-oriented controller (IFOC)) on the system are discussed in this paper. Also, in this study, the objective is to enhance the performance of the controller using HFFC. The purpose and context of this study are outlined in Figure 2.

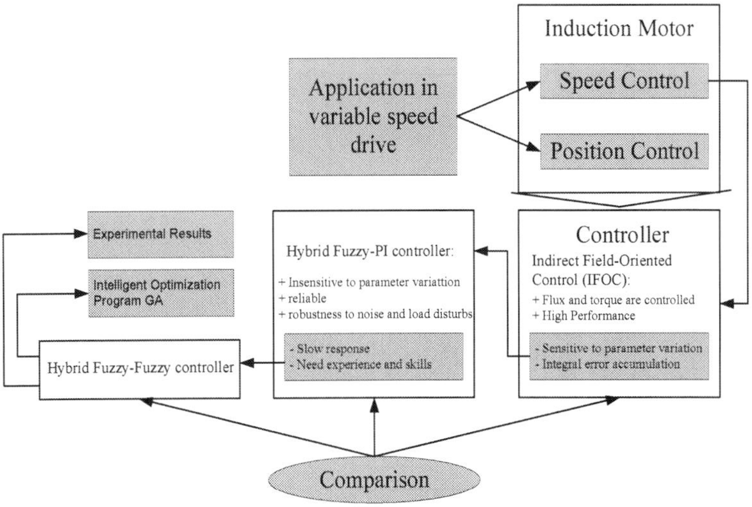

Figure 2. Overview of the study background.

The implementation of a fuzzy current amplitude controller on the induction motor model makes this work unique. This controller possesses the same supply features as FOC and insensitivity to the parameter variation for the motor and system robustness to noise and load disturbances are some of the advantages of this controller. Due to the fact that it provides better performance, the fuzzy current amplitude controller has been selected. While, the common structure of hybrid fuzzy-fuzzy controller is defined in Figure 3.

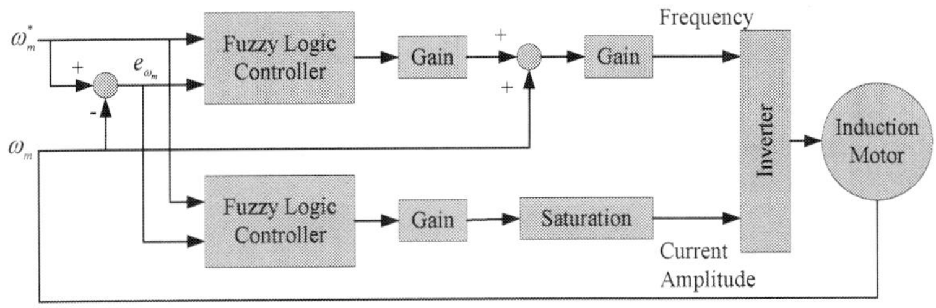

Figure 3. General structure of an HFFC.

2. Mathematical modeling

A higher order of mathematical equations that fall under one of the VSD control classifications can be used to model the dynamics of an induction motor. The steady-state model provides information about the performance of the induction motors in a steady state only. Table 1 provides the related parameters of IM. Figure 4 illustrates the IFOC model block of the proposed system for an IM.

Name	Symbol	Value
Resistance of stator	r_s	$29.4\times10^{-2}\ \Omega$
Resistance of rotor	r_r	$15.6\times10^{-2}\ \Omega$
Self-inductance of stator	L_s	$42.4\times10^{-3}\ H$
Self-inductance of rotor	L_r	$41.7\times10^{-3}\ H$
Magnetizing inductance	L_M	$41\times10^{-3}\ H$
Rotor inertia	J	$40.06\times10^{-2}\ Kg.m^2$
Damping constant	B or f	$0.062\ N.m.s/rad$
Number of pairs of poles	P	3

Table 1. Electrical and mechanical parameters of the IM.

By employing a two-phase motor in a quadrature and direct axis, the dynamic model of the induction motor is developed. The description of the notations is tabulated in Table 2. The state-space model of an induction motor in a stationary reference frame can be derived with the help of the voltage and flux linkage relations of an induction motor in the reference frame that is randomly selected [11, 12].

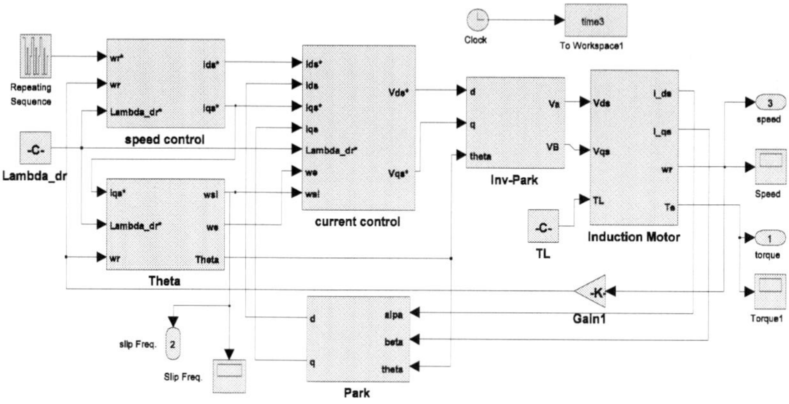

Figure 4. Proposed IFOC model block.

i_{ds}, i_{qs}	d- and q-axis stator current components, respectively, and expressed in stationary reference frame
i_{dr}, i_{qr}	d- and q-axis rotor current components, respectively, and expressed in stationary reference frame
L_M	Magnetizing inductance
L_s, L_r	Self-inductance of the stator and rotor, respectively
r_s, r_r	The resistance of a stator and rotor phase winding, respectively
T_e, T_l	Electromagnetic torque and Load torque reflected on the motor shaft, respectively
V_{ds}, V_{qs}	d- and q-axis stator voltage components, respectively, and expressed in stationary reference frame
L_{ls}, L_{lr}	Leakage resistance of the stator and rotor, respectively
$\lambda_{ds}, \lambda_{qs}$	d- and q-axis stator flux components, respectively, and expressed in stationary reference frame
$\lambda_{dr}, \lambda_{qr}$	d- and q-axis rotor flux components, respectively, and expressed in stationary reference frame
ω_m, ω_r	Mechanical and electrical angular rotor speed, respectively
ω	Synchronous speed or dominant frequency
P	Number of pairs of poles
ρ	Operator $^{d}/_{dt}$
J	The inertia of the rotor kgm^2 or Js^2
B	The damping constant that represents dissipation due to windage and friction
$\gamma=\left[\frac{L_M^2 r_r + L_r^2 r_s}{\sigma L_s L_r^2}\right]$, $\sigma=1-\left[\frac{L_M^2}{L_s L_r}\right]$, $\beta=\left[\frac{L_M}{\sigma L_s L_r}\right]$, $\beta_1=\left[\frac{1}{\sigma L_s}\right]$, $\alpha=\left[\frac{r_r}{L_r}\right]$	

Table 2. Nomenclatures.

The final state-space model of an induction motor with the controlled stator currents in a stationary frame can be written as shown in equations (1)–(6).

$$\frac{di_{ds}^s}{dt} = -\gamma i_{ds}^s + \beta\omega_r\lambda_{qr}^s + \beta\alpha\lambda_{dr}^s + \beta_1 V_{ds}^s \tag{1}$$

$$\frac{di_{qs}^s}{dt} = -\gamma i_{qs}^s - \beta\omega_r\lambda_{qr}^s + \beta\alpha\lambda_{qr}^s + \beta_1 V_{qs}^s \tag{2}$$

http://dx.doi.org/10.5772/61086

$$\frac{d\lambda_{dr}^{s}}{dt}=-\alpha\lambda_{dr}^{s}-\omega_{r}\lambda_{qr}^{s}+\alpha L_{M}i_{ds}^{s} \tag{3}$$

$$\frac{d\lambda_{qr}^{s}}{dt}=-\alpha\lambda_{qr}^{s}+\omega_{r}\lambda_{dr}^{s}+\alpha L_{M}i_{qs}^{s} \tag{4}$$

$$\frac{d\omega_{m}}{dt}=-\frac{B}{J}\omega_{m}+\frac{1}{J}\left(T_{e}-T_{l}\right),\ \omega_{r}=\frac{P}{2}\omega_{m} \tag{5}$$

$$T_{e}=\left(\frac{3}{2}\right)\left(\frac{P}{2}\right)\frac{L_{M}}{Lr}\left(i_{qs}^{s}\lambda_{dr}^{s}-i_{ds}^{s}\lambda_{qr}^{s}\right) \tag{6}$$

Notably, the two features of the FOC that have been used in this research study are shown in Eq. (1) to (6). The first aspect that shows the supply frequency changes with the speed of a rotor [6] is given in Eq. (7) and (8).

$$\omega=\omega_{r}+\frac{P}{2}\omega_{m} \tag{7}$$

where:

$$\omega_{r}=\frac{3r_{r}T_{l}^{*}}{P\lambda_{dr}^{e^{*}2}} \tag{8}$$

If T^{*} is maintained constant during acceleration, then ω_r is also constant. In Eq. (7), as ω_m changes during acceleration, then ω has to be varied, so that Eq. (7) satisfies the first FOC feature. The second feature can be proven by substituting the conditions $\lambda_{qr}=0$ and $\lambda_{dr}=constant$ in Eq. (3) [6, 21].

$$\left|i_{s}\right|=\frac{2}{3}\sqrt{i_{ds}^{2}+i_{qs}^{2}} \tag{9}$$

where:

$$i_{ds}=\frac{\lambda_{dr}^{e}}{L_{M}}=constant \tag{10}$$

$$i_{qs} = \frac{3L_r T^*}{PL_M \lambda_{dr}^{e\,*}} \tag{11}$$

Eq. (9) will be a constant if the torque command T^* is a constant, which satisfies the second feature. The Figure 5 shows that the speed response may be divided into two stages. For the HFFC, Table 3 shows the relationship between the inputs and outputs.

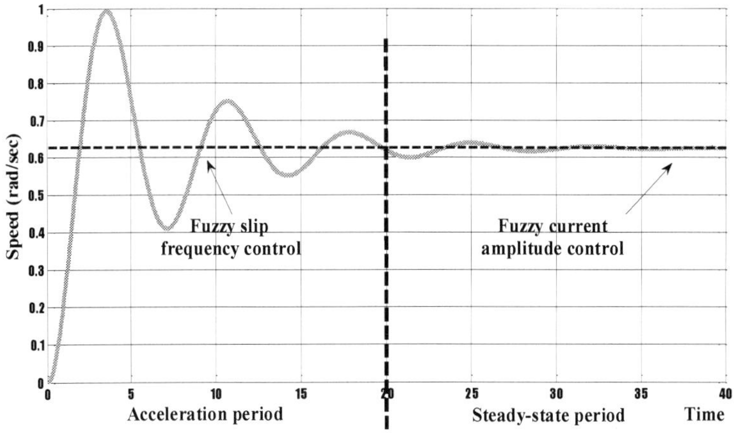

Figure 5. Speed response stages.

3. Design of the controllers

3.1. Fuzzy current amplitude controller

In the stage of acceleration-deceleration, the stator current magnitude is regulated as the system is driven by the maximum permissible values of an inverter. During the final steady-state period, the speed of the rotor is controlled by adjusting the magnitude of the stator current, while the supply frequency is kept constant. In the presence of a constant supply frequency, the relationship between torque-current may be expressed as follows [21].

$$T = \left[\frac{3P}{2}\right]\left[\frac{r_r L_M^2 i_s^2 \left[\omega - \frac{P}{2}\omega_m\right]}{r_r^2 + \left[L_M + L_{lr}\right]^2 \left[\omega - \frac{P}{2}\omega_m\right]}\right] \tag{12}$$

http://dx.doi.org/10.5772/61086

The values for the current amplitude are depicted as follows:

$$|i_s| = \begin{cases} 50.54 & |i_s| \geq 50.54A \quad when \ \Delta\omega_m \neq 0 \\ |i_s| & |i_s| < 50.54A \quad when \ \Delta\omega_m = 0 \end{cases} \tag{13}$$

The main phase from every input-output data couples is to create a fuzzy rule to find out a degree of every data-value in each affiliated area of its corresponding fuzzy domain. Consequently, a variable is assigned to a region having a maximum value.

	Supply Frequency	Current Amplitude	Rotor Speed	Control Objective
Acceleration	Change	Constant	Change	Speed Change
Steady-State	Constant	Change	Constant	Reduce Oscillation

Table 3. HFFC relationship.

A truth or rule degree is assigned to a newly generated fresh rule from the input-output data couples. Therefore, a rule degree can be defined as an extension of assurance to relate the current- and voltage-related functions with an angle. Usually, a degree that is a creation of the membership function degree of every variable in its corresponding area is assigned in a formulated technique.

A compatible fuzzy rule is created by each training data set that is placed in the fuzzy rule base. Consequently, each input-output data couples is preserved and the rules are generated. A two-dimensional form that can be explored by the fuzzy reasoning tool is used to tabulate a knowledge base or fuzzy rule.

Figure 6 demonstrates the general structure of the fuzzy logic control (FLC), a combination of knowledge base, fuzzification, a defuzzification, and inference engine.

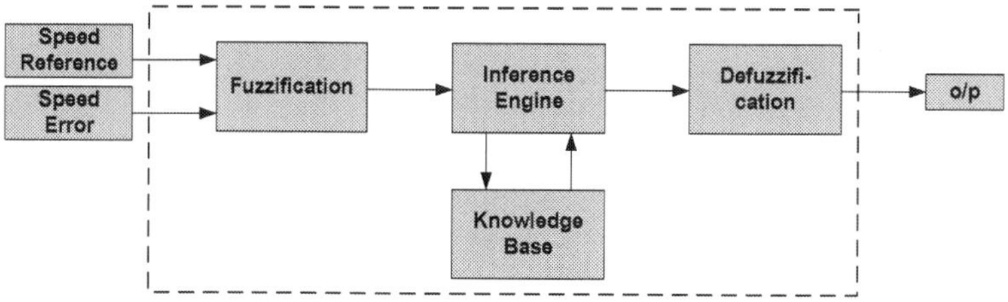

Figure 6. A general block diagram of FLC.

A speed error can be computed by comparing the reference speed and the speed signal feedback. The fuzzy knowledge base consists of membership functions of the inputs of the fuzzy controller including speed reference, error changing, and current amplitude/slip frequency outputs.

Stages	ω_m^*	A	f_A
Deceleration	-	-50.54	NBB
Steady-State	-120	-15.07	NB
-	-80	-12.79	NM
-	-40	-11.214	NS
-	0	0	Z
-	40	11.214	PS
-	80	12.79	PM
-	120	15.07	PB
Acceleration	-	50.54	PBB

Table 4. Speed, current amplitude, and fuzzy linguistic values.

3.2. Membership functions

Primarily, the fuzzy logic controller is used to convert the change of error variables and the crisp error into fuzzy variables, which is then plotted into the linguistic tags. All the membership functions and labels are connected with each other (Figures 7, 8, and 9), which comprises of two inputs and one output. The nine sets are formed by classifying the fuzzy sets that are outlined as follows:

Z: Zero	**PS**: Positive Small	**PM**: Positive Medium
PB: Positive Big	**NS**: Negative Small	**NM**: Negative Medium
NB: Negative Big	**PBB**: Positive Big Big	**NBB**: Negative Big Big

Table 5. Classification of the fuzzy sets

The previously defined seven numbers of linguistics, along with both inputs and outputs, include membership functions. Eq. (9) shows the samples of current amplitude of the two stages as shown in Table 4.

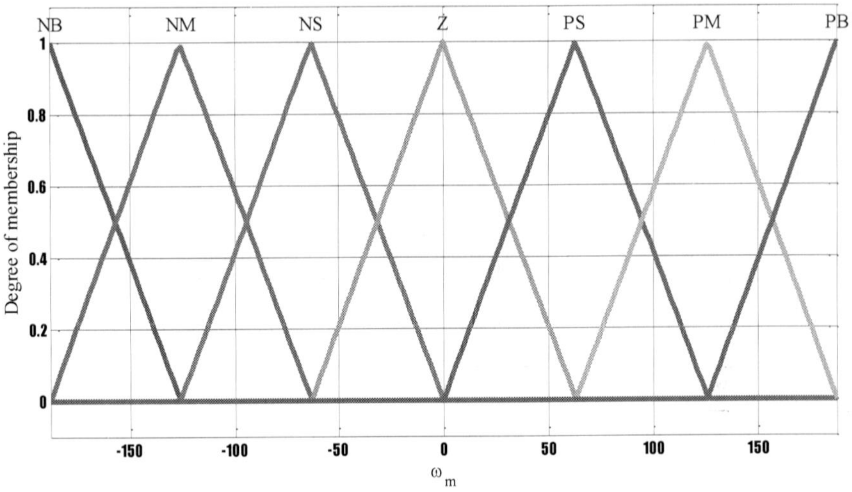

Figure 7. Membership function of speed reference.

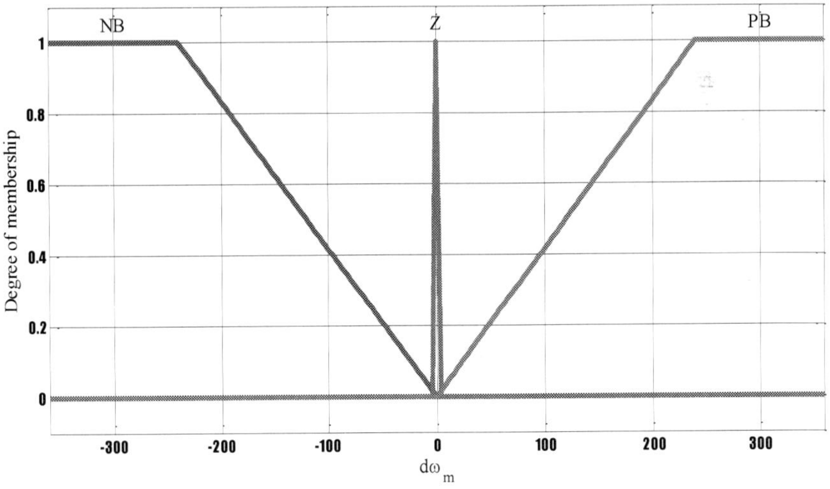

Figure 8. Membership function of speed error.

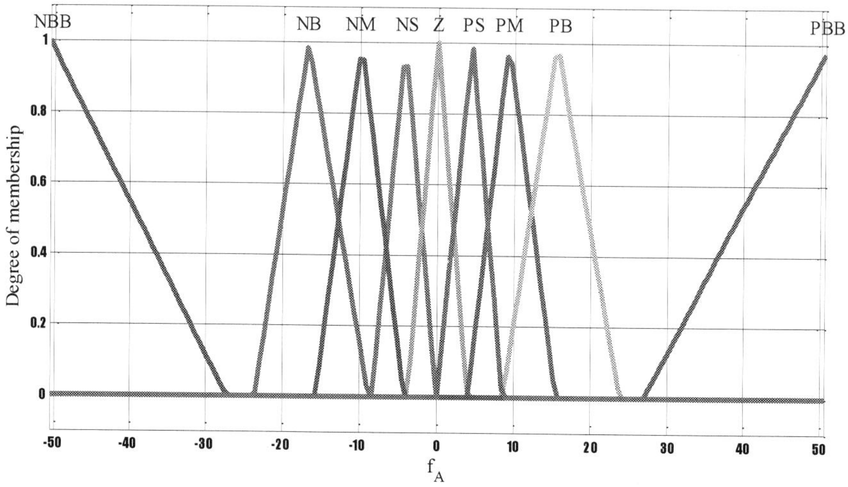

Figure 9. Membership functions of current amplitude.

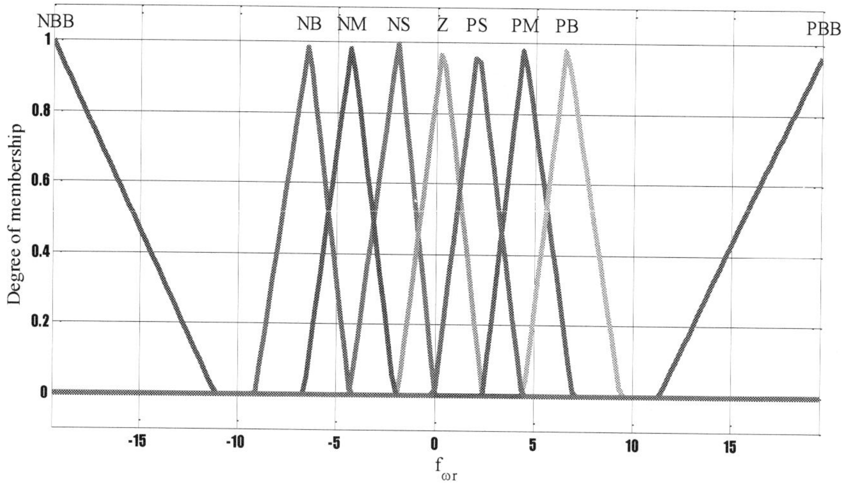

Figure 10. Membership functions of slip frequency.

http://dx.doi.org/10.5772/61086

3.3. Rule base

The fuzzy inputs can be plotted easily into the required output with the help of a useful tool, the rule base, as shown in Table 6.

Speed Error	Speed Reference						
	NB	NM	NS	Z	PS	PM	PB
NB	NBB	NBB	NBB	NBB	NBB	NBB	NBB
Z	NB	NM	NS	Z	PS	PM	PB
PB	PBB	PBB	PBB	PBB	PBB	PBB	PBB

Table 6. Rule matrix for fuzzy amplitude/slip controller.

3.4. Fuzzy frequency controller

A fuzzy frequency control is presented by using a frequency aspect of the field orientation principle. At a steady state phase, the torque command is a smaller value, whereas, the torque command becomes a larger value during the stage of acceleration-deceleration. The speed reference and rotor speed represent these values [4, 6, 11]. The following relations show a slip frequency at a steady-state.

$$\omega_r = \begin{cases} \dfrac{3r_r}{P\lambda_{dr}^{e^*2}} T_{acc} \ \ when\ \Delta\omega_m \neq 0 \\ \dfrac{3r_r}{P\lambda_{dr}^{e^*2}} \mu\omega_m \ \ when\ \Delta\omega_m = 0 \end{cases} \tag{14}$$

So the slip frequency can be written as:

$$\omega_r = f\left(\omega_m, \Delta\omega_m\right) \tag{15}$$

Eq. (14) shows the reference and speed error from the inputs of the fuzzy slip control. The membership functions of the output and input are depicted in Figures 7, 8, and 10. In addition, Table 6 shows the rule matrix. The samples of the slip frequency of the two stages can be obtained by Eq. (14), which is shown in Table 7.

3.5. Defuzzification

The fuzzy control action is executed with the help of simulating the human decision process via the interference engine, from the understanding of the linguistic variable expressions and

Stages	ω_m^*	ω_r	f_{ω_r}
Deceleration	-	-16.80	NBB
Steady-State	-120	-3.6328	NB
-	-80	-2.42192	NM
-	-40	-1.21096	NS
-	0	0	Z
-	40	1.21096	PS
-	80	2.42192	PM
-	120	3.6328	PB
Acceleration	-	16.80	PBB

Table 7. Speed, slip frequency, and fuzzy linguistic values.

the control rules. As a result, the knowledge base along with the inference engine is interconnected during the course of the control process. At first, through substituting the fuzzified inputs into rule base, the active rules are differentiated. Then, by employing one of the fuzzy reasoning methods, these rules are combined. The utmost distinctive fuzzy reasoning methods are the Max-Product and Min–Max. Since the Min-Max interference scheme is commonly used, it is also applied in this research. The fuzzy control actions are then regenerated with the help of defuzzification, which is deduced from the inference engine to a non-fuzzy control action. Thus, using the center of gravity method, the defuzzification is accomplished in the set of Eq. (16) to create a non-fuzzy control signal:

$$y = \frac{\sum_{i=1}^{n} u(i)\mu_{A,\omega_r}(y_i)}{\sum_{i=1}^{n} \mu_{A,\omega_r}(y_i)} \tag{16}$$

where, μ_{A,ω_r} is the degree of membership function.

[NBB=1, NB=2, NM=3, NS=4, Z=5, PS=6, PM=7, PB=8, PBB=9]

Whereas, the linguistic standards of the antecedents relate to the entire values as follows:

[NB=1, NM=2, NS=3, Z=4, PS=5, PM=6, PB=7]

However, it is essential to recognize the values of the precedents, fuzzy operators in the rule base weights, and fuzzy rules to form the rule base of a fuzzy controller; in this situation, the antecedents 1 and 2 and the other stated values are positioned in a matrix inside the MATLAB functions, as can be seen in the matrix below [24]:

$$\begin{bmatrix} Ant_1 Ant_2 Con_1 RwRC & \cdots & Ant_7 Ant_7 Con_7 RwRC \\ \vdots & \ddots & \vdots \\ Ant_3 Ant_3 Con_3 RwRC & \cdots & Ant_{21} Ant_{21} Con_{21} RwRC \end{bmatrix}$$

Where, Ant_n: antecedent n, Ant_n: antecedent n, Con_n: consequent n, Rw: rule weight, and RC: rule connection. Finally, the input/output variables range and the membership function data are in the same MATLAB function. The MATLAB FIS (Fuzzy Inference System) file is the output of this function that relates to a structure in which all the information of fuzzy inference of the system is incorporated, which is utilized as a fuzzy controller in the feedback scheme of the Simulink library. The flowchart procedure is shown in Figure 11.

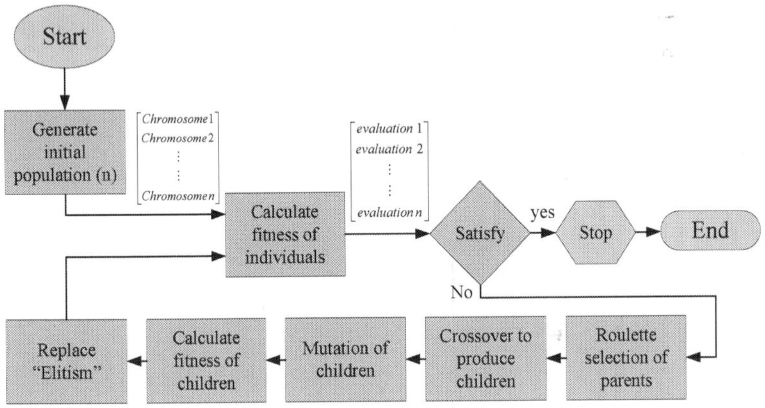

Figure 11. Flowchart procedure.

The parameters of the GA are set as below:

1. Initial population size: 30
2. Maximum number of generations: 30
3. Probability of crossover: 0.8
4. Mutation probability: 0.09
5. Performance measure: IAE

From the feedback scheme, the performance index is entertained and reverted to the genetic algorithm for the stability of the genetic process. Finally, the rule base attained is displayed as below:

$$\begin{bmatrix} 1 & 2 & 2\,2 & 1 & 1\,3 \\ 1 & 2 & 2\,4 & 8 & 8\,8 \\ 9 & 9 & 7\,9 & 7 & 9\,8 \end{bmatrix}$$

4. GA-optimization method

GAs are computational schemes, which on the basis of processes of natural evolution, utilize the operators who understand the process of heuristic search in a search space, in which it is presumed that the perfect solution for the optimization problem is available [22]. The process of GA is revealed in Figure 12 [21, 25].

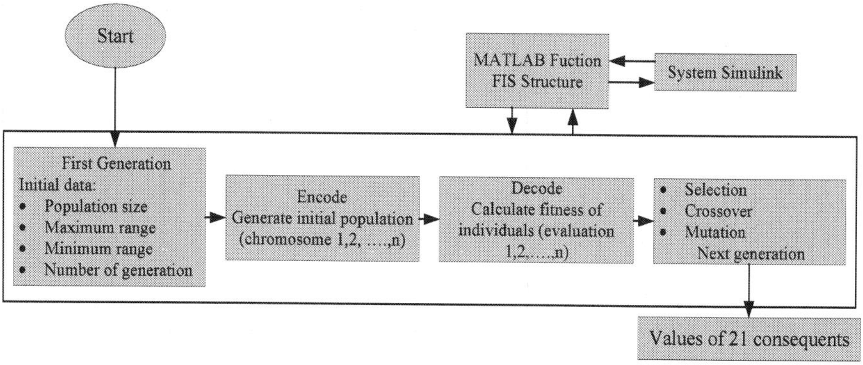

Figure 12. The process flowchart of GA.

The objective function recommended in the optimization problem is the standard *IAE* (Integrated Absolute Error) that is described by [5, 23, 26]:

$$\mathrm{IAE} = \int_0^T |e(t)|\,dt \tag{17}$$

The objective function recommended in Eq. (17) that is reduced in the course of the optimization process reflects a good reaction to set point changes. Among the most significant points, i.e., the decay ratio, the settling time, the rise time, the overshoot, and the steady-state error, this objective function is also considered [23]. However, merely the rule base is attained with the assistance of the optimization technique.

The genetic algorithm is applied to a population of individuals (chromosome) in order to devise the rules where each of them encompasses a certain fuzzy controller. The antecedents conforming to the input linguistic variables are fixed in a group as a fragment of a MATLAB function [24]. During the assessment process, this function is utilized by the genetic algorithm and it gets as arguments in the consequents of a controller; the assessment is done in all the pre-established generations for each of the individuals of the population. Also, the binary code is implemented for simplicity. The values are:

[NBB NB NM NS Z PS PM PB PBB]

And the output language variables are coded in the order:

[0001 0010 0011 0100 0101 0110 0111 1000 1001]

A likely chromosome is shown in Table 5, which would be codified as:

[0001 0010 1001 0001 0100 1001 0001 0101 1001......1000 1001]

With the formation of a population of individuals (generated randomly), the genetic process starts in which every individual comprises the 21 consequents of a fuzzy controller in general. Since 21 consequents could be employed in the MATLAB function, to create the rules, it is essential to change the individual, codified as a binary chain, to complete numbers in the values from 1 to 9.

5. Outcomes and analysis

A combination of the fuzzy frequency controller and fuzzy current amplitude controller is used to form an HFFC. This controller provides similar supply frequency as the FOC and is insensitive to the parameter variation for the motor and system robustness to noise and load disturbances, which are the advantages of this controller. A model of HFFC for an induction motor is produced by using the MATLAB/Simulink software, which is shown in the Figure 13. Table 1 shows the parameters that are chosen to perform the simulation study.

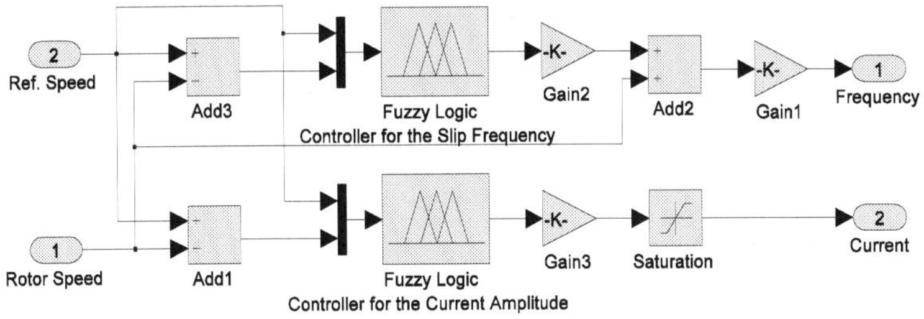

Figure 13. Simulink diagram of the HFFC.

An HFFC can be modeled by combining the fuzzy frequency controller and the fuzzy current amplitude controller. Throughout the final steady-state stage, the fuzzy frequency controller outputs the frequency that relates to the speed command. During the acceleration-deceleration stages, the fuzzy current amplitude controller outputs the maximum permitted current. The model of HFFC for the induction motor is built using MATLAB/Simulink as presented in Figure 14.

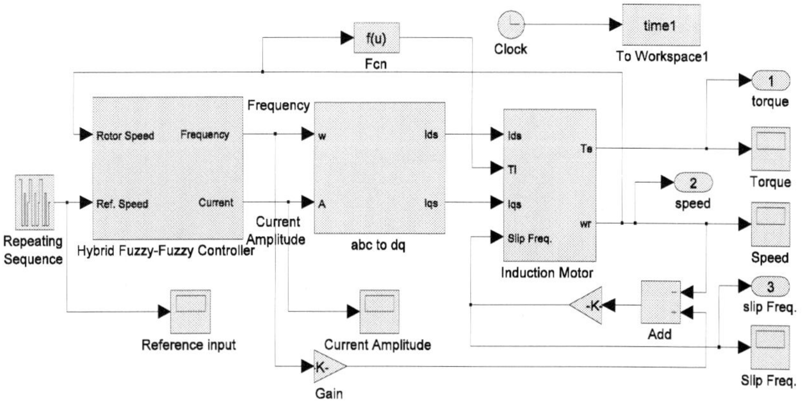

Figure 14. The model diagram of the HFFC.

6. Performance criteria

Generally, following the reference and to cast off the disturbance are the two basic key objectives of control. The following performance criterion is employed for evaluating the efficiency of the controller in achieving the aims of control for complete speed control and the performances are identified for comparisons and investigation [5, 27]. The error (e), the difference between actual and reference value, is generally categorized into a number of quantities. One of the quantities to state the accumulative error magnitude is the IAE, whose formula is given in Eq. (17) [5, 23, 26]. The quadratic measure, i.e., integral of squared error (ISE), provides the error quantitative in quadratic mode. The ISE accrues the squared error. The ISE expression is presented as:

$$\mathrm{ISE} = \int_{0}^{T} e^{2}(\mathrm{t})\mathrm{dt} \tag{18}$$

http://dx.doi.org/10.5772/61086

This criterion's main disadvantage is that it provides large weight if the error is large, such as, a poorly checked system. Other criteria are the integral of time weighted absolute error (ITAE) and integral of the time multiplied by the squared error (ITSE). The expression of ITAE and ITSE are articulated in Eq. (19) and Eq. (20).

$$\mathrm{ITAE} = \int_0^T t\left|e(t)\right|\mathrm{dt} \tag{19}$$

$$\mathrm{ITSE} = \int_0^T te^2(\mathrm{t})\mathrm{dt} \tag{20}$$

The objective of control is to reduce every performance criteria's error. The Simulink model for calculating the performance indices is presented in Figure 15 [28].

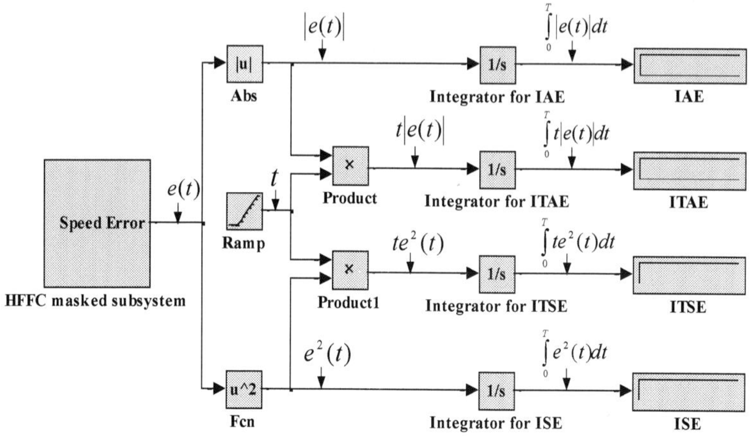

Figure 15. Simulink model for computing the performance indices.

7. Simulation results

By undertaking a simulation of an indirect rotor flux FOC, a new controller is matched with the FOC. As shown in Figure 16, the speed has augmented in 0 s from 0–120 rad/s; in 8 s period, the speed reduced to -120 rad/s; in the 12 s period, the speed has elevated to 50 rad/s; in the 16 s period, the speed has reduced to -120 rad/s; in the 18 s period, the speed has raised to 0 rad/ s; and increased to 120 rad/s in the 20 s.

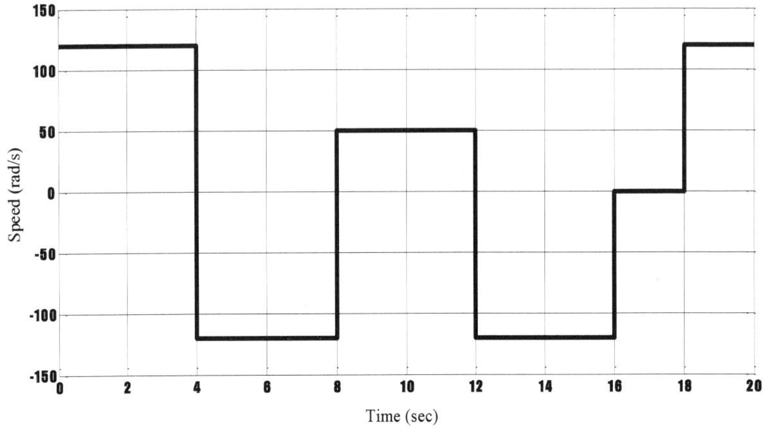

Figure 16. Reference speed step change.

The simulation outcomes of the HFFC are presented in Figures 17, 18, 19 and 20. The two-stage control method provides a very fast speed. Due to the control of the current in a final steady-state stage, the oscillations of speed are completely eliminated at a final operating point. It is evident from the comparison of these results with an IFOC with PI-controller that the HFFC shows the two features of FOC controller, i.e., the current feature at the steady state stage and the frequency feature at accelerate-decelerate stage.

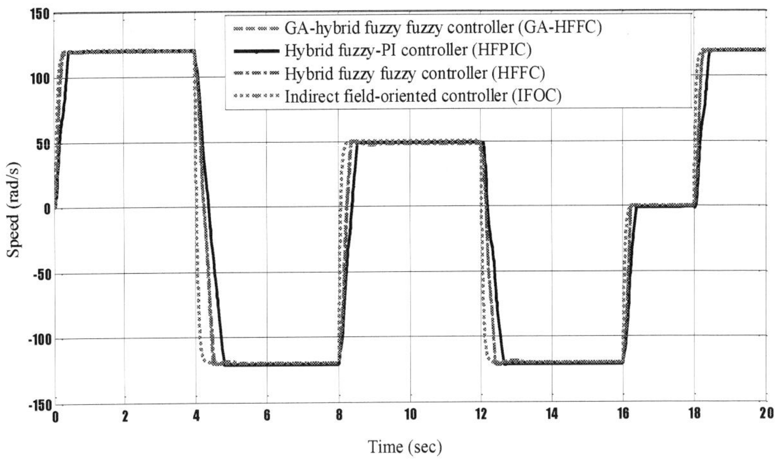

Figure 17. Fuzzy responses of the fuzzy controllers

http://dx.doi.org/10.5772/61086

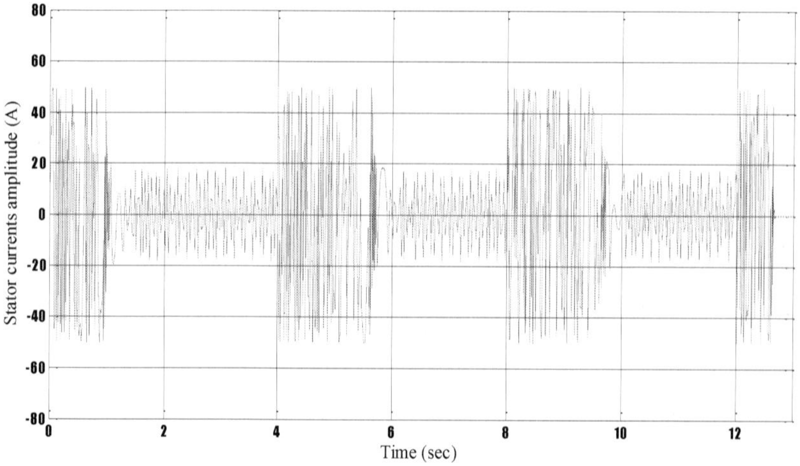

Figure 18. Stator currents amplitude of HFFC.

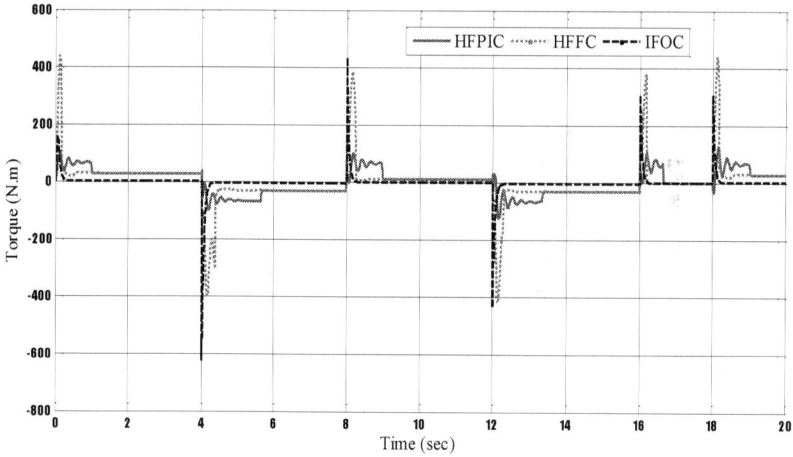

Figure 19. Torque response of HFFC.

The simulation outcomes on the efficiency of the controller centered on the performance measures are revealed in Table 7. A closer look at the overshoot, performance criteria, time rise, IAE, final steady-state value, ISE, ITAE, and ITSE is displayed to show reduced values. This validates the act of the modified controller with a better performance to effectively control the speed.

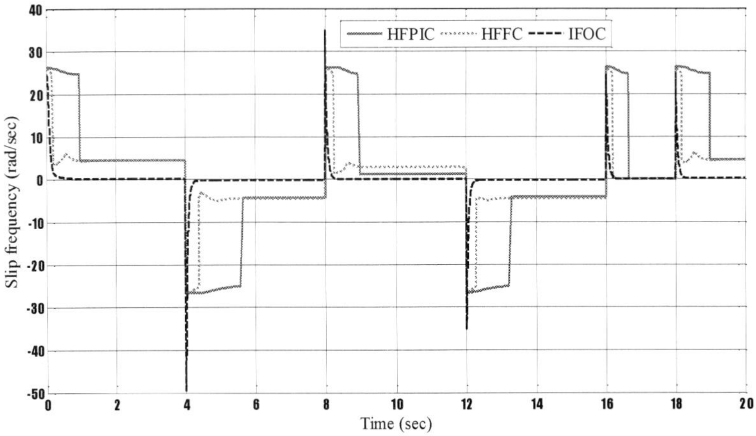

Figure 20. Slip frequency response of the fuzzy controllers.

No.	Performance Index	HFPIC	HFFC	IFOC	GA-HFFC
1	Overshoot (%)	0.000	3.000	0.000	0.000
2	Rise time (sec)	0.4363	0.2803	0.6107	0.2648
3	Final steady-state value	120.0989	120.08483	120.0681	120.08391
4	IAE	289.3	194	158.28	186
5	ISE	3.409e+004	2.311e+004	1.223e+004	2.287e+004
6	ITAE	2544	1656	1457.6	1625
7	ITSE	2.691e+005	1.765e+005	1.31e+005	1.799e+005

Table 8. The performance index comparison of the system.

8. Effects of internal and external disturbances

The mutual inductance and the rotor resistance are expected to be changed to 3 r_r and 0.8 L_M, respectively, in order to demonstrate the insensitivity of the HFFC to the variation of motor parameters. An insensitivity to the parameter variation shown by the speed response of a fuzzy controller is shown in Figure 21.

Additionally, the distributed random noises are added in the input current and the feedback speed to assess the effects of the noise of the input current and the noise of the speed sensor. The speed response to the current noise and with the measured speed illustrates that the HFFC possesses decent disturbance rejection (see Figure 22).

http://dx.doi.org/10.5772/61086

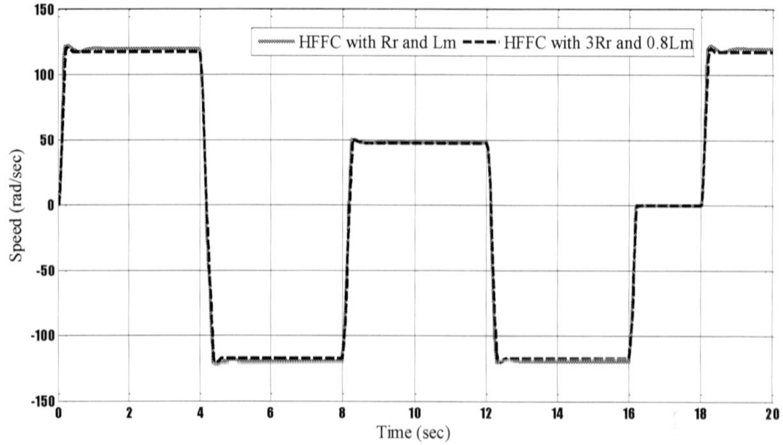

Figure 21. Speed response of HFFC with parameter variations.

Furthermore, for studying the impact of magnetic saturation of the induction motor on the controller performance, in the induction motor model, Figures 23 and 24 demonstrate the simulation results of the torque and rotor speed responses. The flux upsurge is limited, owing to the magnetic saturation so that the torque oscillations are decreased considerably, but during the acceleration-deceleration stage, an extreme magnetic saturation will create a higher temperature rise and larger losses.

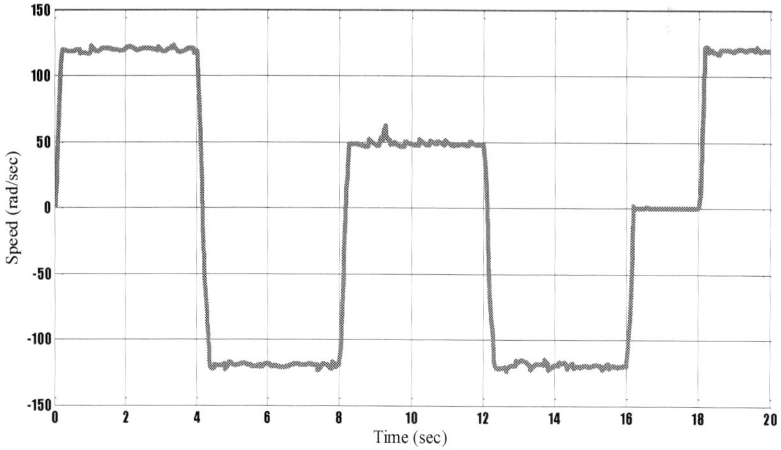

Figure 22. Speed response of HFFC with noise.

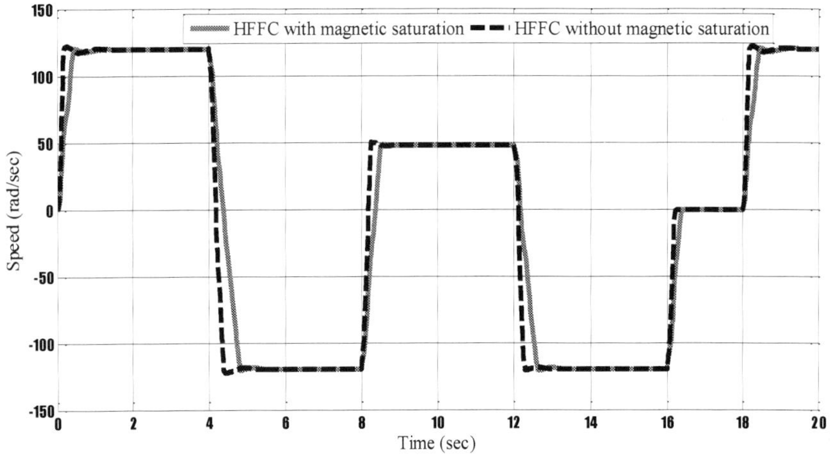

Figure 23. Speed response of HFFC with the effect of magnetic saturation.

Moreover, the effects of load torque variation on the HFFC system are analysed by the simulation. In the simulation, the control system encounters quick variations in the load torque: at $t=2$ s, the load increases from 0% to 100% of the rated torque, T_l, at $t=7$ s, the load decreases to 100% of T_l, and at $t=10$ s and 14 s, the load rises to 100% of T_l again, at $t=19$ s, the load declines to 100% of T_l.

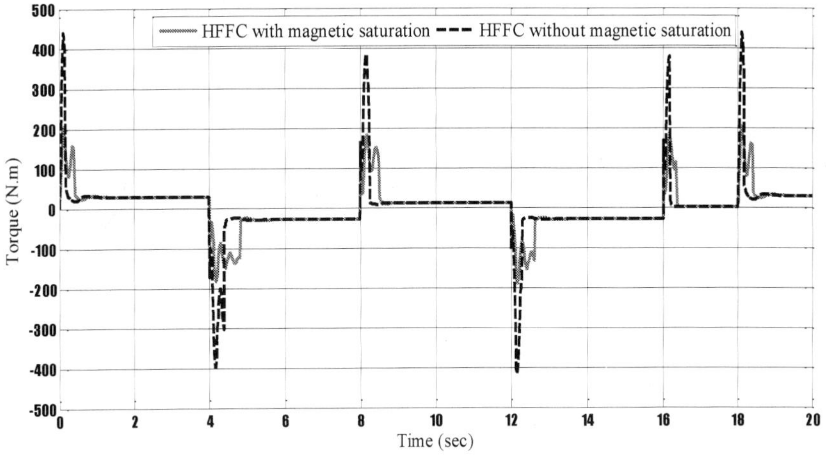

Figure 24. Torque response of HFFC with effect of magnetic saturation.

http://dx.doi.org/10.5772/61086

The torque reaction of the fuzzy-fuzzy control system with a load torque variation is demonstrated in Figure 25. The above-mentioned simulation results illustrate that the greater variations can be created in a load torque by the fuzzy-fuzzy controller. However, the extracted outcomes of performance of the model are found to be corresponding precisely with the anticipations, when it is compared with an IFOC controller. The results also reveal that the HFFC performance is unresponsive to the parameter variation for the motor and the system strength to load and noise disturbances.

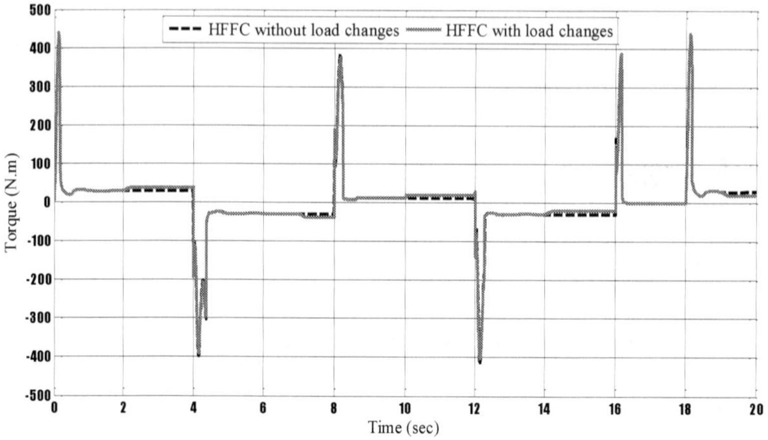

Figure 25. Torque response of HFFC with load changes.

9. The experimental results

The following experiments are conducted to demonstrate and verify the operability of the proposed controller. There are several experiments that can be conducted, however the experimental results presented here are necessary for this purpose. The experimental rig constituting the induction motor speed controller system constructed comprises of the following equipment:

- 3-phase squirrel cage induction motor 240/415 V- 175 W-1360 r/min-0.52 A-50 Hz (Lab-Volt)
- Prime Mover/Dynamometer (Lab-Volt)
- TMS320F28335 (Digital Spectrum Inc.)
- DMC1500 motor controller (Digital Spectrum Inc.)
- Tacho-generator with 500 rpm/volt
- Data acquisition card (DAQ) (Lab-Volt)

- Current sensors
- LVDAM-EMS software for DAQ
- Host PC

Figure 26 (a), (b), and (c) demonstrate the three-phase voltages waveforms V_a, V_b, and V_c, respectively, acquired by DAQ. The actual voltages are 120° different phases from each other.

Figures 27 (a), (b), and (c) show the three-phase sinusoidal PWM for an inverter-fed induction motor in a TMS320F28335 eZdsp control card. These signals are the output from the digital signal processor (DSP) that have been applied to the inverter to control the induction motor. The frequency of switching has been set at 10 kHz. It is worth to point out that for this controller the operating frequency can be applied between 0 to 10 kHz.

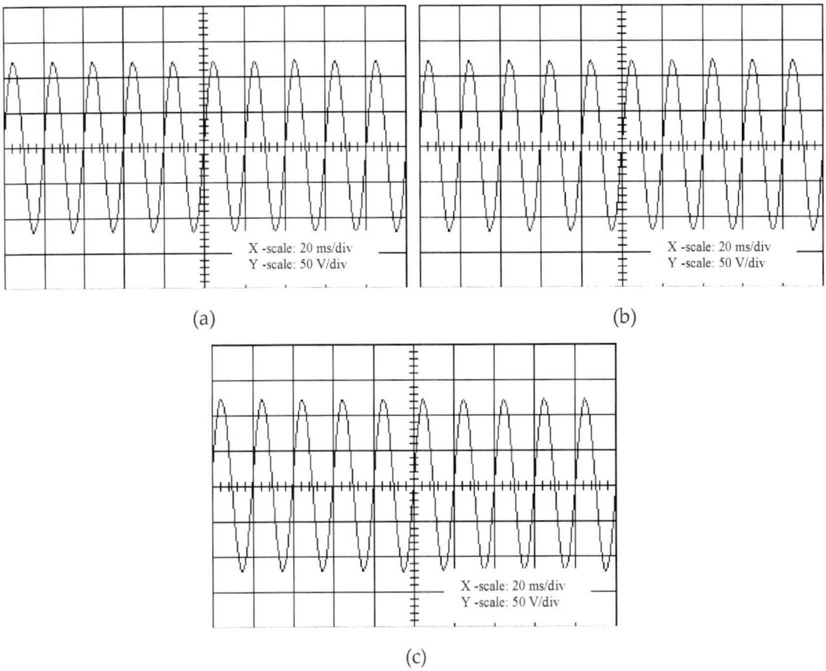

Figure 26. (a) V_a , (b) V_b and (c) V_c waveforms output.

In Figure 28 (a) and (b) are the output signals measured by the encoder at two different sampling rates: 5 kHz and 10 kHz, respectively, which demonstrate the different speeds that can be achieved. The encoder is providing signal to the controller that represents the feedback signals.

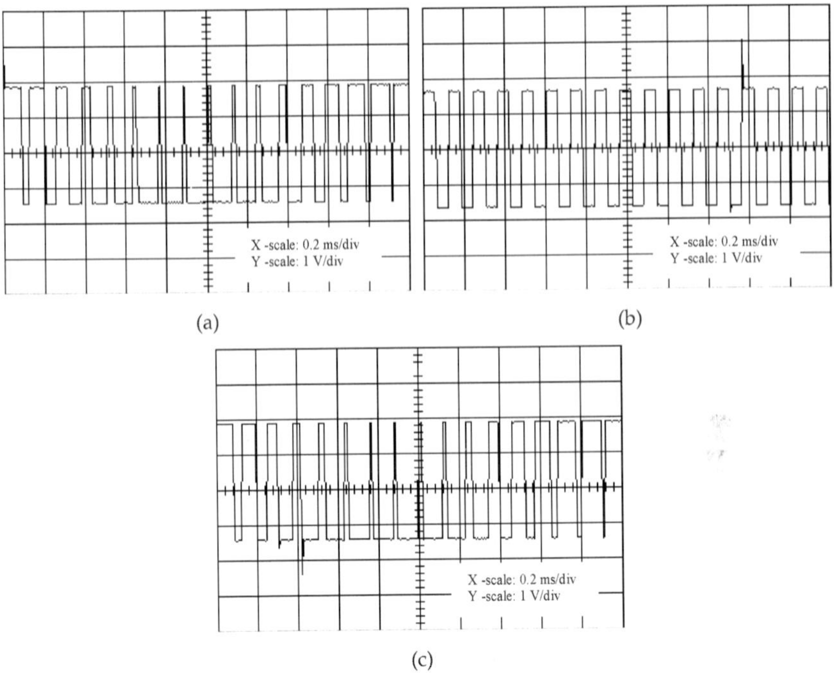

Figure 27. The three-phase PWM. (a) PWM1, (b) PWM2, and (c) PWM3.

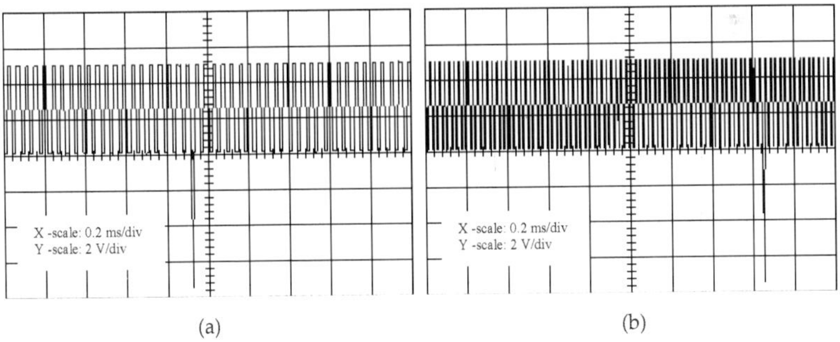

Figure 28. Encoder output signals at different sampling rate (a) 5 kHz and (b) 10 kHz.

Figure 29 (a), (b), and (c) show the stator voltage and current in the acceleration and steady-state stages and speed response in the steady-state stage, respectively, looking at one of the three phases. Figure 29 (a) reveals the stator voltage that increases gradually to a constant value. For the stator current, as shown in Figure 29 (b), the current is initially at maximum value and gradually decreases to a constant value. These verify that the controller is performing as expected during the acceleration and steady-state stages. Figure 29 (c) represents the speed at steady-state stage, which also proves that the controller is performing as expected, though some spikes are observed that can be related to the encoder used in this set-up. The spikes can be eliminated if a more sensitive encoder has been used.

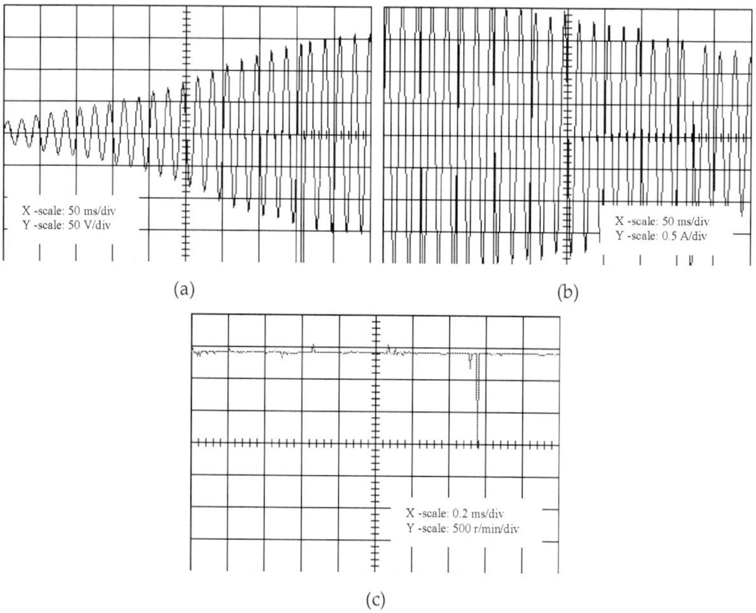

Figure 29. (a) Stator voltage, (b) stator current, and (c) speed response outputs.

10. Conclusions

The objective of this study is to elaborate and elucidate the effects and performance of internal and external disturbances for an established HFFC scheme to accordingly modify the speed of an induction motor. The fuzzy-fuzzy controller has been proven to be more effective as compared to a scalar controller due to the utilisation of the two aspects of the FOC. Besides, one of the key advantages of this controller includes the supply of the same FOC and frequency that is unresponsive to the parameter variation in the motor and system strength to noise and load disturbances. This study, under dynamic settings, produced a comprehensive evaluation and analysis of the three controllers, HFFC, IFOC, and HFPIC. The experimental results verify the performance of the proposed HFFC in controlling the IM variable speed drive. Therefore, for further enhancing the IM-VSD performance, the work to be considered includes the advancement in the augmentation of the controllers to improve the VSD performances.

Acknowledgements

The authors acknowledge the support from the Universiti Teknologi PETRONAS through the award of the Graduate Assistantship scheme and URIF.

Author details

Nordin Saad*, Muawia A. Magzoub, Rosdiazli Ibrahim and Muhammad Irfan

*Address all correspondence to: nordiss@petronas.com.my

Department of Electrical and Electronic Engineering, Universiti Teknologi Petronas, Bandar Sri Iskandar, Perak, Malaysia

References

[1] J. Bocker and S. Mathapati. State of the art of induction motor control. In: IEEE International Electric Machines & Drives Conference; May; 2007. p. 1459-1464.

[2] Muawia Magzoub, Nordin Saad, and Rosdiazli Ibrahim. Analysis and modeling of indirect field-oriented control for PWM-driven induction motor drives. In: 2013 IEEE Conference on Clean Energy and Technology (CEAT); November; IEEE; 2013. p. 488-493.

[3] M. Magzoub, N. Saad, and R. Ibrahim. An intelligent speed controller for indirect field-oriented controlled induction motor drives. In: 2013 IEEE conference on Clean Energy and Technology (CEAT); November; IEEE; 2013. p. 327-331.

[4] M. Magzoub, N. Saad, R. Ibrahim, M. Maharun, and S. Zulkifli. Hybrid fuzzy-fuzzy controller for PWM-driven induction motor drive. In: 2014 IEEE International Conference on Power and Energy (PEcon); IEEE; 2014. p. 260-265.

[5] N. Saad and M. Arrofiq. A PLC-based modified-fuzzy controller for PWM-driven induction motor drive with constant V/Hz ratio control. Robotics and Computer-Integrated Manufacturing. 2012;28(2):95-112.

[6] K. L. Shi, T. F. Chan, and Y. K. Wong. Hybrid fuzzy two-stage controller for an induction motor. In: 1998 IEEE International Conference on Systems, Man, and Cybernetics; October; IEEE; 1998. p. 1898-1903.

[7] K. L. Shi, T. F. Chan, Y. K. Wong, and S. L. Ho. An improved two-stage control scheme for an induction motor. In: Proceedings of the IEEE 1999 International Conference on Power Electronics and Drive Systems; IEEE; 1999. p. 405-410.

[8] K. L. Shi, T. F. Chan, Y. K. Wong, and S. L. Ho. A novel hybrid fuzzy/PI two-stage controller for an induction motor drive. In: IEMDC 2001. IEEE International Electric Machines and Drives Conference; IEEE; 2001. p. 415-421.

[9] D. Asija. Speed control of induction motor using fuzzy-PI controller. In: 2010 2nd International Conference on Mechanical and Electronics Engineering (ICMEE); August; IEEE; 2010. p. V2-460.

[10] N. Tiwary, A. Rathinam, and S. Ajitha. Design of hybrid fuzzy-pi controller for speed control of brushless dc motor. In: 2014 International Conference on Electronics, Communication and Instrumentation (ICECI); January; IEEE; 2014. p. 1-4.

[11] P. Melin and O. Castillo, editors. Hybrid intelligent systems for pattern recognition using soft computing: an evolutionary approach for neural networks and fuzzy systems. Springer Science & Business Media; 2005.

[12] W. S. Oh, Y. T. Kim, C. S. Kim, T. S. Kwon, and H. J. Kim. Speed control of induction motor using genetic algorithm based fuzzy controller. In: The 25th Annual Conference of the IEEE, In Industrial Electronics Society; IEEE; 1999. p. 625-629.

[13] S. V. Wong and A. M. S. Hamouda. Optimization of fuzzy rules design using genetic algorithm. Advances in Engineering Software. 200;31(4):251-262.

[14] I. K. Bousserhane, A. Hazzab, M. Rahli, M. Kamli, and B. Mazari. Adaptive PI controller using fuzzy system optimized by genetic algorithm for induction motor control. In: 10th IEEE, In International Power Electronics Congress; IEEE; 2006. p. 1-8.

[15] J. Y. Zhang and Y. D. Li. Application of genetic algorithm in optimization of fuzzy control rules. In: Proceedings of the Sixth International Conference on Intelligent Systems Design and Applications; IEEE Computer Society; 2006.

[16] H. X. Zhang, B. Zhang, and F. Wang. Automatic fuzzy rules generation using fuzzy genetic algorithm. In: 2009 Sixth International Conference on Fuzzy Systems and Knowledge Discovery; IEEE; 2009. p. 107-112.

[17] M. A. Jaradat, M. I. Awad, and B. S. El-Khasawneh. Genetic-fuzzy sliding mode controller for a dc servomotor system. In: 2012 8th International Symposium on Mechatronics and its Applications (ISMA); IEEE; 2012. p. 1-6.

[18] A. Trzynadlowski. The field orientation principle in control of induction motors. Springer Science & Business Media; 1993.

[19] K. Bimal. Bose. Modern power electronics and AC drives. 2002.

[20] M. Magzoub, N. Saad, and R. Ibrahim. Simulation of a ball on a beam model using a fuzzy-dynamic and a fuzzy-static sliding-mode controller. Research Journal of Applied Sciences, Engineering and Technology. 2014;8(2):288-295.

[21] T. F. Chan and K. Shi. Applied intelligent control of induction motor drives. John Wiley & Sons; 2011. 421 p.

[22] N. Siddique, editor. Intelligent control a hybrid approach based on fuzzy logic, neural networks and genetic algorithms. Springer International Publishing Switzerland 2014; 2014. 282 p. DOI: 10.1007/978-3-319-02135-5

[23] N. Pitalúa Díaz, R. Lagunas Jiménez, and A. González. Tuning fuzzy control rules via genetic algorithms: An experimental evaluation. Research Journal of Recent Sciences. 2013;2(10):81-87.

[24] J. S. R. Jang and and N. Gulley, editors. Fuzzy logic toolbox user's guide. Second printing ed. The MathWorks, Inc.; April 1997. p. 208.

[25] Genetic algorithm and direct search toolbox user's guide. The MathWorks, Inc.; January 2004. p. 210.

[26] M. Nuruzzaman, editor. Modeling and simulation in SIMULINK for engineers and scientists. AuthorHouse; 2005.

[27] O. Wahyunggoro and N. Saad. Development of fuzzy-logic-based self-tuning PI controller for servomotor. In: S. Ehsan Shafiei, editor. Advanced Strategies for Robot Manipulators. Sciyo; August, 2010.

[28] S. Lynch, editor. Dynamical systems with applications using MATLAB. Springer International Publishing; 2014. p. 519.

Chapter 10

DTC-FPGA Drive for Induction Motors

Rafael Rodríguez-Ponce, Fortino Mendoza-Mondragón,
Moisés Martínez-Hernández and Marcelino Gutiérrez-Villalobos

Additional information is available at the end of the chapter

http://dx.doi.org/10.5772/60871

Abstract

Direct torque control, or DTC, is an electrical motor strategy characterized for simplicity and high performance when controlling industrial machines such as induction motors. However, this technique is often accompanied by an unwanted deformation on the torque and flux signals denominated ripple, which can cause audible noise and vibration on the motor. Considerable research has been presented on this topic; nevertheless the original DTC algorithm is often modified to the point that it is as complex as other motor control strategies. To solve this problem, a novel architecture was designed in order to reduce the sampling period to a point where torque ripple is minimal, while maintaining the classical DTC control structure. In this work, the original DTC control strategy was implemented on a Virtex-5 field programmable gate array (FPGA). For the code, a two's complement fixed-point format and a variable word-size approach was followed using very-high-speed integrated circuit hardware description language (VHDL). Results were validated using MATLAB/Simulink simulations and experimental tests on an induction motor. With this new architecture, the authors hope to provide guidelines and insights for future research on DTC drives for induction motors.

Keywords: Direct torque control, AC servo drive, Field programmable gate array

1. Introduction

It is well known that one of the components most commonly found in any industrial or residential machine is the electric motor. Motors are used almost in any application where electricity must be converted to a mechanical motion of some kind. They come in a wide variety of sizes, ranging from very small, as the motors found inside a cellphone for creating vibration, to very large, as the ones used in wind tunnels for aircraft testing.

There are many types of electrical motors but one that has remained the favorite for almost any medium- to large-sized application is the AC induction motor (ACIM). The concept of this "sparkless" motor was first conceived by Nicola Tesla in the late nineteenth century. Although it was first designed as a polyphase structure that consisted of two stator phases in an orthogonal relationship, it has since been modified to a more common three-phase structure, which results in a more balanced operation of the motor voltages and currents [1]. The ACIM is rugged and highly reliable, can be manufactured at a low cost, and is almost maintenance-free, except for bearings and other external mechanical parts.

The ACIM is essentially a fixed-speed machine. However, most industrial applications require a motor in which torque or speed can easily be controlled. Therefore, several high-performance control strategies have been developed for AC motors; two of the most popular motor control methods are field oriented control (FOC) and direct torque control (DTC) [2]. Unlike FOC, DTC is characterized for its simplicity since it does not require PI regulators, coordinate transformations, pulse width modulation (PWM) generators, or position encoders on the motor shaft [3]. In spite of its simplicity, DTC provides fast torque control in the steady state and under transient operating conditions with simple control structure [4].

One major disadvantage of DTC is that it has the distinct characteristic of ripple on torque and flux signals; this ripple is an unwanted deformation or "noise" on motor signals that can lead to audible noise and vibration in the motor [5]. It is possible to reduce ripple by reducing the sampling period [6]; for this reason, recent DTC drives have been implemented by using fast processing devices such as FPGAs [7].

The purpose of this chapter is to present the development of an ACIM drive on an FPGA; the original DTC strategy was implemented by using a fixed-point architecture on a Xilinx Virtex-5 development board and used on real-time experiments. By using an FPGA and a novel architecture, it was possible to reduce processing time to 1.6 μs, therefore reducing torque ripple to a minimum.

This chapter is organized as follows: Section 2 presents a mathematical model for the induction motor. Section 3 presents a simplified description of the DTC strategy. A detailed description of the FPGA-based induction motor drive is described in Section 4. Simulation and experimental results are presented in Section 5. Finally, the conclusion of this work is given in Section 6.

2. Induction motor mathematical model

The cage rotor induction machine is widely used in industrial applications, such as belt conveyors, pumps, fans, cranes, etc. It presents great mechanical sturdiness and there is a good standardization between ACIM manufacturers worldwide. Nevertheless, the relative simplicity of the operation of the motor hides a great complexity, especially when it is aimed at controlling the performed electromechanical conversion [1].

There are a number of ACIM models; the model used for vector control design can be obtained by using the space vector theory. The 3-phase motor quantities, such as currents, voltages, and magnetic fluxes, are expressed in terms of complex space vectors. Such a model is valid for any instantaneous variation of voltage and current and adequately describes the performance of the machine under steady-state and transient operation. The motor is considered to be a 2-phase machine by using two orthogonal axes; with this model, the number of equations is reduced and the control design is simplified [8].

When describing a three-phase IM by a system of equations, the following simplifying assumptions are made [8]:

- The three-phase motor is symmetrical.
- Only the fundamental harmonic is considered, while the higher harmonics of the spatial field distribution and of the magneto-motive force (MMF) in the air gap are disregarded.
- The spatially distributed stator and rotor windings are replaced by a specially formed, so-called concentrated coil.
- The effects of anisotropy, magnetic saturation, iron losses, and eddy currents are neglected.
- The coil resistances and reactance are taken to be constant.
- In many cases, especially when considering steady state, the current and voltages are taken to be sinusoidal.

Taking into consideration the earlier-stated assumptions, the following equations of the instantaneous stator phase voltage values can be written as follows (all variable descriptions are listed in Table 1):

$$V_A = R_s i_A + \frac{d}{dt}\Psi_A \tag{1}$$

$$V_B = R_s i_B + \frac{d}{dt}\Psi_B \tag{2}$$

$$V_C = R_s i_C + \frac{d}{dt}\Psi_C \tag{3}$$

A three-phase variable system can be uniquely described through a space vector, which is a complex term and time-dependent $k(t)$ and a real homopolar component $k_0(t)$ as follows:

$$k(t) = \frac{2}{3}\left[1^* k_A + a^* k_B + a^{2*} k_C\right] \tag{4}$$

$$k_0(t) = \frac{1}{3}\left[k_A + k_B + k_C\right] \tag{5}$$

The real axis direction coincides with that one of phase A. Usually, the neutral connection for a three-phase system is open, so that the homopolar component equals zero.

The ACIM model is given by the space vector form of the voltage equations. The system model defined in a two-phase stationary (α, β) coordinate system attached to the stator is expressed by the following equations:

a. The stator voltage differential equations:

$$V_{s\alpha} = R_s i_{s\alpha} + \frac{d}{dt}\Psi_{s\alpha} \tag{6}$$

$$V_{s\beta} = R_s i_{s\beta} + \frac{d}{dt}\Psi_{s\beta} \tag{7}$$

b. The rotor voltage differential equations:

$$V_{r\alpha} = 0 = R_r i_{r\alpha} + \frac{d}{dt}\Psi_{r\alpha} + \omega\Psi_{r\beta} \tag{8}$$

$$V_{r\beta} = 0 = R_r i_{r\beta} + \frac{d}{dt}\Psi_{r\beta} + \omega\Psi_{r\alpha} \tag{9}$$

c. The stator and rotor flux linkages expressed in terms of the stator and rotor current space vectors:

$$\Psi_{s\alpha} = L_s i_{s\alpha} + L_m i_{r\alpha} \tag{10}$$

$$\Psi_{s\beta} = L_s i_{s\beta} + L_m i_{r\beta} \tag{11}$$

$$\Psi_{r\alpha} = L_r i_{r\alpha} + L_m i_{s\alpha} \tag{12}$$

$$\Psi_{r\beta} = L_r i_{r\beta} + L_m i_{s\beta} \tag{13}$$

http://dx.doi.org/10.5772/60871

d. The electromagnetic torque expressed by utilizing space vector quantities:

$$T_e = \frac{3}{2}P\left(\Psi_{s\alpha} i_{s\beta} - \Psi_{s\beta} i_{s\alpha}\right) \tag{14}$$

The ACIM model is often used in vector control algorithms. The aim of vector control is to implement control schemes that produce high-dynamic performance and are similar to those used to control DC machines [2]. To achieve this, the reference frames may be aligned with the stator flux-linkage space vector, the space vector of the rotor current in the rotor reference frame, the rotor flux-linkage space vector, or the magnetizing space vector. The most popular reference frame is the reference frame attached to the rotor flux linkage space vector with direct axis (d) and quadrature axis (q) [8].

After transformation into d-q coordinates the motor model follows:

$$V_{sd} = R_s i_{sd} + \frac{d}{dt}\Psi_{sd} - \omega_s \Psi_{sq} \tag{15}$$

$$V_{sq} = R_s i_{sq} + \frac{d}{dt}\Psi_{sq} - \omega_s \Psi_{sd} \tag{16}$$

$$V_{rd} = 0 = R_r i_{rd} + \frac{d}{dt}\Psi_{rd} - (\omega_s - \omega)\Psi_{rq} \tag{17}$$

$$V_{rq} = 0 = R_r i_{rq} + \frac{d}{dt}\Psi_{rq} - (\omega_s - \omega)\Psi_{rd} \tag{18}$$

$$\Psi_{sd} = L_s i_{sd} + L_m i_{rd} \tag{19}$$

$$\Psi_{sq} = L_s i_{sq} + L_m i_{rq} \tag{20}$$

$$\Psi_{rd} = L_r i_{rd} + L_m i_{sd} \tag{21}$$

$$\Psi_{rq} = L_r i_{rq} + L_m i_{sq} \tag{22}$$

$$T_e = \frac{3}{2}P\left(\Psi_{sd} i_{sq} - \Psi_{sq} i_{sd}\right) \tag{23}$$

Variable	Description
V_A, V_B, V_C	Instantaneous values of the stator phase voltages
i_A, i_B, i_C	Instantaneous values of the stator phase currents
Ψ_A, Ψ_B, Ψ_C	Flux linkages of the stator phase windings
k_A, k_B, k_C	Arbitrary phase variables
$a,\ a^2$	Spatial operators $a=e^{j2\pi/3}$ and $a^2=e^{j4\pi/3}$
α, β	Stator orthogonal coordinate system
$V_{s\alpha,\beta}$	Stator voltages [V]
$i_{s\alpha,\beta}$	Stator currents [A]
$V_{r\alpha,\beta}$	Rotor voltages [V]
$i_{r\alpha,\beta}$	Rotor currents [A]
$\Psi_{s\alpha,\beta}$	Stator magnetic fluxes [Wb]
$\Psi_{r\alpha,\beta}$	Rotor magnetic fluxes [Wb]
R_s	Stator phase resistance [Ohm]
R_r	Rotor phase resistance [Ohm]
L_s	Stator phase inductance [H]
L_r	Rotor phase inductance [H]
L_m	Mutual (stator to rotor) inductance [H]
ω / ω_s	Electrical rotor speed/synchronous speed [rad/s]
P	Number of pole pairs
T_e	Electromagnetic torque [Nm]

Table 1. Variable description.

3. Classical DTC scheme

The theory for the DTC control strategy was developed by Manfred Depenbrock as direct self-control (DSC) and separately, as direct torque control (DTC) by Isao Takahashi and Toshihiko Noguchi, both in the mid-1980s, although the DTC innovation is usually credited to all three individuals [2]. A block diagram of the DTC strategy is shown in Figure 1.

The main objective of DTC is to estimate instantaneous values of torque and magnetic flux, based on motor current and voltage. Torque and flux vectors are controlled directly and independently by selecting the appropriate inverter voltage vector that will maintain torque and flux errors within the hysteresis comparator limits [3].

http://dx.doi.org/10.5772/60871

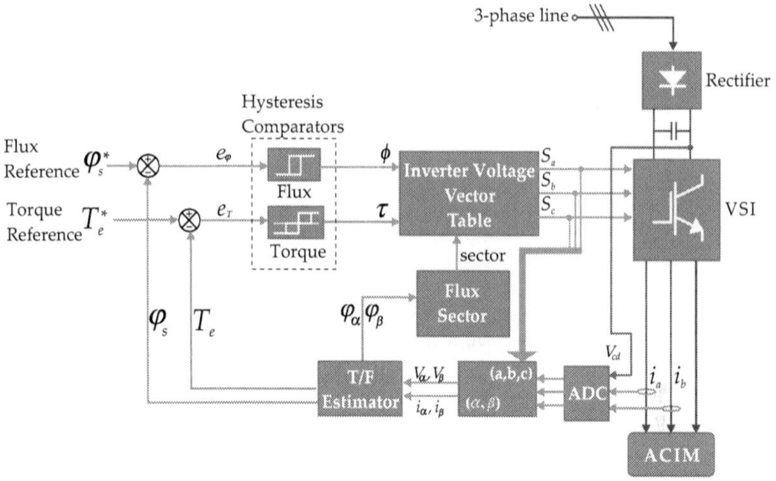

Figure 1. Direct torque control block diagram.

In order to estimate the motor torque and flux values, the instantaneous current (i_a, i_b) and DC bus voltage (V_{cd}) signals are obtained from the ACIM as illustrated in Figure 1. These analog signals are converted to digital values by means of an analog to digital converter (ADC). The current and voltage signals, as well as the current state of the voltage source inverter (VSI) vector (S_a, S_b, S_c), are transformed from a 3-phase reference frame to a 2-phase reference frame (α, β), as follows:

a. The α, β current signals:

$$i_\alpha = i_a \tag{24}$$

$$i_\beta = \frac{\sqrt{3}}{3}\left(i_a + 2i_b\right) \tag{25}$$

b. The α, β voltage signals:

$$V_\alpha = \frac{V_{cd}}{3}\left(2S_a - S_b - S_c\right) \tag{26}$$

$$V_\beta = \frac{\sqrt{3}}{3} V_{cd}\left(S_b - S_c\right) \tag{27}$$

c. The α, β flux components:

$$\varphi_{\alpha} = \varphi_{\alpha 0} + \left(V_{\alpha} - R_{s} i_{\alpha}\right) T_{s} \tag{28}$$

$$\varphi_{\beta} = \varphi_{\beta 0} + \left(V_{\beta} - R_{s} i_{\beta}\right) T_{s} \tag{29}$$

where

T_s – Sampling period [seconds]

R_s – Stator resistance [ohms]

$\varphi_{\alpha 0}$, $\varphi_{\beta 0}$ – Previous flux component value [Wb]

Based on this data, the flux magnitude and the electromagnetic torque are obtained as follows:

a. The stator magnetic flux magnitude:

$$\varphi_{s} = \sqrt{\left(\varphi_{\alpha}^{2}\right) + \left(\varphi_{\beta}^{2}\right)} \tag{30}$$

b. The electromagnetic torque:

$$T_{e} = \frac{3}{2} P\left(\varphi_{\alpha} i_{\beta} - \varphi_{\beta} i_{\alpha}\right) \tag{31}$$

In order to re-orient the flux vector φ_s, first it is necessary to determine where it is localized. For this reason, the flux vector circular trajectory is divided into six symmetrical sectors, as shown in Figure 2.

The angle θ_s can be calculated, based on the α, β flux components as follows:

$$\theta_{s} = tan^{-1} \frac{\varphi_{\beta}}{\varphi_{\alpha}} \tag{32}$$

However, implementing Eq. (32) in an FPGA is complex and time consuming and is usually performed be means of the coordinate rotation digital computer (CORDIC) algorithm [9]. Instead, it is possible to determine the sector in which the flux vector is located, based on the signs of the flux components, as described in [3]. The sector can be determined by using Table 2 and Eq. (33).

http://dx.doi.org/10.5772/60871

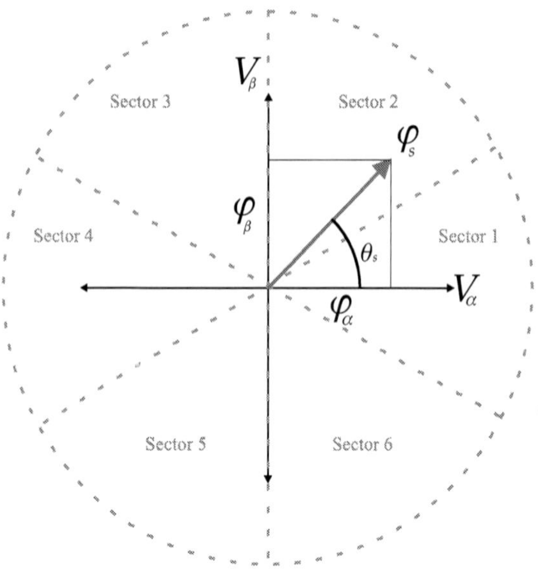

Figure 2. Sectors of the flux vector circular trajectory.

Sign of φ_α	Sign of φ_β	φ_{ref}	Sector
+	+/-	-	1
+	+	+	2
-	+	+	3
-	+/-	-	4
-	-	-	5
+	-	+	6

Table 2. Stator flux space vector's sector.

$$\varphi_{ref} = \sqrt{3}\left|\varphi_\beta\right| - \left|\varphi_\alpha\right| \tag{33}$$

For example, if both flux components are positive and the result of Eq. (33) is also positive, then the flux vector is located in sector 2. Instead, if the result of the equation is negative, the vector is located in sector 1.

The method described previously to determine the sector of the flux vector is easier to implement in a digital device and can be processed faster since it consists of a simple data table.

As shown in Figure 1, the estimated magnetic flux and electromagnetic torque values are compared with the magnetic flux reference and the electromagnetic torque reference, respectively. The flux and torque errors (e_φ, e_T) are delivered to the hysteresis controllers.

A two-level hysteresis controller is used to establish the limits of the flux error. For the torque error, a three-level hysteresis controller is used. The hysteresis controllers are shown in Figure 3.

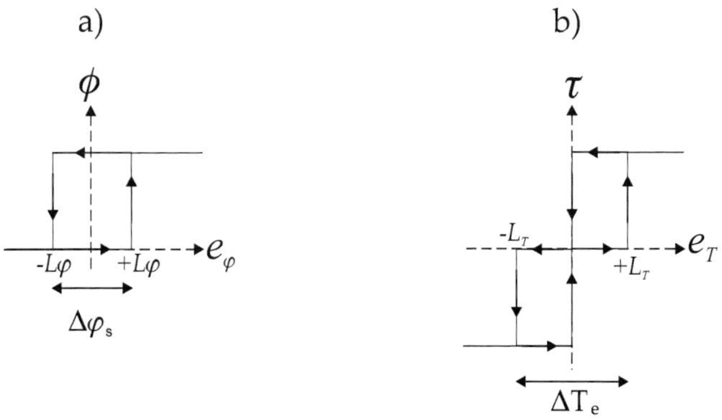

Figure 3. Hysteresis controllers for (a) flux and for (b) torque.

The hysteresis controller output signals ϕ and τ are defined as follows:

$$\phi = 1 \text{ for } e_\varphi > +L_\varphi \tag{34}$$

$$\phi = 0 \text{ for } e_\varphi < -L_\varphi \tag{35}$$

$$\tau = 1 \text{ for } e_T > +L_\tau \tag{36}$$

$$\tau = 0 \text{ for } e_T = 0 \tag{37}$$

$$\tau = -1 \text{ for } e_T < -L_\tau \tag{38}$$

The digitized output variables ϕ, τ and the stator flux sector determine the appropriate voltage vector from the inverter switching table. Thus, the selection table generates pulses S_a, S_b, S_c to control the power switches in the inverter in order to generate six possible active vectors (v1-v6) and two zero vectors (v0, v7), as shown in Figure 4.

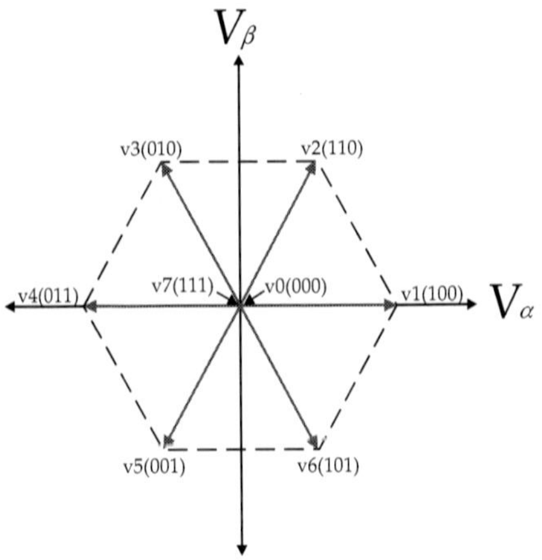

Figure 4. Voltage vectors based on eight possible inverter states.

For the stator flux vector laying in sector 1 (Figure 5), in order to increase its magnitude, voltage vectors v1, v2 or v6 can be selected. Conversely, a decrease can be obtained by selecting v3, v4 or v5. By applying one of the voltage vectors v0 or v7, the stator flux vector is not changed.

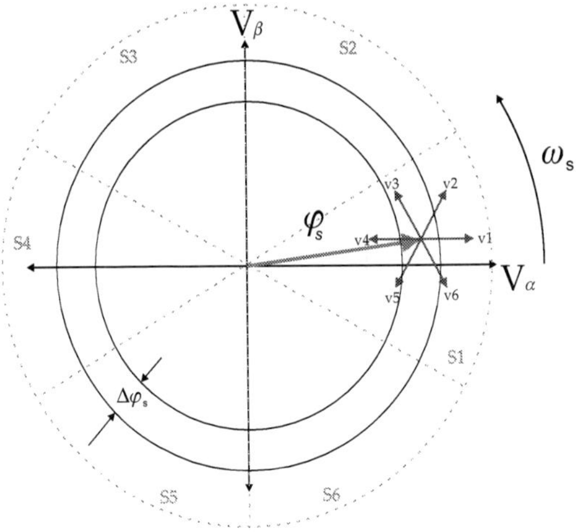

Figure 5. Selection of the optimum voltage vectors for the stator flux vector in sector 1.

For torque control, if the vector is moving as indicated in Figure 5, the torque can be increased by selecting vectors v2, v3 or v4. To decrease torque, vectors v1, v5 or v6 can be selected.

The above considerations allow the construction of the inverter switching table as presented in Table 3.

φ	τ	Sector					
		1	2	3	4	5	6
$\varphi=1$	$\tau=1$	v2(110)	v3(010)	v4(011)	v5(001)	v6(101)	v1(100)
	$\tau=0$	v7(111)	v0(000)	v7(111)	v0(000)	v7(111)	v0(000)
	$\tau=-1$	v6(101)	v1(100)	v2(110)	v3(010)	v4(011)	v5(001)
$\varphi=0$	$\tau=1$	v3(010)	v4(011)	v5(001)	v6(101)	v1(100)	v2(110)
	$\tau=0$	v0(000)	v7(111)	v0(000)	v7(111)	v0(000)	v7(111)
	$\tau=-1$	v5(001)	v6(101)	v1(100)	v2(110)	v3(010)	v4(011)

Table 3. Optimum switching table.

The optimal voltage vector is a vector such that, once applied to the VSI, will maintain the flux and torque signals within the hysteresis comparator limits [4]. The selected voltage vector is applied at the end of the sampling period.

4. DTC Digital Implementation

In this section, a detailed description of a DTC drive for induction motors is presented. The drive was implemented on a Xilinx Virtex-5 FPGA based on two´s complement fixed-point architecture composed of 7 main blocks, which are described as follows and shown in Figure 6:

a. Conversion control block: this block controls 3 external 12-bit serial, ADCs that operate in parallel. The motor current and voltage signals are converted from serial to parallel. Since both signals are scaled versions of the original, in this block both values are restored to their real value.

b. Torque and flux estimator: the real time electromagnetic torque and magnetic flux vectors are estimated based on motor current and voltage signals.

c. Flux sector detection block: the sector for the magnetic flux vector is detected.

d. Reference comparison block: the real torque and flux estimated values are compared with torque and flux references.

e. Hysteresis comparators: two-level and three-level hysteresis comparators are included in this block.

f. Switching table: all the optimal voltage vectors for the inverter are contained in this block.

http://dx.doi.org/10.5772/60871

g. Global control block: a finite state machine (FSM) is included in this block and is in charge of the control of all the other DTC blocks.

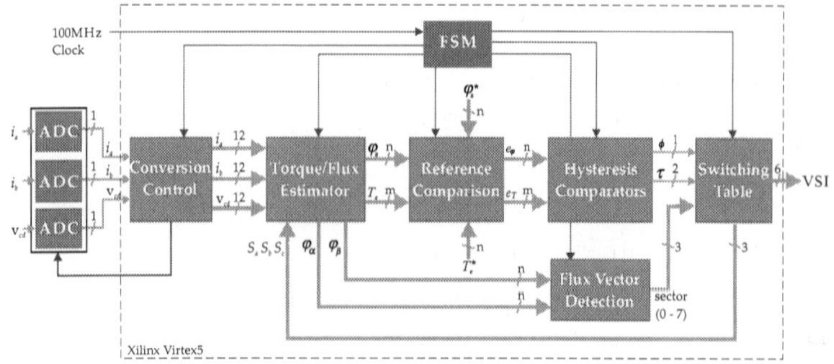

Figure 6. DTC architecture on Xilinx Virtex5 FPGA.

One of the benefits of this DTC architecture presented is that it is completely generic; the data width can be modified depending on the application or the precision required, and all the DTC equations will adjust automatically. The flux data path has n bits while the torque data path can have m bits, as shown in Figure 6. The data paths can be extended for more precision, however this will also extend the sampling time. For this project, the flux data path was left at 20 bits and the torque data path was set to 23 bits in order to achieve a low sampling period of 1.6 μs.

4.1. Conversion control block

The current signals (i_a, i_b) are first obtained by sensing two of the motor lines by means of coil sensors. The output is an AC signal that is amplified and added a DC offset, in order to have a positive only value between 0 and 3.3 V for the ADC. The signals are then converted by the ADC to a serial 12-bit value and then to a 12-bit parallel value. Finally, the offset value is subtracted and multiplied by a scaling factor in order to obtain the original current signal. This process is shown in Figure 7.

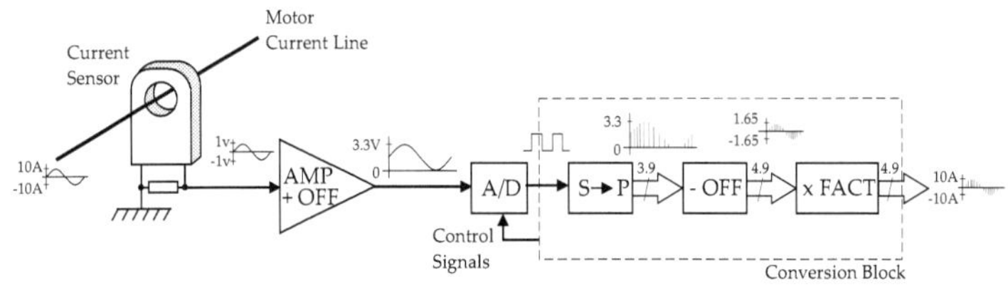

Figure 7. Current signal conversion process from current sensor to conversion control block.

The conversion process for the DC bus voltage signal is similar to the process described for the current signals, except that a current sensor is not used. Instead, by means of a resistive voltage divider, the voltage signal is reduced to a suitable value. Since the signal is always positive, there is no need to add an offset. The signal is only filtered and passed through several operational amplifier (OP AMP) stages, in order to isolate and adjust to a specific value between 0 and 3.3 V.

The DC voltage signal (V_{dc}) is converted to digital using a serial 12-bit ADC; the serial signal is converted to a parallel value and then multiplied by a scaling factor to restore it to the original DC value. The conversion process for the DC voltage signal is shown in Figure 8.

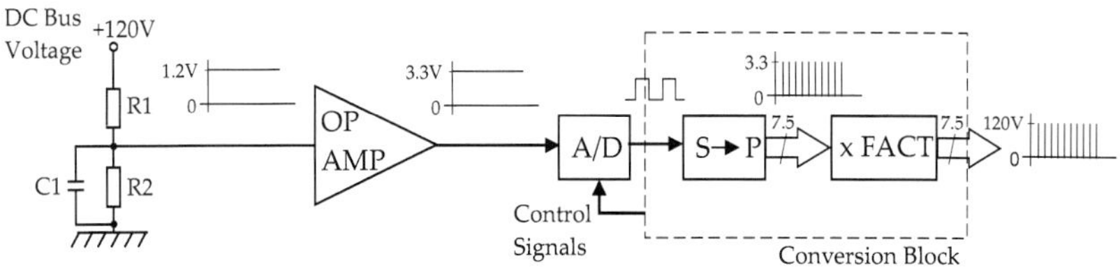

Figure 8. Voltage signal conversion process.

4.2. Torque and flux estimator

The estimation block is the most important part of the DTC process, since the selection of the optimal voltage vector for the VSI depends on the accuracy of the magnetic flux vector [4].

The flux estimator was designed in VHDL to execute Eq. (24-31) presented previously in Section 3, where basically the stator flux is calculated based on stator currents and voltages; once flux stationary components are calculated, the stator flux can be obtained by adding both components squared and applying the square root operation.

Several equations are implemented in parallel, such as the voltage and current transformation to stationary coordinates and later the stator flux stationary components.

In FPGA implementation, word size is critical; a large word size reduces quantization errors but increases area and affects costs. On the contrary, a small word size affects precision, increasing control error and torque ripple [10]. Therefore, a fixed-point format with a variable word size was used in the implementation of the DTC equations.

The DTC architecture was designed for implementation on an FPGA with data words starting at 12 bits and increased according to the mathematical operations to avoid a loss in precision.

The estimator was divided in three stages as follows:

Stage 1 – In the first stage, the values of i_a, i_b and $V_{cd,}$ and the previous inverter vector (S_a, S_b, S_c) are used to calculate the corresponding stationary components i_α, i_β, V_α and V_β. This stage is shown in Figure 9.

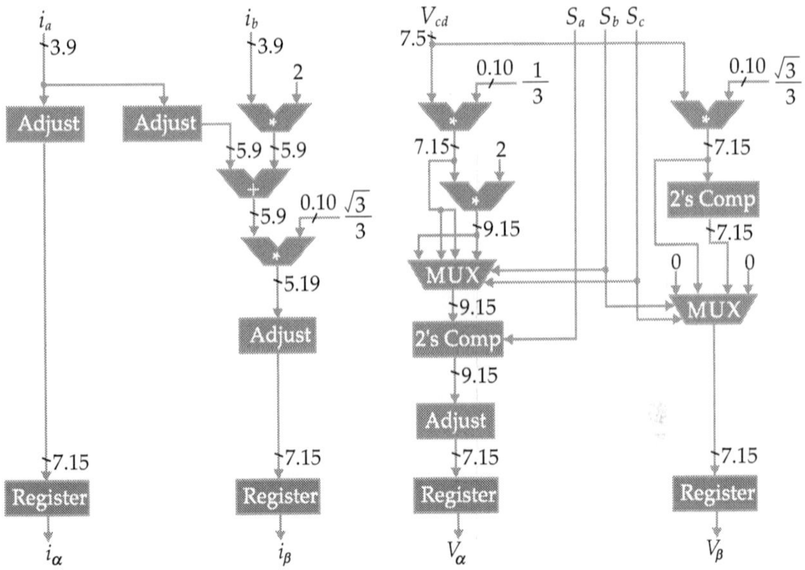

Figure 9. First stage of flux and torque estimator.

In the previous figure, the fixed-point format is indicated in each vertical line and width adjustments are made when required. At the end of this stage, four 22-bit parallel registers restrict data flow until they receive a pulse from the estimator FSM; this assures that all values pass to the next stage at the same time.

Stage 2 – In this stage, flux stationary components φ_α and φ_β are calculated based on data from stage 1 as described in (28) and (29). Both components are calculated based on the same equation, therefore a generic block was designed for this calculation and is used twice in parallel. The architecture for this stage is shown in Figure 10.

In this stage, both flux components are loaded to the register by a pulse from the estimator FSM, which serves as the previous flux value ($\varphi_{\alpha 0}$ or $\varphi_{\beta 0}$) for the next calculation.

Stage 3 – In this last stage, the flux components are squared, added, and then the square root (SQRT) algorithm is applied as in Eq. (30); a special architecture was designed for the SQRT and will be described in detail later in section 4.8. The stator torque is calculated, by means of Eq. (31), in parallel with the flux equation. The architecture for this last stage is shown in Figure 11.

4.3. Flux sector detection block

Based on the value and sign of the stationary flux components, the flux vector sector is determined by means of Eq.(33) and Table 2. The signs of the flux components are used to

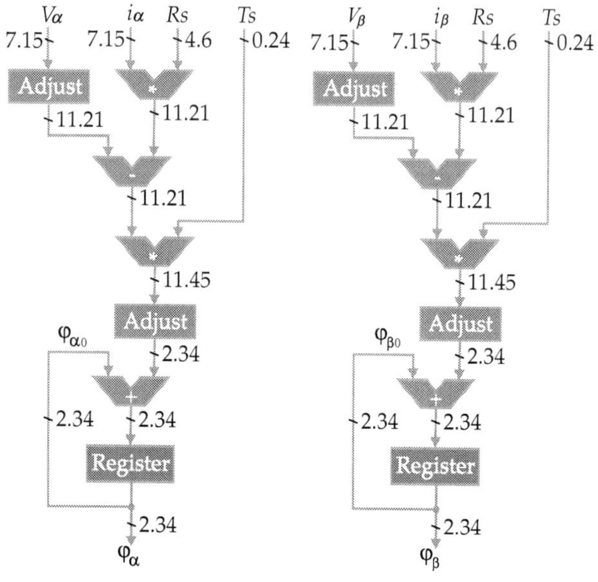

Figure 10. Second stage of flux and torque estimator.

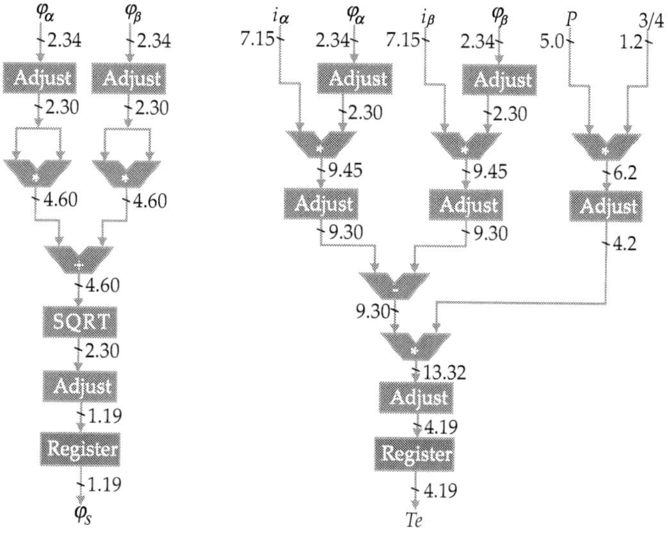

Figure 11. Last stage of flux and torque estimator.

determine the quadrant of the flux vector and the value of φ_{ref} is used for selecting between the upper or lower sectors in that quadrant. The architecture for this block is shown in Figure 12.

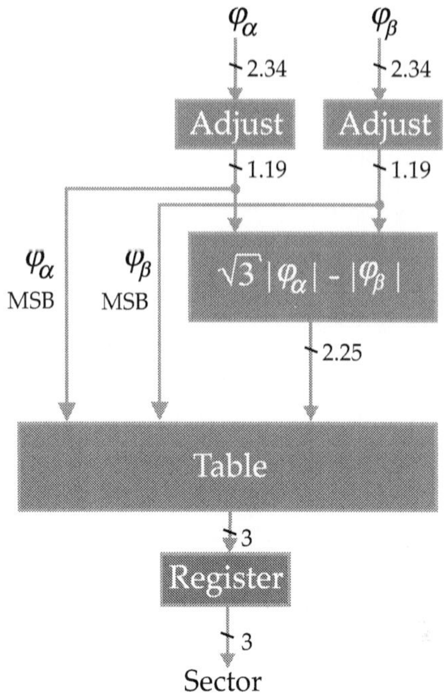

Figure 12. Architecture of the sector detection block.

4.4. Reference comparison block

In this DTC block, the estimated flux and torque values are subtracted from the corresponding reference values. The reference data may be entered by means of external slide switches or it can come from a user interface through the USB port. The structure of the USB interface is not discussed in this document. The structure of the comparison block is shown in Figure 13.

4.5. Hysteresis comparators

A two-level comparator for flux and a three-level comparator for torque are implemented in this block. Both hysteresis comparators were designed as FSMs in order to provide fast transition from one to another state. The FSM for the hysteresis comparators are shown in Figure 14.

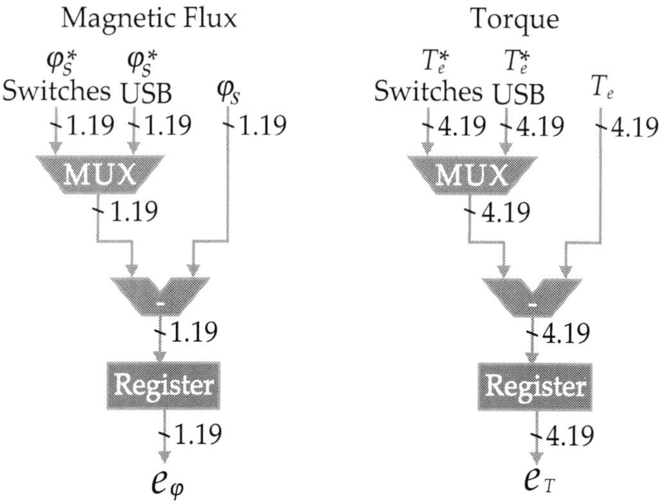

Figure 13. Reference comparison blocks for flux and torque.

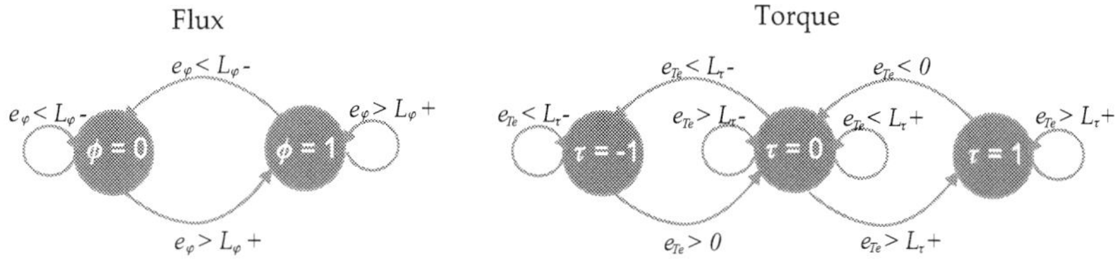

Figure 14. FSMs for flux and torque hysteresis comparators.

4.6. Switching table

The VSI optimal switching vectors listed in Table 3 are included in the switching table. A voltage vector is selected based on the hysteresis comparator values ϕ and τ, and on the flux vector sector. The table output is a 3-bit vector and its complement, which are fed to the VSI. The 3-bit vector is also sent back to the torque-flux estimator to obtain the next torque and flux values. The architecture for the switching table is shown in Figure 15.

4.7. Global control block

In order to have a constant sampling period (T_s), a global FSM was used to control the data flow from one block to the next. Since there is a register at the output of every major block, the FSM sends a timed pulse to each one, depending on the selected width of the data path. The

http://dx.doi.org/10.5772/60871

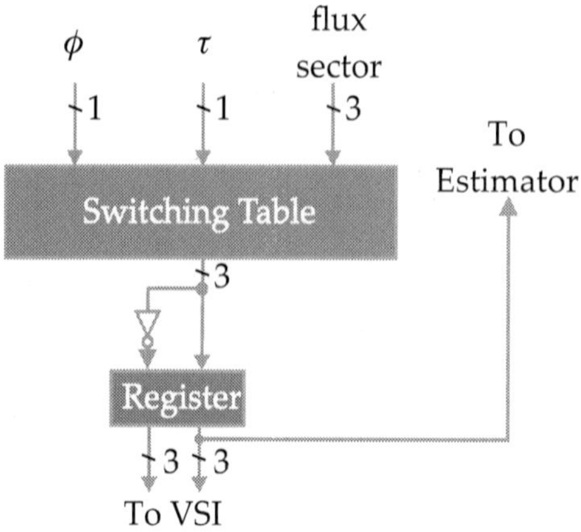

Figure 15. Switching table architecture.

ADC conversion and adjustment take a total of 600 ns and the estimation of torque and flux take 940 ns with the data path set to 20-bits for the flux and 23-bits for torque. The rest of the processes take only 20 ns each, giving a total of 1600 ns for the sampling period. The execution times are shown in Figure 16.

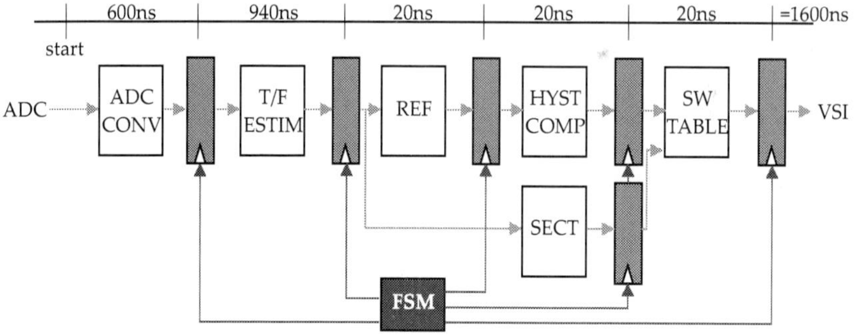

Figure 16. Execution times of each DTC block.

4.8. Square root algorithm

In order to reduce current ripple to a minimum, the DTC algorithm must have a processing period as low as possible, and the square root is the calculation that usually takes the longest [4]. For this reason, a special architecture was devised in order to have an algorithm that could be scaled easily to any number of bits and could execute quickly [11].

A successive approximation register (SAR), similar to the used in commercial ADCs, was designed in order to arrive quickly to the result.

The whole algorithm is based on a square root calculation as follows:

$$y = \sqrt{x} \tag{39}$$

which can also be written as:

$$y^2 - x = 0 \tag{40}$$

In other words, if a certain number y is squared and x is subtracted, and the result is zero or very close to it, then y would be the square root of x. Hence, the main objective is to find y as fast as possible; this is where the SAR proved worthy.

The SAR is basically a register in which each bit is LOW and, bit by bit, is toggled to HIGH, starting from the most significant bit (MSB) down to the least significant bit (LSB) under the flowing conditions; if the result of Eq. (40) is:

- Greater than zero, the HIGH bit is toggled back to LOW and the SAR proceeds to toggle the next bit.
- Less than zero, the HIGH bit is maintained and the SAR proceeds to toggle the next bit.
- Equal to zero or if the SAR ran out of bits to toggle, the current value of y is the square root of x and the process is terminated.

The architecture used for the square root is shown in Figure 17.

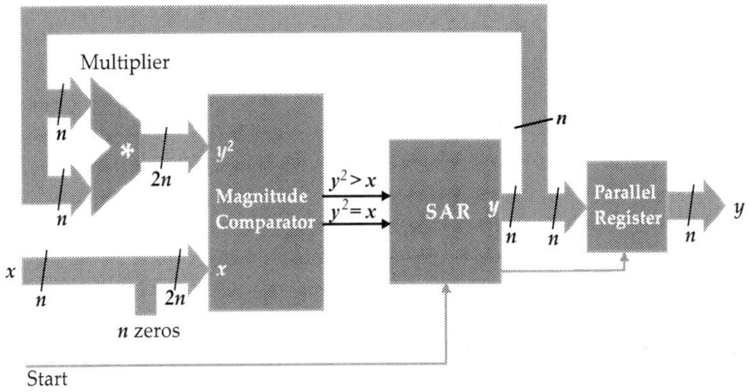

Figure 17. Square root architecture.

http://dx.doi.org/10.5772/60871

The SAR is the main block of the SQRT architecture; a START pulse is received from the estimator FSM to initialize the SAR process. Once the result is found, it is loaded to a parallel register and passed on to the reference comparison block.

Despite the simplicity of the square root architecture presented, it proved to be a fast and precise algorithm that could be scaled easily to adapt to the generic nature of the torque and flux estimator. The only restriction is that it requires x having an even number of bits.

5. Results and discussion

5.1. Simulation results

The DTC architecture presented in Section 4 was first tested in MATLAB/Simulink for simulation studies and later was implemented on an FPGA-based induction motor drive for experimental verification.

The torque dynamic response is shown in Figure 18 with a sampling period of 1.6 μs and the hysteresis band reduced to 0.1 Nm.

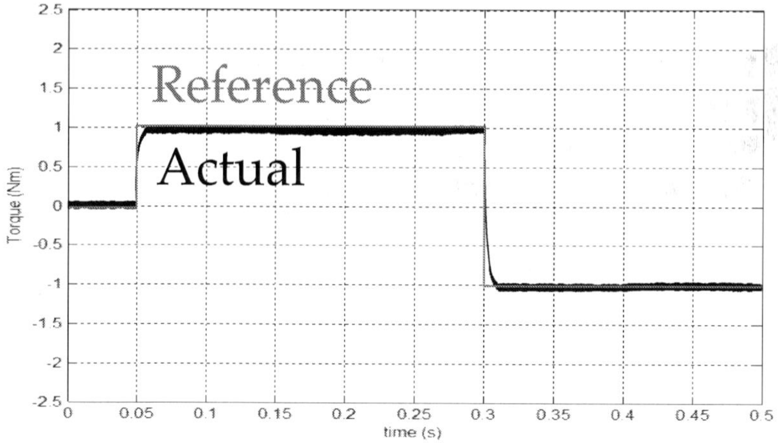

Figure 18. Torque dynamic response in MATLAB/Simulink.

Thanks to the small sampling period, the torque ripple was reduced to a small value.

Similarly, the flux hysteresis band was reduced to 0.06 Wb and as a result, as shown in Figure 19, the flux locus is almost a perfect circle with very small ripple.

Consequently, due to the small sampling period and reduced torque ripple, the stator current signal appears almost as a perfect sinusoidal, as shown in Figure 20.

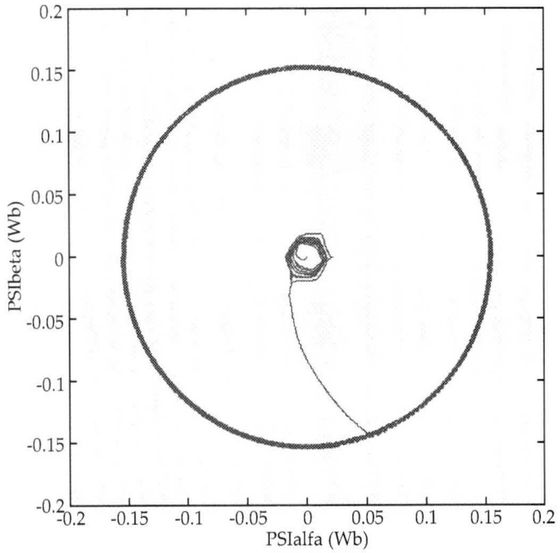

Figure 19. Flux locus in MATLAB/Simulink.

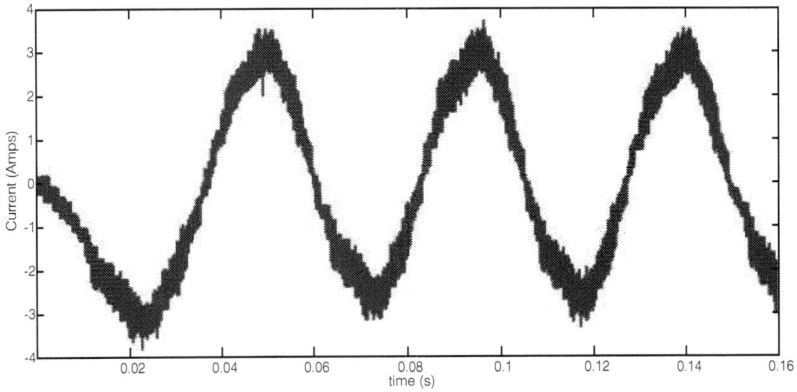

Figure 20. Stator current in MATLAB/Simulink.

5.2. Experimental results

For experimental verification, the DTC strategy was implemented on a Xilinx Virtex-5 development board running at 100 MHz. The current and voltage signal conversions were done using Analog Devices AD7476A ADCs. For the motor power interface, a two-level VSI Fairchild Smart Power Module FNB41560 was used. All motor tests were done using a Texas

Instruments 3-phase induction motor HVACIMTR. The motor parameters shown in Table 4 are the same used in simulations.

Parameter	Value
Poles	4
Rs	11.05Ω
Rr	6.11Ω
Ls	0.316423 mH
Lr	0.316423 mH
Lm	0.293939 mH
Power	0.25KW
V	220VAC

Table 4. Induction motor parameters.

Firstly, the torque dynamic response to a 2 Nm step is shown in Figure 21. A FUTEK torque sensor (TRS300) was used and the output analog signal was scaled, filtered, and displayed on a digital oscilloscope.

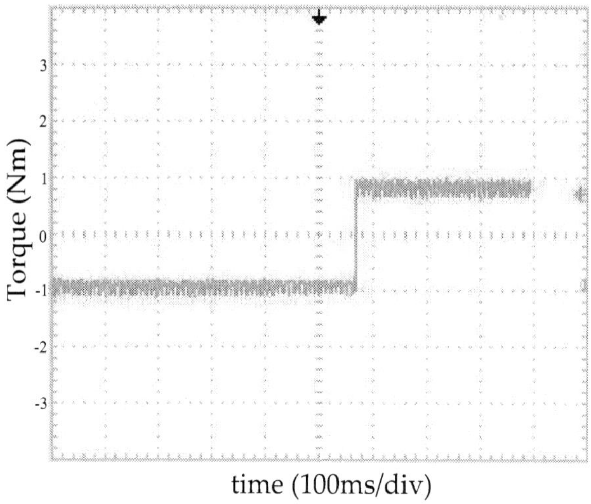

Figure 21. Experimental torque dynamic response.

To visualize the FPGA flux locus on the oscilloscope, two 16-bit digital to analog converters (AD5543) were used. The resulting image is shown in Figure 22, and as the flux simulation, it also appears as a perfect circle with reduced torque ripple.

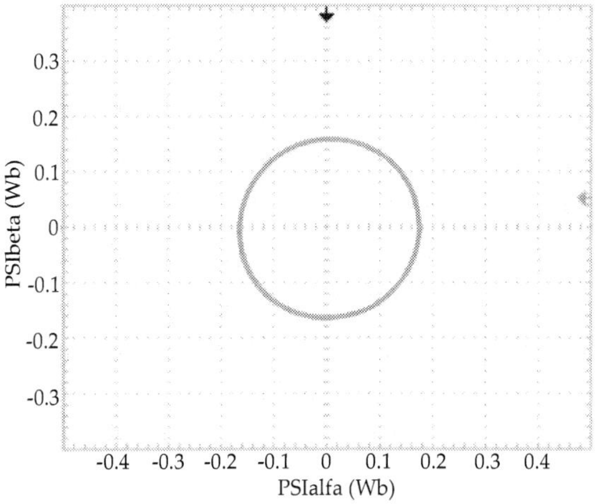

Figure 22. Experimental flux locus.

Finally, the experimental stator current of one of the phases is shown in Figure 23.

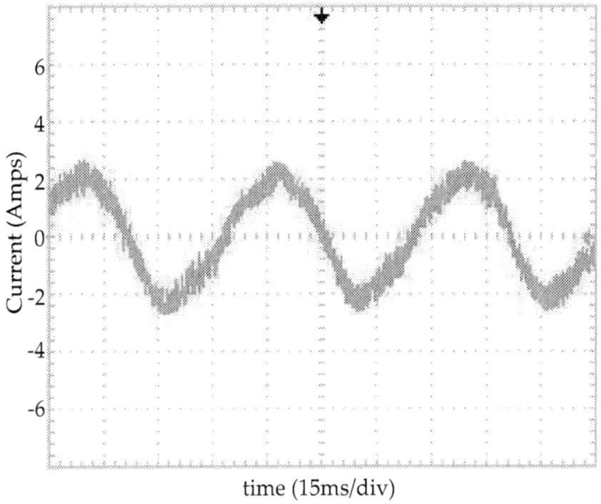

Figure 23. Experimental stator current.

http://dx.doi.org/10.5772/60871

For comparison purposes, the sampling frequency was reduced to 100 KHz and, as expected, a large content of ripple was observed in the torque signal as shown in Figure 24. This increased ripple caused vibration and heating on the motor.

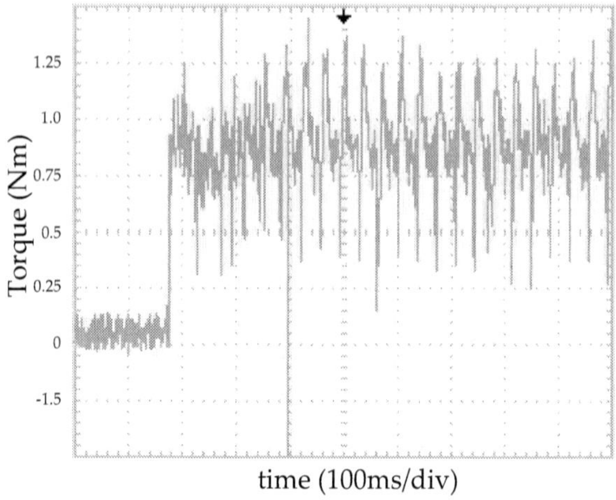

Figure 24. Experimental torque signal at reduced sampling frequency (100 KHz).

6. Conclusion

This chapter has presented an induction motor drive using classical DTC as the main control strategy. This technique was preferred over others due to its simplicity and high performance in motor control. Although DTC is characterized for presenting large ripple on flux and torque signals, it was possible to minimize it to a low value by reducing the sampling period to 1.6 μs. This reduction was achieved by implementation on an FPGA device and the application of a novel architecture for the square root algorithm in the torque/flux estimator. The DTC algorithm was designed based on a structural description and generic VHDL blocks, in order to make the controller easily re-scalable and completely independent of the FPGA technology. A Xilinx Virtex-5 FPGA development board running at 100 MHz was used for this project. The design coded in VHDL uses two´s complement fixed-point format and variable word size for all arithmetic calculations. The complete controller algorithm was simulated using double-precision on MATLAB/Simulink to compare with experimental results. The induction motor presented a smooth, vibration-free operation with a precise torque dynamic, which proves the validity of the presented torque algorithm.

Author details

Rafael Rodríguez-Ponce[1*], Fortino Mendoza-Mondragón[2], Moisés Martínez-Hernández[2] and Marcelino Gutiérrez-Villalobos[2]

*Address all correspondence to: rrodriguez@upgto.edu.mx

1 Polytechnic University of Guanajuato, Robotics Engineering Department, Cortazar, Guanajuato, México

2 Autonomous University of Querétaro, Automation and Control Department, Querétaro, México

References

[1] Chomat M., editor. Electric Machines and Drives. 1st ed. Rijeka, Croatia: InTech; 2011. 274 p. DOI: 10.5772/600

[2] Vas P. Sensorless Vector and Direct Torque Control. 1st ed. New York: Oxford University Press; 1998. 729 p.

[3] Lamchich M.T., editor. Torque Control. 1st ed. Rijeka, Croatia: InTech; 2011. 304 p. DOI: 10.5772/636

[4] Sutikno T., Idris N.R.N., Jidin A. A Review of Direct Torque Control for Induction Motors for Sustainable Reliability and Energy Efficient Drives. Renewable and Sustainable Energy Reviews. 2014; 32:548-558. DOI: 10.1016/j.rser.2014.01.040

[5] Jezernik K., Korelic J., Horvat R. PMSM Sliding Mode FPGA-Based Control for Torque Ripple Reduction. IEEE Transactions on Power Electronics. 2013; 28(7):3549-3556. DOI: 10.1109/TPEL.2012.2222675

[6] Ahmad M., editor. Advances in Motor Torque Control. 1st ed. Rijeka, Croatia: InTech; 2011. 122 p. DOI: 10.5772/862

[7] Bahri I., Idkhajine L., Monmasson E., El Amine Benkhelifa M. Hardware/Software Codesign Guidelines for System on Chip FPGA-Based Sensorless AC Drive Applications. IEEE Transactions on Industrial Informatics. 2013; 9(4):2165-2176. DOI: 10.1109/TII.2013.2245908

[8] Lepka J., Stekl P. 3-Phase AC Induction Motor Vector Control Using 56F80x, 56F8100 or 56F8300 Device. Freescale Application Note. 2005; 1-68.

[9] Lis J., Kowalski C.T., Orlowska-Kowalska T. Sensorless DTC Control of the Induction Motor Using FPGA. IEEE International Symposium on Industrial Electronics; June 30th; Cambridge : IEEE; 2008. p. 1914-1919. DOI: 10.1109/ISIE.2008.4677287

[10] Ferreira S., Haffner F., Pereira L.F., Moraes F. Design and Prototyping of Direct Torque Control of Induction Motors in FPGAs. In: 16th Symposium on Integrated Circuits and System Design; 8-11 Sept.; Sao Paulo, Brazil. IEEE; 2003. p. 105-110. DOI: 10.1109/SBCCI.2003.1232814

[11] Rodriguez R., Gomez R.A., Rodriguez J. Fast Square Root Calculation for DTC Magnetic Flux Estimator. IEEE Latin America Transactions. 2014; 12(2):112-115. DOI: 10.1109/TLA.2014.6749526

Chapter 11

Open-End Winding Induction Motor Drive Based on Indirect Matrix Converter

Javier Riedemann, Rubén Peña and
Ramón Blasco-Giménez

Additional information is available at the end of the chapter

http://dx.doi.org/10.5772/61157

Abstract

Open-end winding induction machines fed from two standard two-level voltage source inverters (VSI) provide an attractive arrangement for AC drives. An alternative approach is to use a dual output indirect matrix converter (IMC). It is well known that IMC provides fully bidirectional power flow operation, with small input size filter requirements. Whilst a standard IMC consists of an AC–DC matrix converter input stage followed by a single VSI output stage, it is possible to replicate the VSI to produce multiple outputs. In this chapter, an open-end winding induction machine fed by an IMC with two output stages is presented. Different modulation strategies for the power converter are analyzed and discussed.

Keywords: Open-end winding, Electrical drive, Matrix converter, Pulse width modulation (PWM)

1. Introduction

An open-end winding induction machine, fed by two 2-level VSIs, offers several advantages when compared to a standard wye or delta connected induction machine drive. The main features of an open-end winding induction machine drive can be summarized as [1, 2]: equal power input from both sides of each winding, thus each VSI is rated at half the machine power rating; each phase stator current can be controlled independently; possibility to have twice the effective switching frequency (depending on the modulation strategy); extensibility to more phases, therefore multiphase induction machines can be considered if current reduction is required; possibility of reducing common-mode voltage; and certain degree of fault tolerance, as there is voltage space vector redundancy.

However, an open-end winding induction machine drive can have some drawbacks, such as [1]: possibility of zero sequence current flowing in the machine because of the occurrence of zero sequence voltage; increased conduction losses; more complex power converter requirements, i.e., more power devices, circuit gate drives, etc.

To supply energy to an open-end winding machine, different power converter topologies have been developed; for instance, [3–6] propose an open-end winding induction machine drive based on two 2-level VSIs fed from isolated DC sources. This topology has the advantage of avoiding the circulation of zero sequence current; however, two isolation transformers are needed. On the other hand, [7–11] present a topology based on two 2-level VSIs fed by a single DC source. In this case, a zero sequence current could circulate in the machine windings (depending on the modulation strategy used), but just one transformer is needed, reducing the volume and cost of the drive.

Multilevel topologies for open-end winding AC drives are presented in [12–16] where different voltage levels can be achieved in the machine phase windings with certain power converter configurations, then reducing the output voltage distortion but increasing the system cost and complexity.

In the past decades, significant research effort has been focused on direct frequency changing power converters, such as the matrix converter (MC) [17] or the indirect matrix converter [18]. It is known that these power converter topologies offer a suitable solution for direct AC–AC conversion, achieving sinusoidal input and output currents, bidirectional power flow capability and controllable input power factor, without using bulky energy storage elements [18]. Matrix converters have been utilized to supply open-end winding AC machines such as reported in [19–21].

In this chapter, the application of an IMC with two output stages to supply energy to an open-end winding induction machine is described [22–24]. For evaluation purposes, simulations and experimental results are presented.

2. Power converter topologies for open-end winding induction machine drives

An open-end winding induction machine drive can be supplied by different configurations of power converters. Some of the most common topologies will be reviewed in this chapter.

2.1. Two 2-level voltage source inverters fed by isolated DC sources

This is the basic power converter for open-end winding AC drives. The circuit configuration is shown in Figure 1 where a standard two-level VSI is connected at each side of the machine stator winding [3]. The VSIs are supplied by isolated DC power sources.

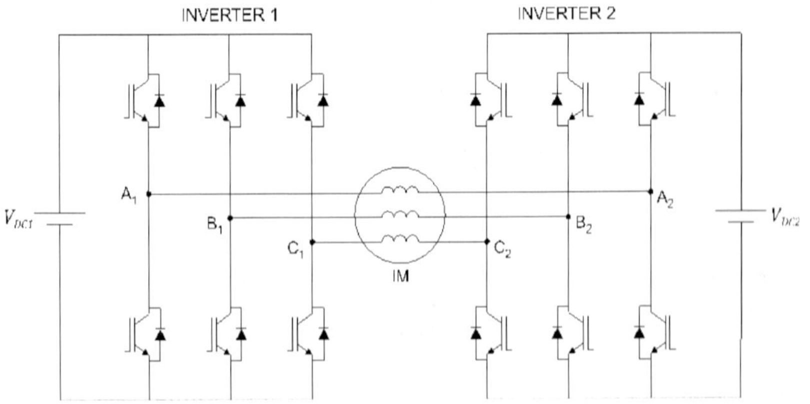

Figure 1. Two 2-level VSIs fed by isolated DC sources for an open-end winding AC machine drive.

The voltage vectors for inverter 1 are shown in Table 1; the same space vectors are valid for inverter 2, but with superscript 2. As each VSI can produce eight voltage space vector locations independent of the other, there are 64 voltage vector combinations of the full converter, resulting in a vector locus similar to a three-level neutral point clamped (NPC) inverter [6].

States of inverter 1 $[S_{A1}S_{B1}S_{C1}]$			
V_1^1=[1 0 0]	V_2^1=[1 1 0]	V_3^1=[0 1 0]	V_7^1=[1 1 1]
V_4^1=[0 1 1]	V_5^1=[0 0 1]	V_6^1=[1 0 1]	V_8^1=[0 0 0]

Table 1. Switching states of the individual inverters

Let $V_{ij}=[V_i^1 V_j^2]$ with i, j=1...8 be the phase voltage vector combination of the dual-inverter output. A representation of the vector locations is shown in Figure 2 [6].

2.2. Two 2-level voltage source inverters fed by a single DC source

This topology is basically the same described in Section 2.1, but now just one DC supply is considered for the drive, as shown in Figure 3. The disadvantage of this converter is that zero sequence current could circulate in the machine windings because of the generation of output zero sequence voltage; however, this issue can be addressed with an appropriate modulation strategy for the inverters [7].

2.3. Multilevel topologies

Several multilevel power converters have been developed for open-end winding induction motor drives. For example, Figure 4a shows a three-level inverter [12] and Figure 4b shows a

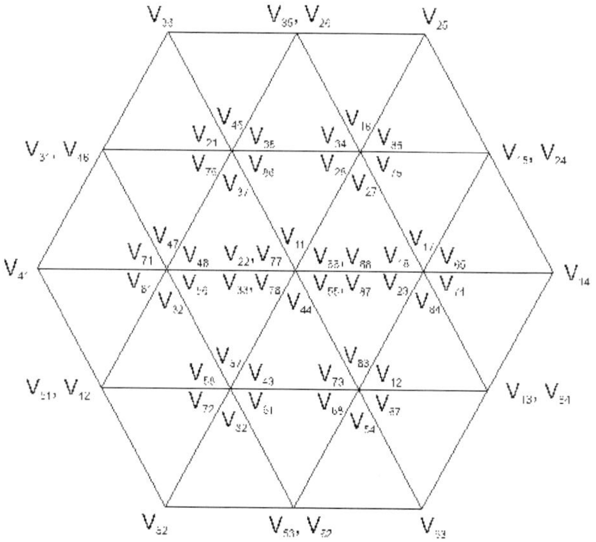

Figure 2. Space vector locations of the dual-inverter scheme.

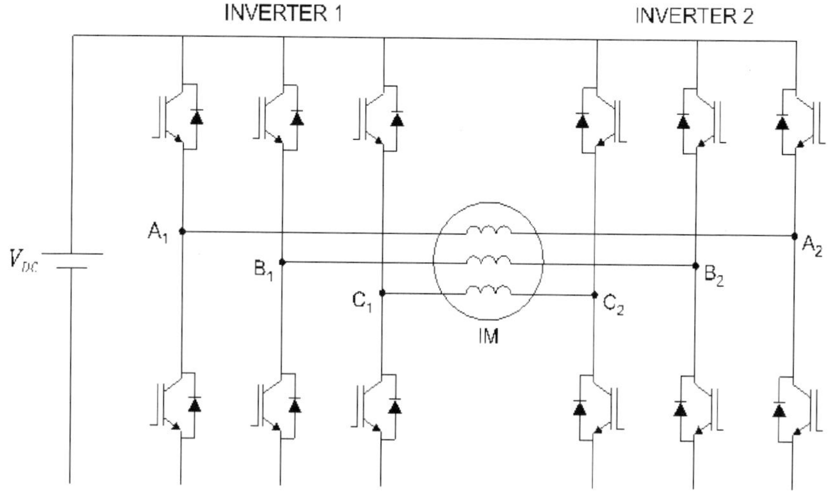

Figure 3. Two 2-level VSIs fed by a single DC source for an open-end winding AC machine drive.

five-level inverter [16]. The main advantage of the multilevel topologies is that the machine phase voltage presents lower voltage distortion, increasing the performance of the drive; but on the other hand, the complexity and cost of the system are also increased.

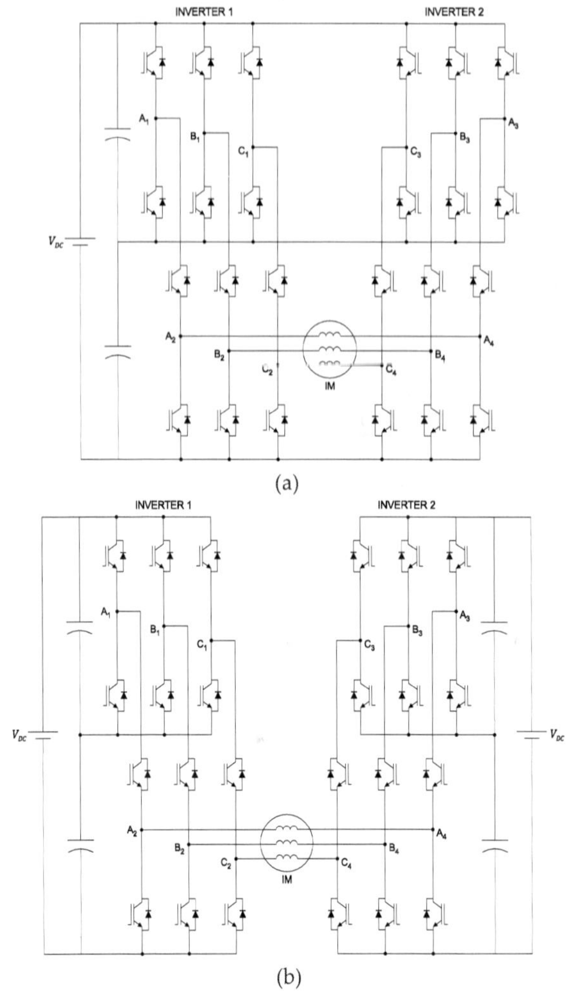

Figure 4. (a) Three-level inverter and (b) five-level inverter for open-end winding AC machine drives.

2.4. Direct power converters

Modern direct power converters consider matrix converter and indirect matrix converters. A matrix converter [17] is a direct frequency converter consisting of nine bidirectional switches (three switches per phase) allowing to connect any of the output terminals to any of the input voltages. For an open-end winding induction motor drive, two MCs are required, connected in the arrangement shown in Figure 5. The main features of a matrix converter are: bidirectional power flow capability, sinusoidal input and output currents without bulky energy storage elements, and controllable input power factor. For a standard matrix converter, a total of 36 IGBTs and diodes are required in this topology.

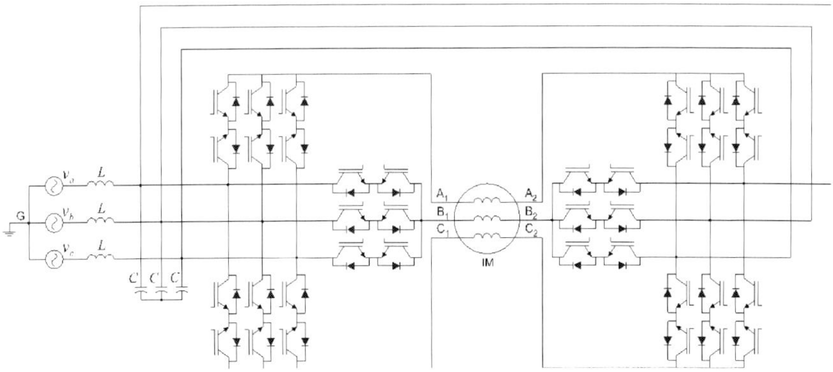

Figure 5. Open-end winding induction motor drive based on matrix converters.

An indirect matrix converter [18] is also a direct frequency converter having the same features of an MC, but now a DC stage is clearly identified in the topology. The IMC consists of an input rectifier, an AC–DC matrix converter, built of six bidirectional switches; this rectifier produces the DC voltage to supply the converter output stage which is a standard two-level VSI. To supply an open-end winding AC machine, two output inverters are required as can be seen in Figure 6. Considering the six bidirectional switches of the input stage and the two output stages, a total of 24 discrete IGBTs and diodes are required in this topology.

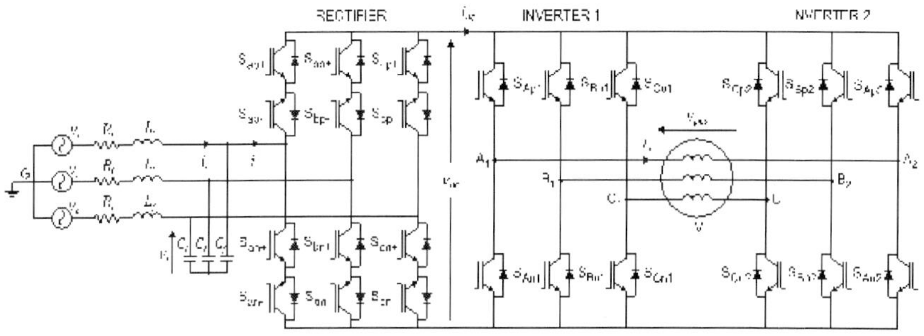

Figure 6. Open-end winding induction motor drive based on indirect matrix converter.

http://dx.doi.org/10.5772/61157

3. Model of the open-end winding induction motor drive based on IMC

The complete drive of Figure 6 can be modeled by state equations which describe the dynamic behavior of the system. The effects of power devices dead-times on zero sequence voltages are neglected. All the equations derived below are written in fixed *abc* coordinates.

The differential equations for the input side are:

$$\boldsymbol{v}_s = R_f \boldsymbol{i}_s + L_f \frac{d\boldsymbol{i}_s}{dt} + \boldsymbol{v}_i \tag{1}$$

$$\boldsymbol{i}_s = C_f \frac{d\boldsymbol{v}_i}{dt} + \boldsymbol{i}_i \tag{2}$$

where

$$\boldsymbol{v}_s = \begin{bmatrix} v_{sa} & v_{sb} & v_{sc} \end{bmatrix}^T, \boldsymbol{i}_s = \begin{bmatrix} i_{sa} & i_{sb} & i_{sc} \end{bmatrix}^T \tag{3}$$

$$\boldsymbol{v}_i = \begin{bmatrix} v_{ia} & v_{ib} & v_{ic} \end{bmatrix}^T, \boldsymbol{i}_i = \begin{bmatrix} i_{ia} & i_{ib} & i_{ic} \end{bmatrix}^T \tag{4}$$

are the source voltage and current (3), and the rectifier input voltage and current (4).

The DC link voltage can be obtained as:

$$v_{DC} = \boldsymbol{S}_r^T \cdot \boldsymbol{v}_i \tag{5}$$

with the rectifier switching matrix:

$$\boldsymbol{S}_r = \begin{bmatrix} S_{ap} - S_{an} \\ S_{bp} - S_{bn} \\ S_{cp} - S_{cn} \end{bmatrix} \tag{6}$$

where S_{xp}, $S_{xn} \in \{0, 1\}$ with $x = a, b, c$. The output pole voltage of Inverter 1 (v_{o1}) and Inverter 2 (v_{o2}), with respect to the negative DC link rail, are defined in (7).

$$\boldsymbol{v}_{o1} = \boldsymbol{S}_{i1} \cdot v_{DC}, \boldsymbol{v}_{o2} = \boldsymbol{S}_{i2} \cdot v_{DC} \tag{7}$$

where the switching matrices of Inverter 1 (S_{i1}) and Inverter 2 (S_{i2}) are:

$$\boldsymbol{S}_{i1}=\begin{bmatrix}S_{A1}\\S_{B1}\\S_{C1}\end{bmatrix}=\begin{bmatrix}S_{Ap1}-S_{An1}\\S_{Bp1}-S_{Bn1}\\S_{Cp1}-S_{Cn1}\end{bmatrix},\boldsymbol{S}_{i2}=\begin{bmatrix}S_{A2}\\S_{B2}\\S_{C2}\end{bmatrix}=\begin{bmatrix}S_{Ap2}-S_{An2}\\S_{Bp2}-S_{Bn2}\\S_{Cp2}-S_{Cn2}\end{bmatrix} \tag{8}$$

and $S_{xpk}=\bar{S}_{xnk}\in\{0,\,1\}$ with $x=a,\,b,\,c\;,\;\;k=1,\,2$. The output phase voltage corresponds to the difference of both inverters pole voltages:

$$\boldsymbol{v}_{ph,o}=\begin{bmatrix}v_{ph,oa} & v_{ph,ob} & v_{ph,oc}\end{bmatrix}^T=\boldsymbol{v}_{o1}-\boldsymbol{v}_{o2}=\left(\boldsymbol{S}_{i1}-\boldsymbol{S}_{i2}\right)v_{DC} \tag{9}$$

Considering a model of the AC machine with R_s the stator resistance, the output phase voltage can be written as:

$$\boldsymbol{v}_{ph,o}=R_s\boldsymbol{i}_o+\frac{d\boldsymbol{\Psi}_s\left(\boldsymbol{i}_o,\boldsymbol{i}_r,\theta_r\right)}{dt} \tag{10}$$

where Ψ_s is the stator flux linkage vector given by:

$$\boldsymbol{\Psi}_s=\begin{bmatrix}\Psi_{sa} & \Psi_{sb} & \Psi_{sc}\end{bmatrix}^T \tag{11}$$

The rotor angle is θ_r and the output (stator) current vector $\boldsymbol{i}_o$ and rotor current vector $\boldsymbol{i}_r$ are given by (12):

$$\boldsymbol{i}_o=\begin{bmatrix}i_{oa} & i_{ob} & i_{oc}\end{bmatrix}^T,\boldsymbol{i}_r=\begin{bmatrix}i_{ra} & i_{rb} & i_{rc}\end{bmatrix}^T \tag{12}$$

As Ψ_s is an implicit function of t, (10) can be rewritten by using the chain rule for the derivative:

$$\boldsymbol{v}_{ph,o}=R_s\boldsymbol{i}_o+\frac{\partial\boldsymbol{\Psi}_s}{\partial\boldsymbol{i}_o}\frac{d\boldsymbol{i}_o}{dt}+\psi_2(t) \tag{13}$$

where

$$\psi_2(t)=\frac{\partial\boldsymbol{\Psi}_s}{\partial\boldsymbol{i}_r}\frac{d\boldsymbol{i}_r}{dt}+\frac{\partial\boldsymbol{\Psi}_s}{\partial\theta_r}\frac{d\theta_r}{dt} \tag{14}$$

http://dx.doi.org/10.5772/61157

Assuming $\partial\Psi_s / \partial i_o$ is a bijective function of t, it can be defined as:

$$\frac{\partial \boldsymbol{\Psi}_s}{\partial \boldsymbol{i}_o} = \psi_1^{-1}(t) \tag{15}$$

and (13) can be redefined by:

$$\boldsymbol{v}_{ph,o} = R_s \boldsymbol{i}_o + \psi_1^{-1}(t)\frac{d\boldsymbol{i}_o}{dt} + \psi_2(t) \tag{16}$$

The DC link current is:

$$i_{DC} = \left(\boldsymbol{S}_{i1} + \boldsymbol{S}_{i2}\right)^T \boldsymbol{i}_o \tag{17}$$

and the rectifier input current:

$$\boldsymbol{i}_i = \boldsymbol{S}_r i_{DC} \tag{18}$$

Taking into account (1) – (18), the state space model of the drive is given by (19) –(21):

$$\frac{d\boldsymbol{i}_s}{dt} = -\frac{R_f}{L_f}\boldsymbol{i}_s - \frac{1}{L_f}\boldsymbol{v}_i + \frac{1}{L_f}\boldsymbol{v}_s \tag{19}$$

$$\frac{d\boldsymbol{v}_i}{dt} = \frac{1}{C_f}\boldsymbol{i}_s - \frac{1}{C_f}\boldsymbol{S}_r\left(\boldsymbol{S}_{i1} + \boldsymbol{S}_{i2}\right)^T \boldsymbol{i}_o \tag{20}$$

$$\frac{d\boldsymbol{i}_o}{dt} = -\psi_1(t) R_s \boldsymbol{i}_o - \psi_1(t)\psi_2(t) + \psi_1(t)\left(\boldsymbol{S}_{i1} - \boldsymbol{S}_{i2}\right)\boldsymbol{S}_r^T \cdot \boldsymbol{v}_i \tag{21}$$

4. Zero sequence voltage

As mentioned before, the dual-inverter fed open-ended winding induction motor drive may suffer from zero sequence current caused by zero sequence voltage. This zero sequence voltage is produced because of the asymmetry of the instantaneous phase voltages applied to the machine windings (due to the voltage space vectors used). In general, zero sequence currents may give rise to increased RMS phase current, thus increasing the system losses, high current/ voltage THD, and machine over-heating and vibrations. The zero sequence voltage is given by [11]:

$$v_{zs} = \frac{v_{A1A2} + v_{B1B2} + v_{C1C2}}{3} \tag{22}$$

The zero sequence voltage contributions from the 64 space vector combinations are shown in Table 2. As can be noted, there are twenty voltage space vectors that do not produce zero sequence voltage; thus in order to avoid the circulation of zero sequence current in the machine windings, only these space vector combinations could be used in the modulation strategy for the dual inverter [9].

V_{zs}	Voltage vector combinations
$-V_{DC}/2$	V_{87}
$-V_{DC}/3$	V_{84}, V_{86}, V_{82}, V_{57}, V_{37}, V_{17}
$-V_{DC}/6$	V_{85}, V_{83}, V_{54}, V_{34}, V_{81}, V_{56}, V_{52}, V_{36} V_{32}, V_{47}, V_{14}, V_{16}, V_{12}, V_{67}, V_{27}
0	V_{88}, V_{55}, V_{53}, V_{35}, V_{33}, V_{44}, V_{51}, V_{31}, V_{46}, V_{42} V_{15}, V_{13}, V_{64}, V_{24}, V_{11}, V_{66}, V_{62}, V_{26}, V_{22}, V_{77}
$+V_{DC}/6$	V_{58}, V_{38}, V_{45}, V_{43}, V_{18}, V_{65}, V_{25}, V_{63} V_{23}, V_{74}, V_{41}, V_{61}, V_{21}, V_{76}, V_{72}
$+V_{DC}/3$	V_{48}, V_{68}, V_{82}, V_{75}, V_{73}, V_{71}
$+V_{DC}/2$	V_{78}

Table 2. Zero sequence voltage contributions from different space vector combinations

Moreover, from Table 2 and Figure 2, it can be noted that there are two different but equivalent sets of active vectors producing null zero sequence voltage (see Table 3), which could be used along with the zero vectors: V_{11}, V_{22}, V_{33}, V_{44}, V_{55}, V_{66}, V_{77}, and V_{88}.

Set 1	V_{15}	V_{35}	V_{31}	V_{51}	V_{53}	V_{13}
Set 2	V_{24}	V_{26}	V_{46}	V_{42}	V_{62}	V_{64}

Table 3. Active space vectors producing null zero sequence voltage

5. Common-mode voltage

Conventional PWM inverters generate alternating common-mode voltages relative to ground which generate currents through the motor parasitic capacitances to the rotor iron [25]. These

http://dx.doi.org/10.5772/61157

currents find their way via the motor bearings back to the grounded stator case. The so-called bearing currents have been found to be a major cause of premature bearing failure in PWM inverter motor drives [26].

One of the main features of an open-end winding induction machine drive is the possibility of reducing the common-mode voltage by using certain space vector combinations of the dual-inverter topology (Figure 2). In general, for an open-end winding machine, the common-mode voltage is given by [8]:

$$v_{cm} = \frac{1}{6}\left(v_{A1} + v_{B1} + v_{C1} + v_{A2} + v_{B2} + v_{C2}\right) \tag{23}$$

where v_{Ai}, v_{Bi}, v_{Ci}, with $i=1, 2$, are the pole voltages of each inverter with respect to a common point of the drive (usually ground).

For the topology depicted in Figure 6, the common-mode voltage is given by:

$$v_{cm} = \frac{1}{6}\left(v_{A1G} + v_{B1G} + v_{C1G} + v_{A2G} + v_{B2G} + v_{C2G}\right) \tag{24}$$

where the common point G is the grounded neutral point of the source. These voltages can also be expressed as:

$$\begin{aligned} v_{AiG} &= S_{Api} v_{pG} + S_{Ani} v_{nG} \\ v_{BiG} &= S_{Bpi} v_{pG} + S_{Bni} v_{nG} \\ v_{CiG} &= S_{Cpi} v_{pG} + S_{Cni} v_{nG} \end{aligned} \tag{25}$$

where v_{pG} and v_{nG} are the voltages of the positive and negative rail of the DC link with respect to the grounded neutral point of the source, respectively; S_{xpi}, $S_{xni} \in \{0, \ 1\}$ with $x = A, \ B, \ C$, $i=1, \ 2$ are the switching functions of the inverter devices (0: switch opened, 1: switch closed) and $S_{xni} = 1 - S_{xpi}$ (because of the complementary operation of the upper and lower switches of each inverter leg). Therefore,

$$\begin{aligned} v_{cm} = \frac{1}{6}\Big[& \left(S_{Ap1} + S_{Bp1} + S_{Cp1} + S_{Ap2} + S_{Bp2} + S_{Cp2}\right) v_{pG} \\ & + \left(S_{An1} + S_{Bn1} + S_{Cn1} + S_{An2} + S_{Bn2} + S_{Cn2}\right) v_{nG} \Big] \end{aligned} \tag{26}$$

Let $N_{sw} = S_{Ap1} + S_{Bp1} + S_{Cp1} + S_{Ap2} + S_{Bp2} + S_{Cp2}$, thus

$$v_{cm} = \frac{1}{6}\left[N_{sw}v_{pG} + (6 - N_{sw})v_{nG}\right] \tag{27}$$

where N_{sw} is the number of upper inverter switches closed. The squared RMS value of the common-mode voltage is:

$$v_{cm_{RMS}}^2 = \frac{1}{36T}\int_0^T \left[N_{sw}v_{pG} + (6 - N_{sw})v_{nG}\right]^2 dt \tag{28}$$

where T is the period of v_{pG} (equals the period of v_{nG}). Further expansion yields:

$$36v_{cm_{RMS}}^2 = N_{sw}^2\frac{1}{T}\int_0^T v_{pG}^2 dt + 2N_{sw}(6 - N_{sw})\frac{1}{T}\int_0^T v_{pG}v_{nG}dt + (6 - N_{sw})^2\frac{1}{T}\int_0^T v_{nG}^2 dt \tag{29}$$

The voltages of the DC link rails are given by:

$$\begin{aligned} v_{pG} &= S_{ap}v_{ra} + S_{bp}v_{rb} + S_{cp}v_{rc} \\ v_{nG} &= S_{an}v_{ra} + S_{bn}v_{rb} + S_{cn}v_{rc} \end{aligned} \tag{30}$$

wherev_{ra}, v_{rb}, and v_{rc} are the converter input phase voltages and S_{xp}, S_{xn} with $x = a, \ b, \ c$ are the switching functions of the rectifier. Accordingly, v_{pG} and v_{nG} will always be segments of different input phase voltages and

$$\left|v_{pG}(t)\right| = \left|v_{nG}(t - t_o)\right|, \ t_o \in \mathrm{R} \tag{31}$$

thus

$$\int_0^T v_{pG}^2 dt = \int_0^T v_{nG}^2 dt \tag{32}$$

Differentiating (29) with respect to N_{sw} and equating to zero, it can be found that $v_{cm_{RMS}}^2$ (and implicitly $v_{cm_{RMS}}$) achieves a minimum value at N_{sw} =3, which means that in order to reduce the RMS common-mode voltage at the machine terminals, only three upper inverter switches should be closed at each switching period.

This can be further investigated by considering a virtual midpoint of the DC link as a reference point (see point 0 in Figure 6). Then, (24) can be rewritten as:

$$v_{cm} = \frac{1}{6}\left(v_{A10} + v_{B10} + v_{C10} + v_{A20} + v_{B20} + v_{C20}\right) + v_{0G} = v_{cm0} + v_{0G} \tag{33}$$

where the contributions of the input and output stages to the overall common-mode voltage have been separated (v_{0G} and v_{cm0}, respectively). The voltage v_{0G} is the voltage between the reference point 0 and the grounded neutral point of the source. This voltage can be calculated as:

$$v_{0G} = \frac{1}{2}\left[\left(S_{ap} + S_{an}\right)v_{ra} + \left(S_{bp} + S_{bn}\right)v_{rb} + \left(S_{cp} + S_{cn}\right)v_{rc}\right] \tag{34}$$

It can be seen in (34) that v_{0G} depends on the modulation of the input stage, which is totally defined by the duty cycles of the rectifier stage. On the other hand, the voltage v_{cm0} can be rewritten as:

$$v_{cm0} = \frac{1}{6}\left[N_{sw}\frac{v_{DC}}{2} + \left(6 - N_{sw}\right)\left(\frac{v_{DC}}{2}\right)\right] = \frac{1}{6}\left[N_{sw}v_{DC} - 3v_{DC}\right] \tag{35}$$

Therefore, it can be seen in (35) that by using $N_{sw}=3$, the contribution of the output inverters to the common-mode voltage is eliminated [8]. Table 4 shows the voltage space vector combinations of the dual-inverter topology which do not produce common-mode voltage.

As can be noted from Table 4 and Figure 2, there are larger and lower active vectors available which produce zero common-mode voltage. Any of them could be considered in the modulation strategy for the dual-inverter system depending on the machine voltage requirement. However, from Table 2 and Table 4, it can be appreciated that the space vectors which reduce the common-mode voltage are not the same vectors which reduce the zero sequence voltage; thus, if a common-mode voltage is required, a compensation should be done for the zero sequence voltage; in other case, large zero sequence current components will circulate in the machine windings; a type of compensation will be reviewed in the following section.

Space vector combinations $[S_{Ap1}S_{Bp1}S_{Cp1}S_{Ap2}\ S_{Bp2}S_{Cp2}]$	
V_{14}=[1 0 0 0 1 1]	V_{56}=[0 0 1 1 0 1]
V_{25}=[1 1 0 0 0 1]	V_{61}=[1 0 1 1 0 0]
V_{36}=[0 1 0 1 0 1]	V_{43}=[0 1 1 0 1 0]
V_{41}=[0 1 1 1 0 0]	V_{12}=[1 0 0 1 1 0]
V_{52}=[0 0 1 1 1 0]	V_{16}=[1 0 0 1 0 1]
V_{63}=[1 0 1 0 1 0]	V_{65}=[1 0 1 0 0 1]

Space vector combinations $[S_{Ap1}S_{Bp1}S_{Cp1}S_{Ap2}\ S_{Bp2}S_{Cp2}]$	
V_{23}=[1 1 0 0 1 0]	V_{54}=[0 0 1 1 1 0]
V_{34}=[0 1 0 0 1 1]	V_{32}=[0 1 0 1 1 0]
V_{45}=[0 1 1 0 0 1]	V_{21}=[1 1 0 1 0 0]
V_{78}=[1 1 1 0 0 0]	V_{87}=[0 0 0 1 1 1]

Table 4. Space vectors with zero common mode

6. Pulse width modulation strategies

In this section, PWM strategies for an open-end winding induction motor drive based on a two-output IMC are described.

6.1. Modulation strategies for the input stage of the IMC

The modulation for the input (rectifier) stage of the converter aims to obtain a positive DC link voltage in each sampling period and unity displacement factor at the input [27]. Two different space vector modulation (SVM) strategies can be used for the rectifier [28]. One modulation maximizes the DC voltage by commutating between the largest and the second largest positive line input voltage (Figure 7). The other modulation produces a reduced DC voltage commutating between the lowest and the second lowest input line voltage (Figure 8).

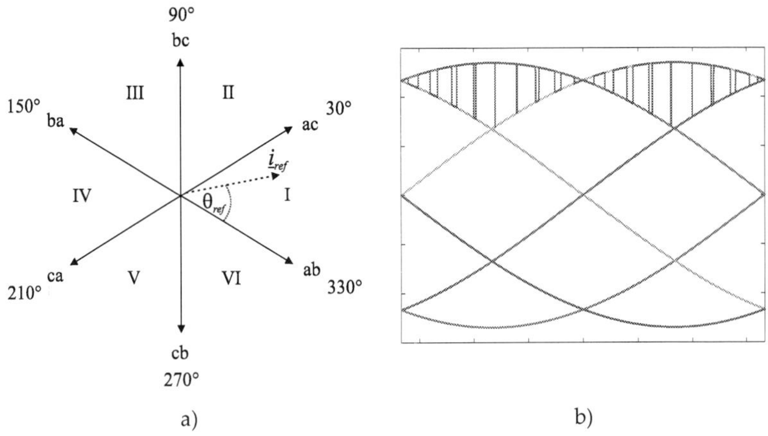

Figure 7. (a) Space vector locations for maximum DC voltage. (b) Maximum DC voltage

http://dx.doi.org/10.5772/61157

If $\theta_{ref,i}$ is the angle of the voltage reference vector; the $\gamma-\delta$ duty cycles for both modulation strategies are given in (36)–(38). Further details about the modulations for the input stage can be found in [27, 28].

$$d_\gamma^R = \frac{d_\gamma}{d_\gamma + d_\delta}, \; d_\delta^R = \frac{d_\delta}{d_\gamma + d_\delta} \tag{36}$$

with

$$d_\gamma = \sin\left(\pi/3 - \theta_{ref,i}\right), \; d_\delta = \sin\left(\theta_{ref,i}\right) \tag{37}$$

the duty cycles for maximum DC voltage and

$$d_\gamma = \cos\left(\theta_{ref,i}\right), \; d_\delta = \cos\left(\pi/3 - \theta_{ref,i}\right) \tag{38}$$

the duty cycles for reduced DC voltage. Figure 9a shows the transition from reduced DC voltage to maximum DC voltage and Figure 9b shows the opposite situation.

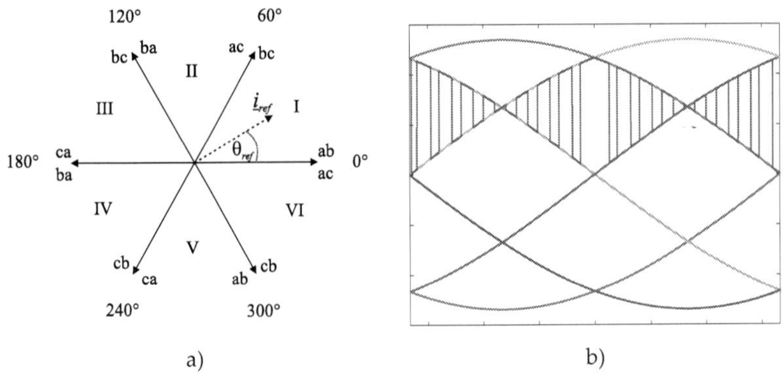

Figure 8. (a) Space vector locations for reduced DC voltage. (b) Reduced DC voltage.

The rectifier SVM for reduced DC link voltage decreases the voltage gain by $\sqrt{3}$. Thus, the transition between reduced and maximum DC link voltage should take place when the output voltage reference is higher than $\frac{1.5}{\sqrt{3}} \cdot V_{ph,\,input} = 0.866 \cdot V_{ph,\,input}$.

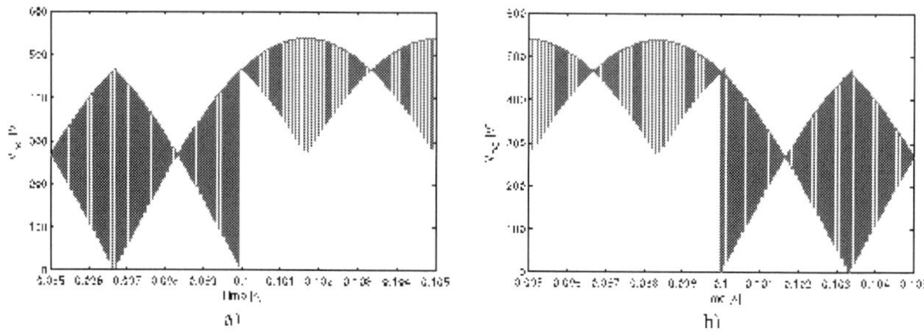

Figure 9. Transition between both rectifier modulation strategies.

6.2. Modulation strategies for the output stages of the IMC

6.2.1. Carrier-based modulation strategy

In a PWM strategy based on a triangular carrier (SPWM), the duty cycles for each leg of inverter 1 are [29]:

$$d_a = \frac{1}{2}\left(m(t)\cos\left(\frac{2\pi}{m_f}k\right) + 1\right) \tag{39}$$

$$d_b = \frac{1}{2}\left(m(t)\cos\left(\frac{2\pi}{m_f}k - \frac{2\pi}{3}\right) + 1\right) \tag{40}$$

$$d_c = \frac{1}{2}\left(m(t)\cos\left(\frac{2\pi}{m_f}k + \frac{2\pi}{3}\right) + 1\right) \tag{41}$$

In this case, It is necessary a variable modulation index given by $m(t)=m_o(d_\gamma + d_\delta)$ to compensate the fluctuations of the DC link voltage, m_o is the final modulation index ($0 \le m_o \le 1$), $m_f = f_s / f_o$is the frequency index (f_o: output frequency, f_s: switching frequency) and $0 \le k \le m_f$. Duty cycles of inverter 2 are obtained using (39)–(41), but considering a phase shift of 180° for the cosine functions. For implementation purposes, the duty cycles d_a, d_b, and d_c are transformed into equivalent $\alpha-\beta-0$ duty cycles. Thus, considering Figure 10, the $\alpha-\beta-0$ duty cycles are:

$$d_0 = 1 - d_{max} \tag{42}$$

$$d_\alpha = d_{max} - d_{mid} \tag{43}$$

$$d_\beta = d_{mid} - d_{min} \tag{44}$$

$$d_7 = d_{min} \tag{45}$$

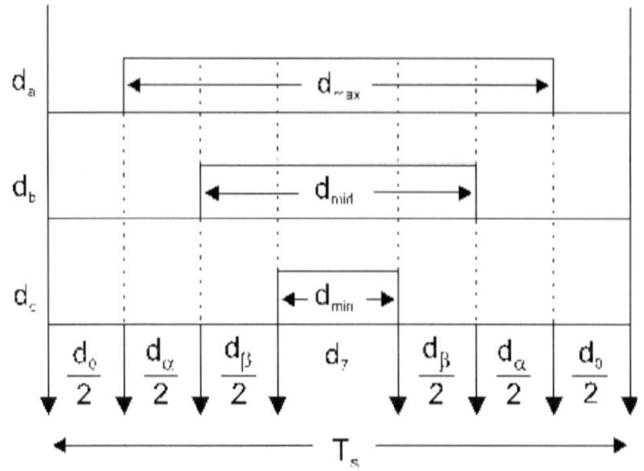

Figure 10. A single inverter stage duty cycles.

To obtain a correct balance of the input currents and the output voltages in a switching period, the modulation pattern should produce all combinations of the rectification and the inversion switching states [27], resulting in the following duty cycles for the active vectors:

$$d_{\alpha\gamma} = d_\alpha d_\gamma^R, d_{\beta\gamma} = d_\beta d_\gamma^R, d_{\alpha\delta} = d_\alpha d_\delta^R, \ d_{\beta\delta} = d_\beta d_\delta^R \tag{46}$$

The duty cycle corresponding to the switching state $[S_{Ap} S_{Bp} S_{Cp}]$=[0 0 0] is:

$$d_{00} = d_{0,tot} - d_{min} \tag{47}$$

and the combined zero vectors duty cycles are:

$$d_{0\gamma} = d_{00} d_\gamma^R, d_{0\delta} = d_{00} d_\delta^R \tag{48}$$

The output stages duty cycles are different for each inverter and are represented in Figure 11 [23] for inverter i (i=1, 2).

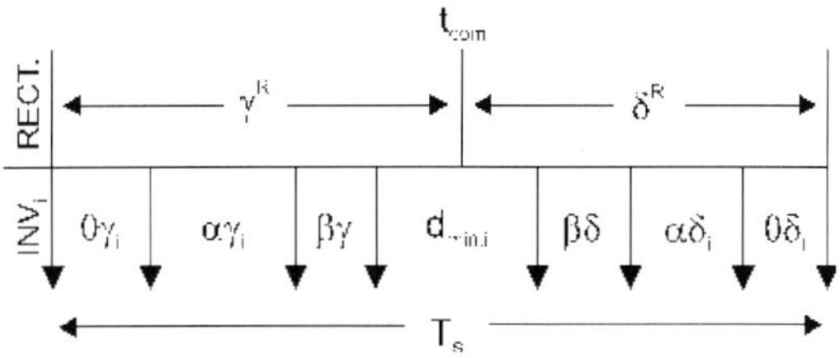

Figure 11. Inverters duty cycles.

6.2.2. Space vector modulation strategy for zero sequence reduction

As mentioned in Section 4, in an open-end winding induction machine drive, the zero sequence voltage can be reduced by using the active space vectors given in Table 3. The zero vectors are mapped depending on the sector information [9]. The mapping is shown in Table 5.

Sector	I	II	III	IV	V	VI
Set 1 zero vectors	V_{55}	V_{33}	V_{11}	V_{55}	V_{33}	V_{11}
Set 2 zero vectors	V_{22}	V_{66}	V_{44}	V_{22}	V_{66}	V_{44}

Table 5. Mapping of zero vectors

The duty cycles for the output stages are calculated as:

$$d_\alpha = m(t)\sin\left(\pi/3 - \theta_{ref,o}\right), d_\beta = m(t)\sin\left(\theta_{ref,o}\right) and\, d_0 = 1 - d_\alpha - d_\beta \tag{49}$$

where $m(t)=m_0(d_\gamma + d_\delta)$ and $0 \le m_0 \le 1$. $\theta_{ref,o}$ is the angle of the output reference voltage space vector. As in the carrier-based modulation strategy, the duty cycles of the rectifier and the inverters should be combined; thus the active vector duty cycles are given in (46).

The combined zero vectors duty cycles are:

$$d_{0\gamma} = d_0 d_\gamma^R, d_{0\delta} = d_0 d_\delta^R \tag{50}$$

Thus, the switching sequence, which is the same for both output stages, is shown in Figure 12, commutating the input stage with zero DC current [27].

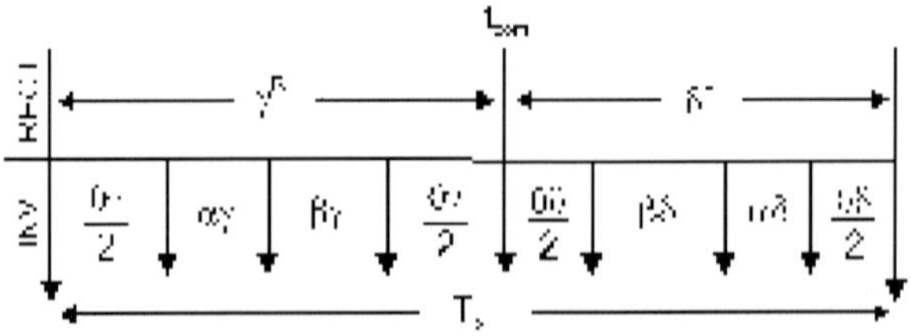

Figure 12. Standard switching sequence for IMCs.

6.2.3. Space vector modulation strategy for common-mode voltage reduction

If only the voltage space vectors shown in Table 4 are used in the modulation strategy for the output stages of the IMC, the contribution of the dual-inverter to the overall common-mode voltage can be eliminated. The duty cycles for the IMC outputs are calculated as in (49) – (50). However, as mentioned in Section 5, the space vectors producing null v_{cm0} are not the same vectors producing null zero sequence voltage. Hence, as in this SVM strategy, the vectors used to modulate the converter output stages will eliminate the common-mode voltage; a compensation must be performed to avoid the circulation of zero sequence current in the machine.

The average zero sequence voltage within a sampling interval can be eliminated by forcing the zero sequence volt-seconds to zero [11] by applying the null voltage vectors with unequal times. Accordingly, the standard switching sequence used in the IMC [27] is modified in order to reduce/eliminate the average zero sequence voltage within a sampling period. For the modulation strategy presented, the duty cycles for both output VSIs are the same, which can be noted in Figure 13.

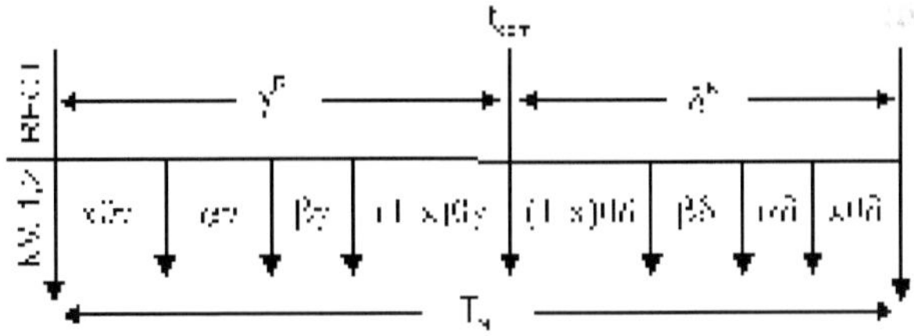

Figure 13. Modified switching sequence for the IMC with two outputs.

Taking into account that the same space vectors sequence applied in γ^R interval is applied in the δ^R interval but in reverse order, then the value of x, which causes the cancellation of the zero sequence volt-seconds, is calculated at every sampling period to satisfy [11]:

$$v_{zs1}x(0\gamma + 0\delta) + v_{zs2}(\alpha\gamma + \alpha\delta) + v_{zs3}(\beta\gamma + \beta\delta) + v_{zs4}(1 - x)(0\gamma + 0\delta) = 0 \quad (51)$$

where v_{zsk} with $k=1,\ 2,\ 3,\ 4$, is the zero sequence voltage value at intervals $x0\gamma$, $\alpha\gamma$, $\beta\gamma$, and $(1-x)0\gamma$, respectively. The value of v_{zsk} can be calculated by using (22) and considering the space vectors used in the modulation strategy.

7. Open-end winding induction machine drive

In this section, the application of a two-output IMC supplying an open-end winding induction machine will be presented. The performance of the drive by using some of the modulation strategies discussed above will be shown and analyzed via simulation and experimental results. The simulations are carried out in a PSIM/MATLAB simulation platform. On the other hand, experimental results are obtained using the setup shown in Figure 14. The IMC has been designed and built at the Power Electronics, Machines and Control lab facilities, University of Nottingham, UK. A six-pole induction machine rated at 7.5 kW is used. A DSP board, based on the TMS320C6713 processor, is used as the control platform. The calculation of duty cycles is carried out on the DSP among several other tasks. An FPGA interface board, designed at Nottingham University, is used to implement the modulation strategies and data acquisition. Communication between the DSP and a PC is achieved using a DSK6713HPI (Host Port Interface) daughter card. The converter input stage uses SK60GM123 modules and the output stages use SK35GD126 modules. The switching frequency is 10 kHz and voltages and currents are also sampled at 10 kHz. The load used in the experimental system is a DC generator, supplying a resistive load, coupled with the induction motor shaft.

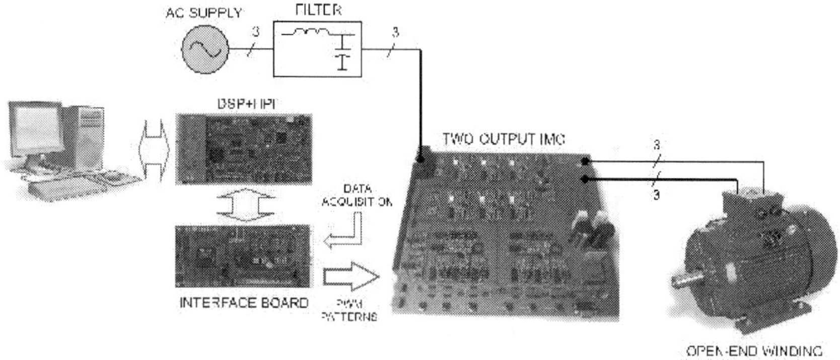

Figure 14. Experimental setup.

7.1. SVM strategy for zero sequence voltage reduction and changing DC voltage

The modulation strategies, presented in Section 6.1 and 6.2.2, are used to implement a standard feedforward vector control scheme [30] for the machine currents (Figure 15). Simulation and experimental results are shown in Sections 7.1.1 and 7.1.2, respectively.

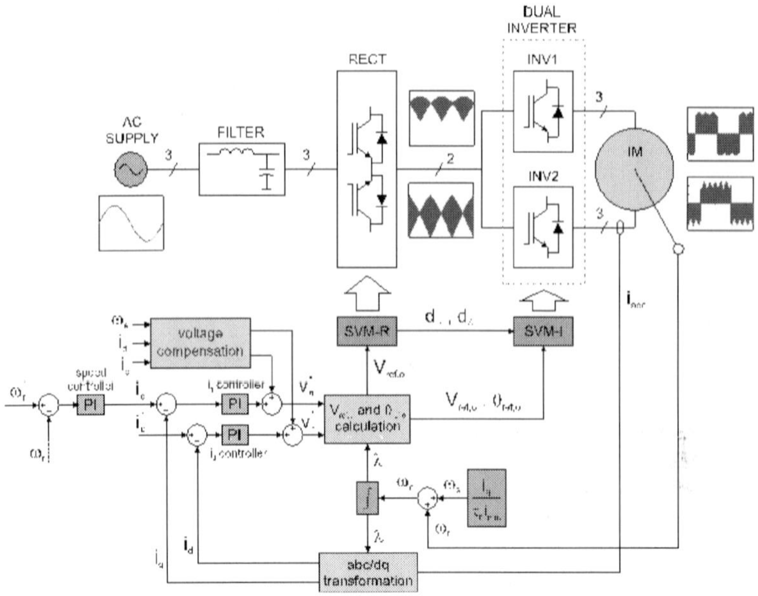

Figure 15. Feedforward vector control scheme of induction machine.

7.1.1. *Simulation results*

The performance of the vector control scheme is verified by applying step changes in the dq-axis reference currents. Figure 16 shows the waveforms obtained by applying a step change in the q-axis reference current from 7.7 to 10 A at $t = 0.16$ s, while d-axis current is kept constant at 6 A. Figure 16a shows dq reference currents and their corresponding responses. A good tracking of the reference currents is obtained. The motor currents and phase a voltage are shown in Figure 16b. The step change in q-axis current is followed with changes in the magnitude and phase of the instantaneous machine currents. Moreover, a transition between reduced virtual DC voltage to maximum virtual DC voltage can be noticed in the output phase voltage of Figure 16b (bottom).

Figure 17a shows the performance of the control scheme when a step change from 6 to 8 A is applied in d-axis reference current while q-axis current is kept constant at 7.7 A. The motor currents and phase- a voltage are shown in Figure 17b. Again, the transition between both modulation strategies for the input rectifier can be noticed.

Figure 18 (top) shows the converter input phase voltage and current; the unity input displacement factor can be noted. Figure 18 (bottom) shows the output zero sequence voltage that has been obtained from the voltage across the load in each phase and then applying (22). As can be seen, the zero sequence voltage has been eliminated due to the modulation strategy used for the output stages.

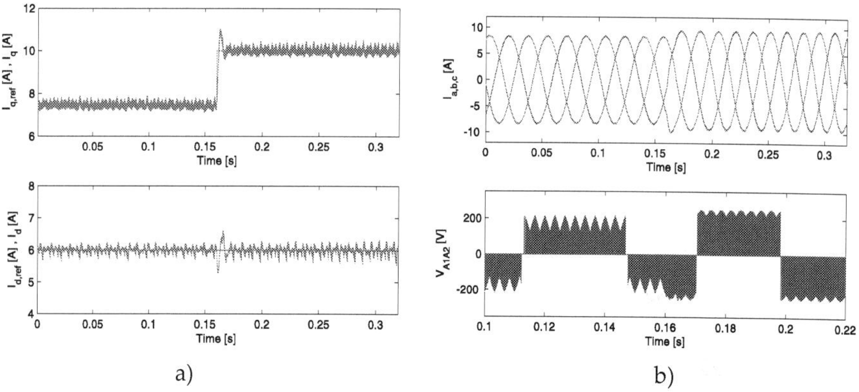

Figure 16. q-axis current step change. (a) Motor q-axis current (top) and d-axis current (bottom). (b) Motor currents (top) and phase voltage (bottom).

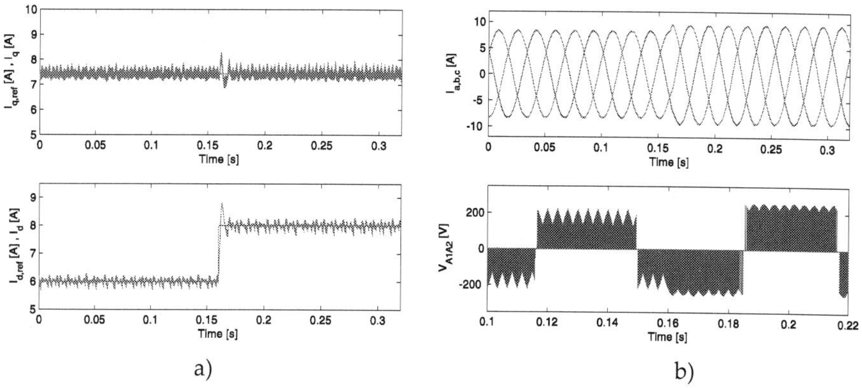

Figure 17. d-axis current step change. (a) Motor q-axis current (top) and d-axis current (bottom). (b) Motor currents (top) and phase voltage (bottom).

7.1.2. Experimental results

The dq-axis currents are shown in Figure 19a. As the speed controller saturates when a step change in the speed reference takes place, a step change in q-axis current reference is applied. The d-axis current reference is kept constant at 6 A. A good performance of the control scheme can be appreciated agreeing with the simulated results. The instantaneous motor currents and phase a voltage are shown in Figure 19b. Again, the transition between both rectifier modulation strategies can be noted in the output phase voltage (Figure 19b bottom) when the change in q-axis reference current is applied.

http://dx.doi.org/10.5772/61157

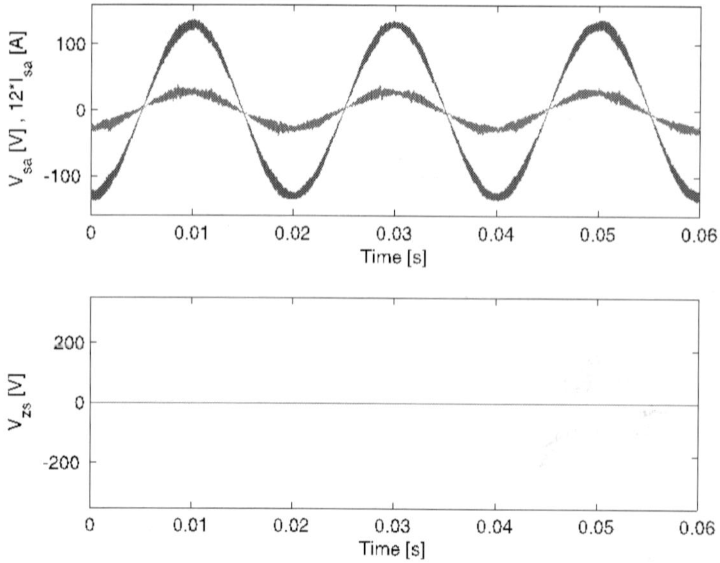

Figure 18. Input phase voltage and current (top) and zero sequence voltage (bottom).

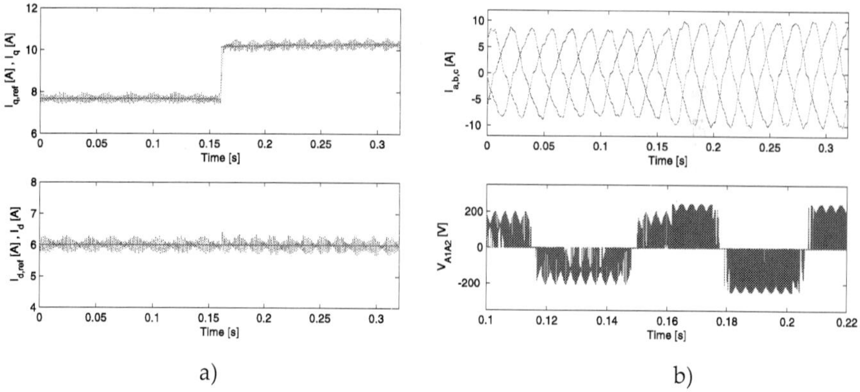

Figure 19. q-axis current step change. (a) Motor q-axis current (top) and d-axis current (bottom). (b) Motor currents (top) and phase voltage (bottom).

Figure 20a shows the dq machine currents when a step change in d-axis reference current is applied while q-axis current is kept constant at 7.7 A. The motor currents and phase- a voltage are shown in Figure 20b. A good correspondence with the simulation results can be noted.

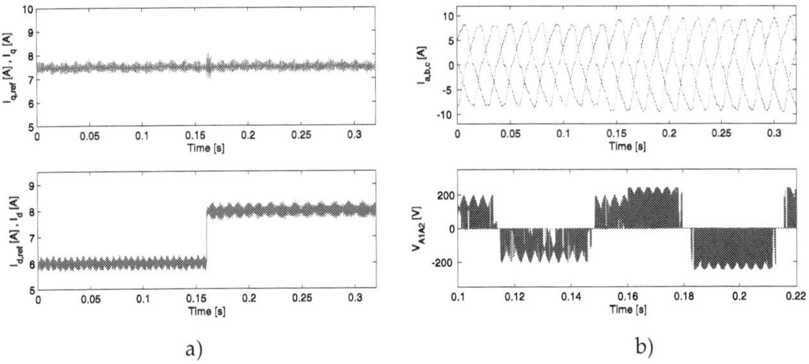

Figure 20. q-axis current step change. (a) Motor q-axis current (top) and d-axis current (bottom). (b) Motor currents (top) and phase voltage (bottom).

Finally, Figure 21 (top) shows the input phase voltage and current. The zero sequence voltage shown in Figure 21 (bottom) is not exactly zero, but this is probably due to the measurement procedure because not all of the channels are sampled at the same time and because in Figure 18 the input switches are ideal.

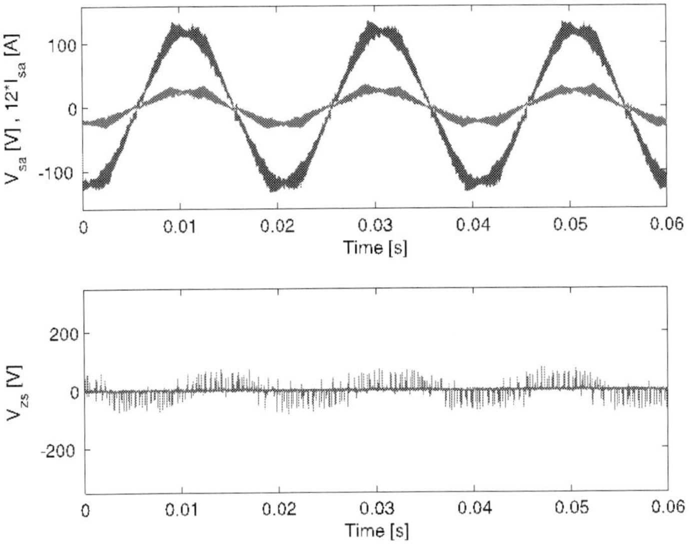

Figure 21. Input rectifier voltage and current (top) and zero sequence voltage (bottom).

http://dx.doi.org/10.5772/61157

7.2. SVM strategy for common-mode voltage reduction and zero sequence voltage compensation

Simulation and experimental results for the modulation strategy presented in Section 6.2.2 will be shown in this section. The rectifier is modulated to maximize the DC link voltage (Figure 7) and the full drive is tested in open-loop *v/f* operation.

7.2.1. Simulation results

The DC link voltage and phase- *a* machine voltage are shown in Figure 22a, top and bottom, respectively. The reference output voltage and frequency were set to 150 V and 50 Hz, respectively. The machine currents for 25 Hz operation are shown in Figure 22b (top), while Figure 22b (bottom) shows machine currents for 50 Hz operation.

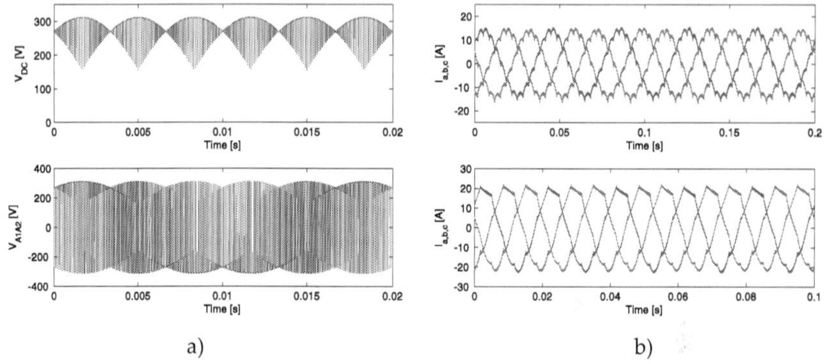

Figure 22. (a) DC link voltage (top) and output phase voltage (bottom). (b) Machine currents for 25 Hz output (top) and 50 Hz output (bottom).

Small disturbances, occurring every 60°, can be noted in the motor currents shown in Figure 22b. These current disturbances are due to the application of zero voltage vectors to machine windings, see PWM pattern in Figure 13, aiming to reduce the zero sequence voltage. During the application of zero voltage vectors, each machine phase winding is supplied with a voltage of $-V_{DC}$ or $+V_{DC}$. When $-V_{DC}$ voltage is applied to the machine windings, the current decreases according to the zero vector duty cycle. Figure 23 shows the current disturbance along with the corresponding DC link voltage and output phase voltage.

The input (supply) currents are shown in Figure 24 (top) while Figure 24 (bottom) shows the converter input phase voltage (blue) and current (green) for an output reference of 150 V and 50 Hz. The unity displacement factor is evident in Figure 24 (bottom).

The low-order harmonics of the machine currents are presented in Table 6.

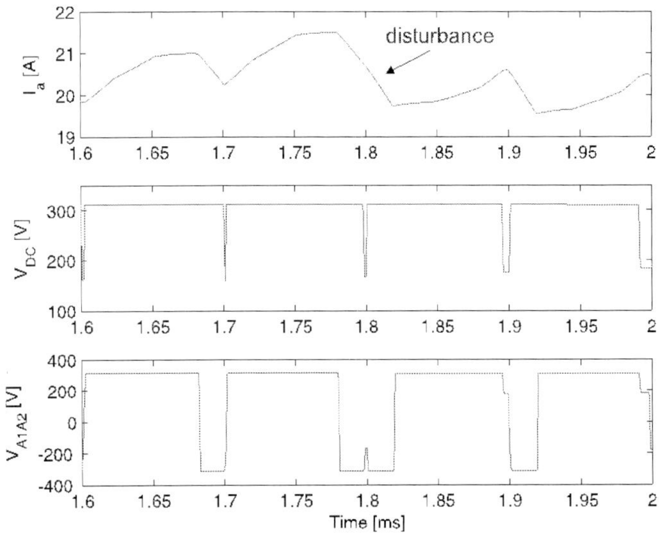

Figure 23. Phase a machine current (top), DC link voltage (middle), and machine phase- a voltage (bottom).

Current Harmonic	RMS Value (A)
Fundamental	14.400
2nd	0.250
3rd	0.184
4th	0.112
5th	0.025
6th	0.815

Table 6. Harmonic content of the machine currents

Figure 25a shows the common-mode voltage separated into v_{0G} and v_{cm0} as defined in (33)–(35). Due to absence of the reference point 0 in the real (and also simulated) converter, the common-mode voltages v_{0G} and v_{cm0} shown in Figure 25a top and bottom, respectively, are obtained as:

$$\begin{aligned} v_{0G} &= v_{nG} + \frac{v_{DC}}{2} \\ v_{cm0} &= \frac{1}{6}\left(v_{A1n} + v_{B1n} + v_{C1n} + v_{A2n} + v_{B2n} + v_{C2n}\right) - \frac{v_{DC}}{2} \end{aligned} \tag{52}$$

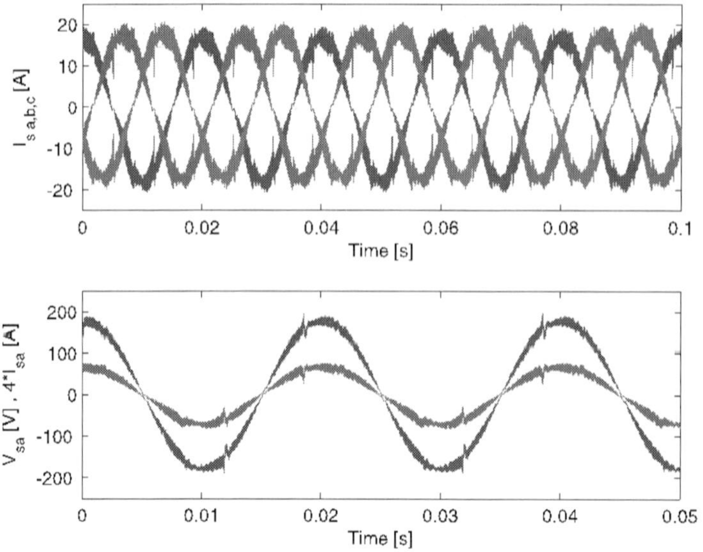

Figure 24. Input currents (top) and input phase voltage and current (bottom).

where n is the negative rail of the DC link.

It can be seen in the simulation results that the contribution of the output inverters to the common-mode voltage is completely eliminated due to the modulation strategy used. Figure 25b shows the zero sequence voltage (top) and its frequency spectrum (bottom). It can be noted that the low-order zero sequence harmonics are reduced because of the asymmetry of the null vector duty cycles used in the switching sequence for each output stage.

7.2.2. Experimental results

Figure 26a (top) shows the DC link voltage while Figure 26a (bottom) shows the voltage across the machine phase- a winding. The output phase voltage presents a fundamental component of 141 V, 50 Hz, slightly less than the voltage reference (150 V) because of the device voltage drops. As can be seen, the modulation strategy used results in a bipolar pulse width modulated voltage at the converter output. The machine currents for 25 Hz and 50 Hz operation are shown in Figure 26b (top and bottom), respectively. The reference output voltages are set to 75 and 150 V, respectively. Correspondence between the simulation and the experimental result can be observed.

In Figure 26b, the effect of the zero voltage vectors in the PWM pattern shown in Figure 13 is also observed. The supply currents are shown in Figure 27 (top), again with good correspondence with the simulation study. Figure 27 (bottom) shows the input phase voltage and current.

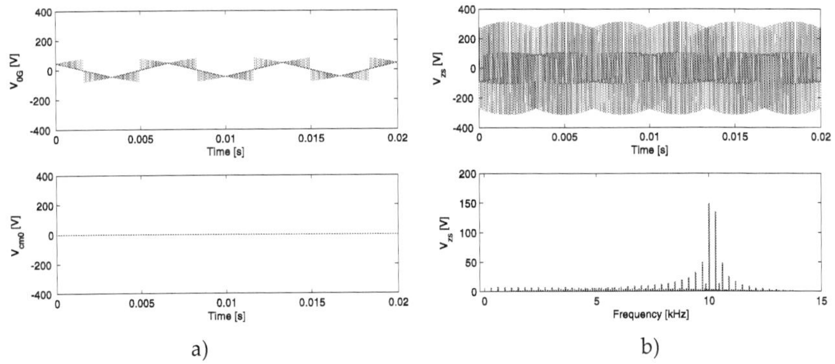

Figure 25. (a) Common-mode voltages v_{0G} (top) and v_{cm0} (bottom). (b) Zero sequence voltage (top) and its frequency spectrum (bottom).

Figure 28a shows the common-mode voltages v_{0G} (top) and v_{cm0} (bottom). The voltage v_{0G} follows very closely the simulation results shown in Figure 25. The voltage v_{cm0} is not exactly zero, but this is probably due to the measurement procedure because not all of the channels are sampled at the same time and because in Figure 25 the input switches are ideal. Finally, Figure 28b shows the zero sequence voltage (top) and its frequency spectrum (bottom), agreeing closely with the simulation results.

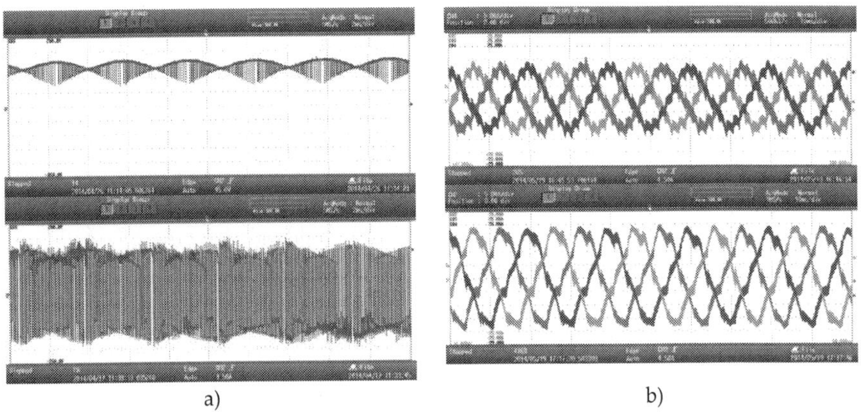

Figure 26. (a) DC link voltage (top) and output phase voltage (bottom). (b) Machine currents for 25 Hz output (top) and 50 Hz output (bottom).

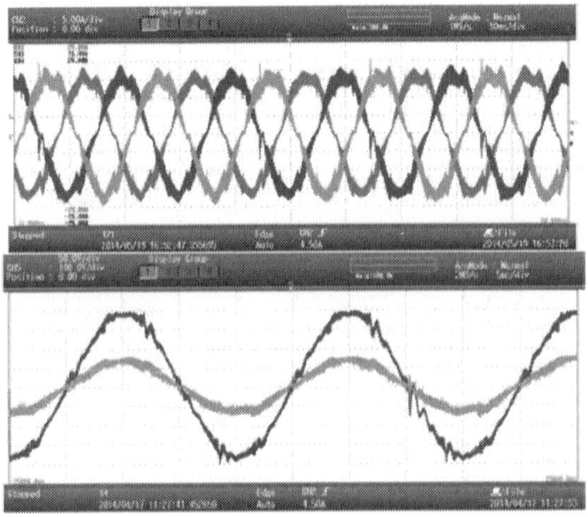

Figure 27. Input currents (top) and input phase voltage and current (bottom).

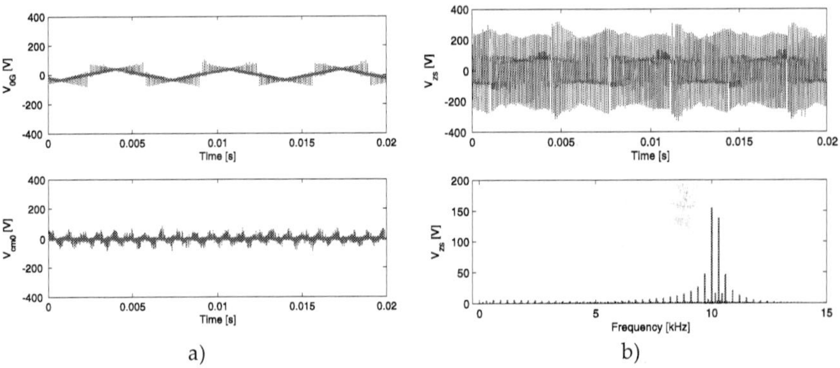

Figure 28. a) Common-mode voltages v_{0G} (top) and v_{cm0}. (bottom). (b) Zero sequence voltage (top) and its frequency spectrum (bottom).

8. Conclusion

The open-end winding induction machine has been presented in this chapter. Different power converter topologies for this type of machine connection have been shown and briefly discussed. Because of the advantages of indirect matrix converter, emphasis has been done in the application of this converter to control the machine. Therefore, an indirect matrix converter with two outputs stages has been proposed to supply energy to the open-end winding induction motor. This topology has been thoroughly modeled and pulse width modulation strategies for the input and output stages of the proposed topology have been shown. Issues such as zero sequence voltage and common-mode voltage, presented in the dual-inverter configuration, have been analyzed and strategies to eliminate and/or reduce such effects have been presented. The control and modulation strategies have been simulated and experimentally tested in a prototype rig. Results for open and closed-loop operation of the open-end winding topology based on IMC have been shown.

Acknowledgements

This work was funded by Fondecyt Chile under Grant 1151325. The financial support of CONICYT/FONDAP/15110019 is also acknowledged.

Author details

Javier Riedemann[1], Rubén Peña[2] and Ramón Blasco-Giménez[3]

1 Department of Electrical and Electronic Engineering, University of Bío-Bío, Concepción, Chile

2 Department of Electrical Engineering, University of Concepción, Concepción, Chile

3 Department of System Engineering and Control, Polytechnic University of Valencia, Valencia, Spain

References

[1] Wang Y, Panda D, Lipo,TA, Pan D. Open-winding power conversion systems fed by half-controlled converters. Power Electron, IEEE Transac 2013;28(5):2427,2436.

[2] Wang Y, Lipo TA, Pan D. Robust operation of double-output AC machine drive. Power Electron ECCE Asia (ICPE & ECCE), 2011 IEEE 8th Int Conf 2011;1:140,144.

[3] Ramachandrasekhar K, Mohan S, Srinivas S. An improved PWM for a dual two-level inverter fed open-end winding induction motor drive. Electric Machines (ICEM), 2010 XIX Int Conf 2010;1(1):1–6, 6–8.

[4] Mohapatra KK, Gopakumar K, Somasekhar VT, Umanand L. A harmonic elimination and suppression scheme for an open-end winding induction motor drive. Ind Electron, IEEE Transac 2003;50(6):1187–98.

[5] Reddy BV, Somasekhar VT. A decoupled SVPWM technique for a dual inverter four-level open-end winding induction motor drive with voltage balancing of Dc-link capacitors. Power Electron, Drives Energy Syst (PEDES) & 2010 Power India, 2010 Joint Int Conf 2010;1:1–6, 20–23.

[6] George DS, Baiju MR. Decoupled random modulation technique for an open-end winding induction motor based 3-level inverter. Ind Electron Applic 2009. ISIEA 2009. IEEE Symp 2009;2:1022–7.

[7] Somasekhar VT, Gopakumar K, Shivakumar EG, Sinha SK. A space vector modulation scheme for a dual two level inverter fed open end winding induction motor drive for the elimination of zero sequence currents. Eur Power Electron Drives J 2002;12(2):26-36.

[8] Srinivasan P, Reddy BV, Somasekhar VT. PWM switching strategy for the elimination of common mode voltage of a two-level inverter drive with an open-end winding induction motor configuration. Power Electron Drives Energy Syst (PEDES) & 2010 Power India, 2010 Joint Int Conf 2010;1:1–6, 20–23.

[9] Somani A, Gupta RK, Mohapatra KK, Basu K, Mohan N. Modulation strategies for direct-link drive for open-end winding AC machines Electric Machines Drives Conf 2009. IEMDC '09. IEEE Int 2009;1863–1868.

[10] Somani A, Gupta RK, Mohapatra KK, Mohan N. A minimum-switch direct-link drive with common-mode voltage suppression and active filtering for open-end winding AC machines Power Electron Specialists Conf 2008. PESC 2008. IEEE 2008;1:2889–93.

[11] Somasekhar VT, Srinivas S, Prakash Reddy B, Nagarjuna Reddy C, Sivakumar K. Pulse width-modulated switching strategy for the dynamic balancing of zero-sequence current for a dual-inverter fed open-end winding induction motor drive. Electric Power Applic, IET 2007;1(4):591 –600.

[12] Kanchan RS, Tekwani PN, Gopakumar K. Three-level inverter scheme with common mode voltage elimination and DC-link capacitor voltage balancing for an open end winding induction motor drive. Electric Machines Drives, IEEE Trans Power Electron 2005;21(6):1676-1683.

[13] Figarado S, Gopakumar K, Mondal G, Sivakumar K, Dinesh NS. Three-level inverter fed open-end winding IM drive with common mode voltage elimination and reduced power device count. Ind Electron Soc 2007. IECON 2007. 33rd Ann Conf IEEE 2007;1:1238–43.

[14] Somasekhar VT, Baiju MR, Mohapatra KK, Gopakumar K. A multilevel inverter system for an induction motor with open-end windings. IECON 02 [Ind Electron Soc IEEE 2002 28th Ann Conf 2002;2:973 –8.

[15] Lakshminarayanan S, Mondal G, Gopakumar K. Multilevel inverter with 18-sided polygonal voltage space vector for an open-end winding induction motor drive. EUROCON, 2007. Int Conf Computer as a Tool. 2007;1:1810–7.

[16] Baiju MR, Gopakumar K, Mohapatra KK, Somasekhar VT, Umanand L. Five-level inverter voltage-space phasor generation for an open-end winding induction motor drive. Electric Power Applic IEE Proc 2003;150(5):531,538.

[17] Wheeler PW, Rodriguez J, Clare JC, Empringham L, Weinstein A. Matrix converters: a technology review. Ind Electron IEEE Transac 2002;49(2):276,288.

[18] Kolar JW, Friedli T, Rodriguez J, Wheeler PW. Review of three-phase PWM AC–AC converter topologies. Ind Electron IEEE Transac 2011;58(11):4988,5006.

[19] Mohapatra KK, Mohan N,. Open-end winding induction motor driven With matrix converter for common-mode elimination. Power Electron Drives Energy Syst2006. PEDES '06. Int Conf 2006;1:1–6.

[20] Gupta RK, Somani A, Mohapatra KK, Mohan N. Space vector PWM for a direct matrix converter based open-end winding ac drives with enhanced capabilities. Appl Power Electron Conf Exposition (APEC), 2010 Twenty-Fifth Ann IEEE 2010;1:901–8.

[21] Pitic CI, Klumpner C. A new matrix converter-voltage source inverter hybrid arrangement for an adjustable speed-open winding induction motor drive with improved performance. Power Electron Machines Drives, 2008. PEMD 2008. 4th IET Conf 2008;1:60–65.

[22] Riedemann J, Pena R, Cardenas R, Clare J, Wheeler P, Rivera M. Common mode voltage and zero sequence current reduction in an open-end load fed by a two output indirect matrix converter. Power Electron Applic (EPE), 2013 15th Eur Conf 2013;1:1,9.

[23] Javier R, Ruben P, Roberto C, Jon C, Pat W, Marco R. Switching strategies for an indirect matrix converter fed open-end load. Ind Electron (ISIE), 2013 IEEE Int Symp 2013;1:1,6.

[24] Riedemann J, Pena R, Cardenas R, Clare J, Wheeler P, Blasco-Gimenez R. Control strategy of a dual-inverter system for an open-end winding induction machine based on indirect matrix converter. Power Electron Applic (EPE'14-ECCE Europe), 2014 16th Eur Conf 2014;1:1,8.

[25] Chen S. Bearing current, EMI and soft switching in induction motor drives: a systematic analysis, design and evaluation, Ph.D. dissertation, Univ. Wisconsin, Madison, 1995.

http://dx.doi.org/10.5772/61157

[26] Erdman M, Kerkman RJ, Schlegel DW, Skibinski GL. Effect of PWM inverters on ac motor bearing currents and shaft voltages. IEEE Trans Ind Applicat, 1996;32(2):250–9.

[27] Klumpner C, Blaabjerg F. Modulation method for a multiple drive system based on a two-stage direct power conversion topology with reduced input current ripple. IEEE Trans Power Electron 2005;20:922–9.

[28] Kolar JW, Schafmeister F. Novel modulation schemes minimizing the switching losses of sparse matrix converters. The 29th Annual Conference of the IEEE Ind. Electronics Society IECON '03, 2003;1:2085–90.

[29] Hava AM, Kerkman RJ, Lipo TA. Simple analytical and graphical methods for carrier-based PWM-VSI drives. Power Electron IEEE Transac 1999;14(1):49,61.

[30] Leonhard W. Control of Electrical Drives. Berlin: Springer, 2001.

Section 4

Multiphase Induction Motors

Chapter 12

Open-Phase Fault Operation on Multiphase Induction Motor Drives

Hugo Guzman, Ignacio Gonzalez, Federico Barrero and Mario Durán

Additional information is available at the end of the chapter

http://dx.doi.org/10.5772/60810

Abstract

Multiphase machines have been recognized in the last few years like an attractive alternative to conventional three-phase ones. This is due to their usefulness in a niche of applications where the reduction in the total power per phase and, mainly, the high overall system reliability and the ability of using the multiphase machine in faulty conditions are required. Electric vehicle and railway traction, all-electric ships, more-electric aircraft or wind power generation systems are examples of up-to-date real applications using multiphase machines, most of them taking advantage of the ability of continuing the operation in faulty conditions. Between the available multiphase machines, symmetrical five-phase induction machines are probably one of the most frequently considered multiphase machines in recent research. However, other multiphase machines have also been used in the last few years due to the development of more powerful microprocessors. This chapter analyzes the behavior of generic n-phase machines (being n any odd number higher than 3) in faulty operation (considering the most common faulty operation, i.e. the open-phase fault). The obtained results will be then particularized to the 5-phase case, where some simulation and experimental results will be presented to show the behavior of the entire system in healthy and faulty conditions. The chapter will be organized as follows: First, the different faults in a multiphase machine are analyzed. Fault conditions are detailed and explained, and the interest of a multiphase machine in the management of faults is stated. The effect of the open-phase fault operation in the machine model is then studied. A generic n-phase machine is considered, being n any odd number greater than three. The analysis is afterwards particularized to the 5-phase machine, where the open-phase fault condition is managed using different control methods and the obtained results are compared. Finally, the conclusions are presented in the last section of the chapter.

Keywords: Multiphase drives, Fault-tolerance, Predictive control techniques, Resonant controllers

1. Introduction

Multiphase machines have been recognized in the last few years as an attractive alternative to conventional three-phase ones. This is due to their usefulness in a niche of applications where the reduction in the total power per phase and, the high overall system reliability and the ability of using the multiphase machine in faulty conditions are required. Electric vehicle and railway traction, all-electric ships, more-electric aircraft or wind power generation systems are examples of up-to-date real applications using multiphase machines, most of them taking advantage of the ability of continuing the operation in faulty conditions. Among the available multiphase machines, symmetrical five-phase induction machines are probably one of the most frequently considered multiphase machines in recent research. However, other multiphase machines have also been used in the last few years due to the development of more powerful microprocessors. This chapter analyzes the behavior of generic n-phase machines (n being any odd number higher than 3) in faulty operation (considering the most common faulty operation, i.e. the open-phase fault). The obtained results will be then particularized to the 5-phase case, where some simulation and experimental results will be presented to show the behavior of the entire system in healthy and faulty conditions.

The chapter will be organized as follows:

First, the different faults in a multiphase machine are analyzed. Fault conditions are detailed and explained, and the interest of a multiphase machine in the management of faults is stated. The effect of the open-phase fault operation in the machine model is then studied. A generic n-phase machine is considered, n being any odd number greater than three. The analysis is afterwards particularized to the 5-phase machine, where the open-phase fault condition is managed using different control methods and the obtained results are compared. Finally, the conclusions are presented in the last section of the chapter.

2. Faults in electromechanical multiphase drives

An electrical drive is an electromagnetic equipment subject to different electrical and mechanical faults which, depending on its nature and on the special characteristics of the system, may result in abnormal operation or shut down. In order to increase the use of electrical drives in safety-critical and high-demand applications, the development of cost-effective, robust and reliable systems is imperative. This issue has recently become one of the latest challenges in the field of electrical drives design [1]. Therefore, fault-tolerance, which can be defined as the ability to ensure proper speed or torque reference tracking in the electrical drive under abnormal conditions, has been considered in three-phase electrical drives taking into account different design and research approaches, including redundant equipment and over-dimensioned designs, leading to effective and viable fault standing but costly solutions. Fault-tolerance in three-phase drives for different types of faults is a viable and mature research field, where the drive performance and the control capability is ensured at the expense of extra equipment [1]. However, this is not the case in the multiphase drives area in spite of the higher

http://dx.doi.org/10.5772/60810

number of phases that the multiphase machine possesses, which favors its higher fault-tolerance capability compared with conventional three-phase drives. Multiphase drives do not need extra electrical equipment to manage post-fault operation, requiring only proper post-fault control techniques in order to continue operating [2]. Therefore, they are ideal for traction and aerospace applications for security reasons or in offshore wind farms where corrective maintenance can be difficult under bad weather conditions [3-6].

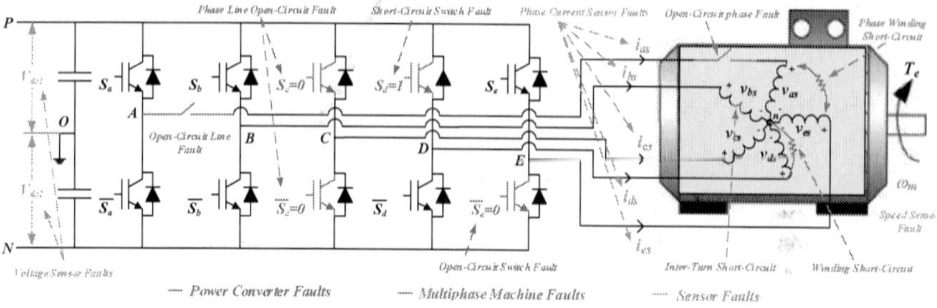

Figure 1. Types of faults on a five-phase drive.

Faults in an electromechanical drive can be also classified depending on the nature (electrical or mechanical), the location or the effect they have on the overall system (notice that different types of faults can result in the same abnormal machine behavior). The most common classification of faults in electrical drives defines three main groups of faults that can appear in the electrical drive. The power converter, electronic sensors (current, temperature, speed and voltage) and the electrical machine focus the main faults in an electrical drives, as shown in Figure 1. These faults are detailed hereafter.

1. Electrical machine faults that can be caused by either electrical or mechanical problems/ stress and are further divided as follows [7-11]:

2. Stator faults: open-circuit or short-circuit of one or more stator phase windings. These kinds of faults appear due to mechanically damaged connections caused by insulation failure, extreme electrical operating conditions (high temperatures in the stator core or winding coils, starting stresses, over- or under-voltage operation, electrical discharges, unbalanced stator voltages) or nonappropriate ambient conditions (dirt, oil and moisture contamination) [10-12], leading to inter-turn [12], stator-winding [13], and different phase winding short circuits [10- 11], which may result in further open-phase faults of one or more phase windings [14-15].

3. Rotor faults: shorted rotor field winding, broken rotor bars and cracked rotor end-rings. They are caused by electrical (shorted rotor windings) or mechanical (broken bars and cracked rotor rings) problems. These types of faults appear due to thermal stress (drive operation under overload and unbalanced load conditions), electromagnetic stress,

manufacturing problems, dynamic stress from shaft torque, environmental conditions and fatigued mechanical parts [10-11].

4. Air-gap irregularities due to static or dynamic eccentricity problems. Eccentricity is caused by manufacturing and constructive errors that generate an unequal air gap between the stator and the rotor, leading to unbalanced radial forces and possible rotor-stator contact [10]. Static eccentricity appears when the position of the air gap inequality is fixed, whereas dynamic eccentricity happens when the rotor center is not properly aligned at the rotation center and the position of the air gap does not rotate equally.

5. Bearing faults, which are mainly caused by assembling errors (misalignment of bearings) that result in bearing vibration forced into the shaft [16].

6. Bent shaft faults, which are similar to dynamic eccentricity faults [10]. These faults appear when force unbalance or machine-load misalignment happens, resulting in machine vibration and further machine failure [11].

Statistically, the most common faults in electrical machines are the bearing failures, stator winding faults, broken rotor bar, shaft and coupling faults, cracked rotor end-rings, and air-gap eccentricity [7-9], leading to unbalanced stator currents and voltages, the appearance of specific harmonics in the phase currents, overall torque oscillation and reduction, machines vibration, noise, overheating and efficiency reduction [10- 11].

1. Sensor Faults. Electrical drives commonly include speed, voltage and current sensors for control and protection purposes (Figure 1). In the multiphase drive case, standard FOC and predictive control techniques require a speed and at least $n-1$ (for an n-phase drive) current measurements in order to ensure proper control behavior. In case of abnormal sensor operation, inexistent or nonaccurate signals can downgrade the system performance or result in a complete drive failure [17-20]. Sensor faults have been mainly analyzed for three-phase drives, and recent works have also addressed this type of faults for the multiphase case [21-23]. Notice that depending on the faulty sensor (i.e. DC-link voltage, current or speed), the effect in conventional three-phase or multiphase drive is mainly the same. In any case, the analysis of these kinds of faults mainly focuses on handling only one faulty sensor due to the small probability of fault in more than one sensor [24], which would include current and speed sensor faults, which are the most critical in electrical drive applications. The main reason for this is that high-performance drives are based on speed and current closed-loop controllers and consequently on speed and current sensors. Any variation or systematic error on the measured quantities may result in instantaneous power demanding control actions, subjecting the whole system to possible electrical stress [17].

2. Power Converter Faults. The most common types of faults in electrical drives are those associated to the power converter [25]. Power converter faults are presented graphically in Figure 1, and can be further classified as single short-circuit switch fault, single open-circuit switch fault, phase-leg short-circuit fault, phase-leg open-circuit fault or open-circuit line fault [2]. These types of faults are mainly due to the burn out of the semiconductor or due to the semiconductor driver failure, forcing the semiconductor to

remain in a constant ON or OFF state. As a result, the power converter may either lose a complete phase (also termed open-phase fault) or may physically maintain the number of phases and current flow but lose specific control capabilities on either one or both of the semiconductor of a certain phase. Thus, the configuration of the electrical drive varies, and the post-fault electrical drive may be regarded as an entire different system [26].

The phase redundancy that multiphase drives possess allows managing faulty operation without the need of extra equipment, depending on the specific electrical machine configuration. Postfault control techniques exploit extra degrees of freedom of the multiphase system to maintain a circular Magneto-Motive Force (MMF) and achieve the desired speed or torque references. Depending on the type of fault and the electrical drive characteristics, different postfault control strategies, drive configuration and electrical machine winding connections are adopted under postfault operation. For instance, in the case of short-circuit faults, the proposed fault management strategies are based on controlling the available four healthy phases in a five-phase drive, maintaining operation at the expense of higher stator phase winding losses and torque ripple [27]. Nonetheless, this increase in torque ripple is managed in a dual three-phase drive [13] maintaining postfault operation with one three-phase drive in short-circuit and compensating the braking torque with the healthy three-phase drive [28]. The inclusion of auxiliary semiconductors in the electrical machine windings, in order to change from a short-circuit fault to an open-circuit or open-phase fault, was also addressed in [29], where ripple-free output torque was obtained with the appropriate control of the remaining four healthy phases. As a result, the multiphase electrical drive is able to manage different types of faults but at the expense of extra electronic equipment, like in the conventional three-phase case. Different winding connections have also been considered for single and phase short circuit faults for a dual three-phase machine, assessing the effect of the harmonics obtained in the machines losses and torque, and evaluating its performance under different working conditions [30]. A similar approach has also been followed for open-phase and open-line faults, where different drive topologies or machine winding connections have been considered. In one study [3], a six-phase drive was designed in order to independently control each phase of a three-phase machine under different types of faults and its viability was stated emulating an open-circuit line fault. Five-phase machines considering penta- and star-type winding connections are also compared in another study [30], where fundamental and third-harmonic components are used to control the post-fault operation of the electrical drive. The available torque is increased, while torque ripple and losses are reduced. It is concluded that penta-winding connection results in improved fault-tolerance capabilities due to the higher number of open-circuit phases it can withstand (three open-phase faults in a five-phase drive).

But fault management does not include only postfault control techniques. It is divided in four different states namely, fault occurrence, fault detection, fault isolation and, finally, postfault control or the fault-tolerant control operation. Different fault detection and fault isolation techniques have been proposed based on the specific characteristics of the electrical drive to ensure proper postfault behavior. Then, a proper postfault control method is implemented to maintain correct reference tracking. This book chapter will be only focused on the postfault controller, and fault-detection and fault-isolation techniques will not be addressed.

3. Analysis of an open-phase fault in multiphase drives with odd number of phases

The most common fault, the open-phase fault, is studied in this section. The ability of a multiphase machine managing the fault operation lies in the greater number of phases and in the greater number of independent variables that model the system. The model of the multiphase machine is analyzed. The analysis is done for a generic multiphase machine. Then, the modeling equations of an n-phase multiphase drive under an open-phase fault operation is presented, emphasizing their effect in the healthy model to understand the imposed constraints for the design of postfault control techniques.

In the first place, the n-phase one neutral induction machine model is studied. The machine can be modeled by a set of stator and rotor phase voltage equilibrium equations referred to a fixed reference frame linked to the stator as follows:

$$[V_s]=[R_s]\cdot[I_s]+\frac{d}{dt}[\lambda_s]=[R_s]\cdot[I_s]+p\cdot\left([L_{ss}]\cdot[I_s]+[L_{sr}(\theta)]\cdot[I_r]\right) \tag{1}$$

$$[V_r]=[R_r]\cdot[I_r]+\frac{d}{dt}[\lambda_r]=[R_r]\cdot[I_r]+p\cdot\left([L_{rr}]\cdot[I_r]+[L_{rs}(\theta)]\cdot[I_s]\right) \tag{2}$$

Where θ represents the rotor electrical angular position with respect to the stator, and rotates at the rotor electrical velocity ω_r. The voltage, current and flux matrices are given by (3)-(8). Notice that the voltage rotor components (4) are equal to zero.

$$[V_s]=[v_{as}v_{bs}v_{cs}v_{ds}v_{es}\cdots v_{ns}]^T \tag{3}$$

$$[V_r]=[v_{ar}v_{br}v_{cr}v_{dr}v_{er}\cdots v_{nr}]^T \tag{4}$$

$$[\lambda_s]=[\lambda_{as}\lambda_{bs}\lambda_{cs}\lambda_{ds}\lambda_{es}\cdots\lambda_{ns}]^T \tag{5}$$

$$[\lambda_r]=[\lambda_{ar}\lambda_{br}\lambda_{cr}\lambda_{dr}\lambda_{er}\cdots\lambda_{nr}]^T \tag{6}$$

$$[I_s]=[i_{as}i_{bs}i_{cs}i_{ds}i_{es}\cdots i_{ns}]^T \tag{7}$$

$$[I_r]=[i_{ar}i_{br}i_{cr}i_{dr}i_{er}\cdots i_{nr}]^T \tag{8}$$

The rotor and stator resistance and inductance matrices are defined as follows:

http://dx.doi.org/10.5772/60810

$$[R_s]=\begin{bmatrix} R_s & 0 & 0 & 0 & \cdots & 0 \\ 0 & R_s & 0 & 0 & \cdots & 0 \\ 0 & 0 & R_s & 0 & \cdots & 0 \\ 0 & 0 & 0 & R_s & \cdots & 0 \\ \vdots & \vdots & \vdots & \vdots & \ddots & \vdots \\ 0 & 0 & 0 & 0 & \cdots & R_s \end{bmatrix} \quad [R_r]=\begin{bmatrix} R_r & 0 & 0 & 0 & \cdots & 0 \\ 0 & R_r & 0 & 0 & \cdots & 0 \\ 0 & 0 & R_r & 0 & \cdots & 0 \\ 0 & 0 & 0 & R_r & \cdots & 0 \\ \vdots & \vdots & \vdots & \vdots & \ddots & \vdots \\ 0 & 0 & 0 & 0 & \cdots & R_r \end{bmatrix} \tag{9}$$

$$\left[L_{ss}\right]=L_{ls}\cdot\left[I_n\right]+L_{ms}\cdot\left[\Lambda(\vartheta)_n\right] \tag{10}$$

$$\left[L_{rr}\right]=L_{lr}\cdot\left[I_n\right]+L_{mr}\cdot\left[\Lambda(\vartheta)_n\right] \tag{11}$$

$$\left[\Lambda(\vartheta)_n\right]=\begin{bmatrix} 1 & \cos(\vartheta) & \cos(2\vartheta) & \cos(3\vartheta) & \cdots & \cos((n-1)\vartheta) \\ \cos((n-1)\vartheta) & 1 & \cos(\vartheta) & \cos(2\vartheta) & \cdots & \cos((n-2)\vartheta) \\ \cos((n-2)\vartheta) & \cos((n-1)\vartheta) & 1 & \cos(\vartheta) & \cdots & \cos((n-3)\vartheta) \\ \cos((n-3)\vartheta) & \cos((n-2)\vartheta) & \cos((n-1)\vartheta) & 1 & \cdots & \cos((n-4)\vartheta) \\ \vdots & \vdots & \vdots & \vdots & \ddots & \vdots \\ \cos(2\vartheta) & \cos(3\vartheta) & \cos(4\vartheta) & \cos(5\vartheta) & \cdots & \cos(\vartheta) \\ \cos(\vartheta) & \cos(2\vartheta) & \cos(3\vartheta) & \cos(4\vartheta) & \cdots & 1 \end{bmatrix} \tag{12}$$

Due to the machine symmetry, the stator-rotor (L_{msr}) and rotor-stator (L_{mrs}) mutual inductances are given by:

$$L_{msr}=L_{mrs}=L_{ms}\rightarrow L_{ms}=k_w^2\cdot L_{mr} \tag{13}$$

Making possible to conclude that:

$$\left[L_{sr}(\theta)\right]=\left[L_{rs}(\theta)\right]^T\rightarrow\left[L_{sr}(\theta)\right]=L_{ms}\cdot\left[\Psi(\theta)\right] \tag{14}$$

$$\left[\Psi(\theta)\right]=\begin{bmatrix} \cos(\Delta_1) & \cos(\Delta_2) & \cos(\Delta_3) & \cos(\Delta_4) & \cdots & \cos(\Delta_n) \\ \cos((\Delta_n) & \cos(\Delta_1) & \cos(\Delta_2) & \cos(\Delta_3) & \cdots & \cos(\Delta_{(n-1)}) \\ \cos(\Delta_{(n-1)}) & \cos(\Delta_n) & \cos(\Delta_1) & \cos(\Delta_2) & \cdots & \cos(\Delta_{(n-2)}) \\ \cos(\Delta_{(n-2)}) & \cos(\Delta_{(n-1)}) & \cos(\Delta_n) & \cos(\Delta_1) & \cdots & \cos(\Delta_{(n-3)}) \\ \vdots & \vdots & \vdots & \vdots & \ddots & \vdots \\ \cos(\Delta_2) & \cos(\Delta_3) & \cos(\Delta_4) & \cos(\Delta_5) & \cdots & \cos(\Delta_1) \end{bmatrix} \tag{15}$$

Notice that $[I_n]$ is the identity matrix of order n, Δ_i angles are defined as: $\Delta_i=\theta + (i-1)\vartheta$, being $i=\{1,2,3, ,n\}$, L_{ls} and L_{lr} are the stator and rotor leakage inductances, and ϑ is the angle between phase windings.

Depending on the working state of the electrical drive and the number of phases it possesses, different transformation matrices can be used in order to describe the machine's electrical parameters in an α - β - x - y - z reference frame. For instance, for normal operation the traditional Clarke transformation (16) is used. (16)

$$[T_n]=\frac{2}{n}\begin{bmatrix} 1 & \cos(\vartheta) & \cos(2\vartheta) & \cos(3\vartheta) & \cdots & \cos((n-1)\vartheta) \\ 0 & \sin(\vartheta) & \sin(2\vartheta) & \sin(3\vartheta) & \cdots & \sin((n-1)\vartheta) \\ 1 & \cos(2\vartheta) & \cos(4\vartheta) & \cos(6\vartheta) & \cdots & \cos(2(n-1)\vartheta) \\ 0 & \sin(2\vartheta) & \sin(4\vartheta) & \sin(6\vartheta) & \cdots & \sin(2(n-1)\vartheta) \\ 1 & \cos(3\vartheta) & \cos(6\vartheta) & \cos(9\vartheta) & \cdots & \cos(3(n-1)\vartheta) \\ 0 & \sin(3\vartheta) & \sin(6\vartheta) & \sin(9\vartheta) & \cdots & \sin(3(n-1)\vartheta) \\ 1 & \cos(4\vartheta) & \cos(8\vartheta) & \cos(12\vartheta) & \cdots & \cos(4(n-1)\vartheta) \\ 0 & \sin(4\vartheta) & \sin(8\vartheta) & \sin(12\vartheta) & \cdots & \sin(4(n-1)\vartheta) \\ \vdots & \vdots & \vdots & \vdots & \ddots & \vdots \\ 1 & \cos\left(\frac{n-1}{2}\vartheta\right) & \cos\left(2\frac{n-1}{2}\vartheta\right) & \cos\left(3\frac{n-1}{2}\vartheta\right) & \cdots & \cos\left((n-1)\frac{n-1}{2}\vartheta\right) \\ 0 & \sin\left(\frac{n-1}{2}\vartheta\right) & \sin\left(2\frac{n-1}{2}\vartheta\right) & \sin\left(3\frac{n-1}{2}\vartheta\right) & \cdots & \sin\left((n-1)\frac{n-1}{2}\vartheta\right) \\ \frac{1}{2} & \frac{1}{2} & \frac{1}{2} & \frac{1}{2} & \cdots & \frac{1}{2} \end{bmatrix} \tag{16}$$

In order to eliminate the time dependence of the coupling inductances and divide the model in a set of different independent-orthogonal equations, the Clarke transformation is applied to the machine model. The stator and rotor voltage, current and flux components in the α_1 - β_1 - α_2 - β_2 - ... - z_n reference frame can be calculated by:

$$\begin{bmatrix} v_{s\alpha_1} \\ v_{s\beta_1} \\ v_{s\alpha_2} \\ v_{s\beta_2} \\ \vdots \\ v_{sz_n} \end{bmatrix} = [T_n]*[V_s] \quad \begin{bmatrix} i_{s\alpha_1} \\ i_{s\beta_1} \\ i_{s\alpha_2} \\ i_{s\beta_2} \\ \vdots \\ i_{sz_n} \end{bmatrix} = [T_n]*[I_s] \quad \begin{bmatrix} \lambda_{s\alpha_1} \\ \lambda_{s\beta_1} \\ \lambda_{s\alpha_2} \\ \lambda_{s\beta_2} \\ \vdots \\ \lambda_{sz_n} \end{bmatrix} = [T_n]*[\lambda_s] \tag{17}$$

http://dx.doi.org/10.5772/60810

$$\begin{bmatrix} v'_{r\alpha_1} \\ v'_{r\beta_1} \\ v'_{r\alpha_2} \\ v'_{r\beta_2} \\ \vdots \\ v'_{rz_n} \end{bmatrix} = [T_n]*[V_r] \quad \begin{bmatrix} i'_{r\alpha_1} \\ i'_{r\beta_1} \\ i'_{r\alpha_2} \\ i'_{r\beta_2} \\ \vdots \\ i'_{rz_n} \end{bmatrix} = [T_n]*[I_r] \quad \begin{bmatrix} \lambda'_{r\alpha_1} \\ \lambda'_{r\beta_1} \\ \lambda'_{r\alpha_2} \\ \lambda'_{r\beta_2} \\ \vdots \\ \lambda'_{rz_n} \end{bmatrix} = [T_n]*[\lambda_r] \tag{18}$$

Multiplying the transformation matrix T_n with the stator and rotor phase voltage equations, (1) and (2), we get:

$$\begin{aligned} [T_n]\cdot[V_s] &= [T_n]\cdot[R_s]\cdot[T_n]^{-1}\cdot[T_n]\cdot[I_s] \\ &+ p\cdot[T_n]\cdot[L_{ss}]\cdot[T_n]^{-1}\cdot[T_n]\cdot[I_s] \\ &+ p\cdot[T_n]\cdot[\mathrm{L}_{\mathrm{sr}}(\theta)]\cdot[T_n]^{-1}\cdot[T_n]\cdot[I_r] \end{aligned} \tag{19}$$

$$\begin{aligned} [0] &= [T_n]\cdot[R_r]\cdot[T_n]^{-1}\cdot[T_n]\cdot[I_r] \\ &+ p\cdot[T_n]\cdot[L_{rr}]\cdot[T_n]^{-1}\cdot[T_n]\cdot[I_r] \\ &+ p\cdot[T_n]\cdot[\mathrm{L}_{\mathrm{rs}}(\theta)]\cdot[T_n]^{-1}\cdot[T_n]\cdot[I_s] \end{aligned} \tag{20}$$

When an open-phase fault occurs in phase "i", the stator windings become an unbalanced system, the faulty phase current is now zero (i_{is}=0), leading to a modification in the machine equations.

Due to the fact that the machine has no longer symmetrical stator windings, the back-emf terms are no longer mutually canceled, consequently the sum of the phase voltages are no longer zero.

$$\sum[Vs] \neq 0 \tag{21}$$

Even though the faulty phase stator current will be zero, the corresponding phase voltage with respect to the neutral machine point will have an equivalent voltage value equal to the back-emf (22).

$$v_{is} = R_s i_{is} + p\cdot\lambda_{is} = p\cdot\lambda_{is} = BackEmf_i \tag{22}$$

Taking this into account the new voltage, current and flux matrices are:

$$\left[V_s\right]=\left[v_{as}v_{bs}v_{cs}\cdots-BackEMF_i\cdots v_{ns}\right]^T \tag{23}$$

$$\left[0\right]=\left[v_{ar}v_{br}v_{cr}v_{dr}v_{er}\cdots v_{nr}\right]^T \tag{24}$$

$$\left[\lambda_s\right]=\left[\lambda_{as}\lambda_{bs}\lambda_{cs}\lambda_{ds}\lambda_{es}\cdots\lambda_{ns}\right]^T \tag{25}$$

$$\left[\lambda_r\right]=\left[\lambda_{ar}\lambda_{br}\lambda_{cr}\lambda_{dr}\lambda_{er}\cdots\lambda_{nr}\right]^T \tag{26}$$

$$\left[I_s\right]=\left[i_{as}\,i_{bs}i_{cs}\cdots 0\cdots i_{ns}\right]^T \tag{27}$$

$$\left[I_r\right]=[i_{ar}i_{br}i_{cr}i_{dr}i_{er}\cdots i_{nr}]^T \tag{28}$$

Notice that the rotor components remain the same as in normal operation, due to the fact that in postfault operation the machine rotor maintains a symmetrical winding distribution.

The absence of the stator phase results in a loss in one degree of freedom. Depending on the position of the faulty phase the transformation matrix (16) is modified, making it no longer possible to generate the same number of orthogonal sub-systems, leading to the removal of one or more of the generated components (29).

$$\left[T_{nUF}\right]=\frac{2}{n}\begin{bmatrix}
1 & \cos(\vartheta) & \cos(2\vartheta) & \cos(3\vartheta) & \cdots & 0 & \cdots & \cos((n-1)\vartheta)\\
0 & \sin(\vartheta) & \sin(2\vartheta) & \sin(3\vartheta) & \cdots & 0 & \cdots & \sin((n-1)\vartheta)\\
1 & \cos(2\vartheta) & \cos(4\vartheta) & \cos(6\vartheta) & \cdots & 0 & \cdots & \cos(2(n-1)\vartheta)\\
0 & \sin(2\vartheta) & \sin(4\vartheta) & \sin(6\vartheta) & \cdots & 0 & \cdots & \sin(2(n-1)\vartheta)\\
1 & \cos(3\vartheta) & \cos(6\vartheta) & \cos(9\vartheta) & \cdots & 0 & \cdots & \cos(3(n-1)\vartheta)\\
0 & \sin(3\vartheta) & \sin(6\vartheta) & \sin(9\vartheta) & \cdots & 0 & \cdots & \sin(3(n-1)\vartheta)\\
\vdots & \vdots & \vdots & \vdots & \vdots & 0 & \cdots & \vdots\\
0 & 0 & 0 & 0 & \cdots & 0 & \cdots & 0\\
\vdots & \vdots & \vdots & \vdots & \vdots & 0 & \ddots & \vdots\\
1 & \cos\left(\frac{n-1}{2}\vartheta\right) & \cos\left(2\frac{n-1}{2}\vartheta\right) & \cos\left(3\frac{n-1}{2}\vartheta\right) & \cdots & 0 & \cdots & \cos\left((n-1)\frac{n-1}{2}\vartheta\right)\\
0 & \sin\left(\frac{n-1}{2}\vartheta\right) & \sin\left(2\frac{n-1}{2}\vartheta\right) & \sin\left(3\frac{n-1}{2}\vartheta\right) & \cdots & 0 & \cdots & \sin\left((n-1)\frac{n-1}{2}\vartheta\right)\\
\frac{1}{2} & \frac{1}{2} & \frac{1}{2} & \frac{1}{2} & \cdots & 0 & \cdots & \frac{1}{2}
\end{bmatrix} \tag{29}$$

Consequently, the machine coupling inductance matrices (10-14) for the stator-rotor components need to be arranged considering the absence of the faulty phase.

$$\left[\Lambda(\vartheta)_{nUF}\right]=\begin{bmatrix} 1 & \cos(\vartheta) & \cos(2\vartheta) & \cdots & 0 & \cdots & \cos((n-1)\vartheta) \\ \cos((n-1)\vartheta) & 1 & \cos(\vartheta) & \cdots & 0 & \cdots & \cos((n-2)\vartheta) \\ \cos((n-2)\vartheta) & \cos((n-1)\vartheta) & 1 & \cdots & 0 & \cdots & \cos((n-3)\vartheta) \\ \cos((n-3)\vartheta) & \cos((n-2)\vartheta) & \cos((n-1)\vartheta) & \cdots & 0 & \cdots & \cos((n-4)\vartheta) \\ \vdots & \vdots & \vdots & \vdots & 0 & \ddots & \vdots \\ 0 & 0 & 0 & 0 & 0 & \cdots & 0 \\ \vdots & \vdots & \vdots & \vdots & 0 & \vdots & \vdots \\ \cos(\vartheta) & \cos(2\vartheta) & \cos(3\vartheta) & \cdots & 0 & \cdots & 1 \end{bmatrix} \tag{30}$$

The equations (1) and (2) need to be multiplied by the new Clarke transformation matrix $[T_{nuf}]$, to express the stator and rotor voltage, current and flux components in the α_1 - β_1 - α_2 - β_2 - ... - z_n reference frame in post-fault situation:

$$\begin{aligned}\left[T_{nuf}\right]\left[V_s\right]&=\left[T_{nuf}\right]\left[R_s\right]\cdot\left[T_{nuf}\right]^{-1}\cdot\left[T_{nuf}\right]\cdot\left[I_s\right]\\ &+p\cdot\left[T_{nuf}\right]\cdot\left[L_{ss}\right]\cdot\left[T_{nuf}\right]^{-1}\cdot\left[T_{nuf}\right]\cdot\left[I_s\right]\\ &+p\cdot\left[T_{nuf}\right]\cdot\left[L_{sr}(\theta)\right]\cdot\left[T_{nuf}\right]^{-1}\cdot\left[T_{nuf}\right]\cdot\left[I_r\right]\end{aligned} \tag{31}$$

$$\begin{aligned}\left[0\right]&=\left[T_{nuf}\right]\left[R_r\right]\cdot\left[T_{nuf}\right]^{-1}\cdot\left[T_{nuf}\right]\cdot\left[I_r\right]\\ &+p\cdot\left[T_{nuf}\right]\cdot\left[L_{rr}\right]\cdot\left[T_{nuf}\right]^{-1}\cdot\left[T_{nuf}\right]\cdot\left[I_r\right]\\ &+p\cdot\left[T_{nuf}\right]\cdot\left[L_{rs}(\theta)\right]\cdot\left[T_{nuf}\right]^{-1}\cdot\left[T_{nuf}\right]\cdot\left[I_s\right]\end{aligned} \tag{32}$$

The equations (31) and (32) depict the stator and rotor voltage vector equations in the α_1 - β_1 - α_2 - β_2 - ... - z_n reference frame in postfault situation, when an open-phase fault occurs in a multiphase drive with odd number of phases. In the next section these equations are particularized for a five-phase machine.

4. Open-phase fault operation in five-phase drives

The case study presented in this chapter is a 5-phase induction machine with symmetrical and distributed windings. The n-phase mathematical model presented in the previous section must be first particularized for the 5-phase case to understand the system behavior in the faulty situation and to predict the effect of the selected control actions on the post-fault controlled system. Two different post-fault control strategies for the open-phase fault management will be presented. The first one is based on linear Proportional Resonant (PR) current controllers and the field oriented control technique. The second one is also based on the field oriented control method but combined with a Predictive Current Control (PCC) technique. Both control methods can be applied during postfault operation, and will be described in this section, along with the criteria that can be used to generate the current references in the drive during the fault. These criteria differ from those established in healthy operation, and constitute one of the bases of the postfault operation of the drive.

4.1. A. Five-phase induction machine modeling under an open phase fault

The general n-phase induction machine model introduced before can be particularized for the 5-phase case. Taking also into account that the faulty phase is 'a', which can be made without any lack of generality, from now on it can be assumed that i_{as} = 0. The stator/rotor resistance, inductance and coupling general matrices can be obtained as follows:

$$\left[R_s\right] = R_s \cdot [I_4] \tag{33}$$

$$\left[R_r\right] = R_r \cdot [I_4] \tag{34}$$

$$L_{ss} = L_{ls} \cdot \left[I_4\right] + L_m \cdot \left[\Lambda(\vartheta)\right] \tag{35}$$

$$L_{rr} = L_{lr} \cdot \left[I_4\right] + L_m \cdot \left[\Lambda(\vartheta)\right] \tag{36}$$

$$\left[\Lambda(\vartheta)_n\right] = \begin{bmatrix} 1 & \cos(\vartheta) & \cos(2\vartheta) & \cos(3\vartheta) & \cos(4\vartheta) \\ \cos(4\vartheta) & 1 & \cos(\vartheta) & \cos(2\vartheta) & \cos(3\vartheta) \\ \cos(3\vartheta) & \cos(4\vartheta) & 1 & \cos(\vartheta) & \cos(2\vartheta) \\ \cos(2\vartheta) & \cos(3\vartheta) & \cos(4\vartheta) & 1 & \cos(\vartheta) \\ \cos(\vartheta) & \cos(2\vartheta) & \cos(3\vartheta) & \cos(4\vartheta) & 1 \end{bmatrix} \tag{37}$$

$$\left[\Psi(\theta)\right] = \begin{bmatrix} \cos(\Delta_1) & \cos(\Delta_2) & \cos(\Delta_3) & \cos(\Delta_4) & \cos(\Delta_5) \\ \cos(\Delta_5) & \cos(\Delta_1) & \cos(\Delta_2) & \cos(\Delta_3) & \cos(\Delta_4) \\ \cos(\Delta_4) & \cos(\Delta_5) & \cos(\Delta_1) & \cos(\Delta_2) & \cos(\Delta_3) \\ \cos(\Delta_3) & \cos(\Delta_4) & \cos(\Delta_5) & \cos(\Delta_1) & \cos(\Delta_2) \\ \cos(\Delta_2) & \cos(\Delta_3) & \cos(\Delta_4) & \cos(\Delta_5) & \cos(\Delta_1) \end{bmatrix} \tag{38}$$

where L_{ls} and L_{lr} are the stator and rotor leakage inductances, M is the mutual inductance of the machine $M = 5L_m / 2$, and the stator and rotor inductances are defined as $L_s = M + L_{ls}$ and $L_r = M + L_{lr}$, respectively.

The 5-phase case is characterized by the transformation matrix (T_5). The stator and rotor phase variables can be mapped to a set of four independent variables divided in two orthogonal stationary planes (namely α-β and x-y subspaces) and a zero sequence component (z component). Notice that the distributed windings' characteristic of the five-phase machine determines that the torque production is only dependent of the α - β components, while x - y components only generate motor losses. This particularization for the 5-phase case can be summarized in the following equations:

$$\begin{bmatrix} i_{\alpha s} & i_{\beta s} & i_{xs} & i_{ys} & i_{zs} \end{bmatrix}^T = \begin{bmatrix} T_5 \end{bmatrix} \cdot \begin{bmatrix} i_{as} & i_{bs} & i_{cs} & i_{ds} & i_{es} \end{bmatrix}^T \tag{39}$$

$$\begin{bmatrix} v_{\alpha s} & v_{\beta s} & v_{xs} & v_{ys} & v_{zs} \end{bmatrix}^T = \begin{bmatrix} T_5 \end{bmatrix} \cdot \begin{bmatrix} v_{as} & v_{bs} & v_{cs} & v_{ds} & v_{es} \end{bmatrix}^T \tag{40}$$

$$\begin{bmatrix} T_5 \end{bmatrix} = \frac{2}{5} \cdot \begin{bmatrix} 1 & \cos(\vartheta) & \cos(2\vartheta) & \cos(3\vartheta) & \cos(4\vartheta) \\ 0 & \sin(\vartheta) & \sin(2\vartheta) & \sin(3\vartheta) & \sin(4\vartheta) \\ 1 & \cos(2\vartheta) & \cos(4\vartheta) & \cos(\vartheta) & \cos(3\vartheta) \\ 0 & \sin(2\vartheta) & \sin(4\vartheta) & \sin(\vartheta) & \sin(3\vartheta) \\ 1/2 & 1/2 & 1/2 & 1/2 & 1/2 \end{bmatrix} \tag{41}$$

While the traditional Clarke transformation matrix (T_5) is applied in healthy state, a modified matrix can be used under open-phase postfault operation in order to have a reduced-order subset of equations. If the reduced-order Clarke transformation matrix remains orthogonal as in (T_5), the asymmetries lead to noncircular α - β current components. In order to compensate for the stator/rotor impedance asymmetries appearing in postfault situation, a new nonorthogonal transformation matrix that will be named (T_{PCC}) is used here [31]. When the fault appears, it is no longer possible to define four independent variables in the system because a fixed relationship exists between α and x current components, being $i_{sx} = -i_s$.

$$\begin{bmatrix} T_{PCC} \end{bmatrix} = \frac{2}{5} \cdot \begin{bmatrix} \cos(\vartheta)-1 & \cos(2\vartheta)-1 & \cos(3\vartheta)-1 & \cos(4\vartheta)-1 \\ \sin(\vartheta) & \sin 2\vartheta & \sin(3\vartheta) & \sin(4\vartheta) \\ \sin(2\vartheta) & \sin(4\vartheta) & \sin(6\vartheta) & \sin(8\vartheta) \\ 1 & 1 & 1 & 1 \end{bmatrix} \tag{42}$$

In a similar way, the coordinate transformation can be applied to the machine voltage equations. The stator phase voltages in normal operation (V_{pre}) depend on the switching state of every leg of the power converter (S_i), being S_i=0 if the lower switch is ON and the upper switch is OFF, and S_i=1 if the opposite occurs.

$$\begin{bmatrix} v_{as} \\ v_{bs} \\ v_{cs} \\ v_{ds} \\ v_{es} \end{bmatrix} = \frac{V_{DC}}{5} \cdot \begin{bmatrix} 4 & -1 & -1 & -1 & -1 \\ -1 & 4 & -1 & -1 & -1 \\ -1 & -1 & 4 & -1 & -1 \\ -1 & -1 & -1 & 4 & -1 \\ -1 & -1 & -1 & -1 & 4 \end{bmatrix} \cdot \begin{bmatrix} S_a \\ S_b \\ S_c \\ S_d \\ v_e \end{bmatrix} = V_{pre} \tag{43}$$

During prefault operation the five-phase drive possesses 2^5=32 switching states and the sum of the healthy phase voltages is zero $\left(\sum v_{in}=0\right)$. However, if an open-phase fault appears, the available switching states are reduced to 2^4=16 and the faulty phase current is zero. Nonetheless, the faulty phase voltage is not null since there is a back-emf induced in the faulty phase, leading to an asymmetric effect in the machine modeling [31]. Taking this into account, the phase voltage of the faulty phase ('*a*') is given by:

$$v_{as} = R_s \cdot i_{as} + \frac{d}{dt}\lambda_{as} = \frac{d}{dt}\lambda_{as} = BackEmf_a \tag{44}$$

Consequently, the back-emf term can be expressed as (45), estimating the stator flux term in (44) and considering the transformation matrix (T_5) and $i_{sx} = -i_s$.

$$\begin{aligned} \lambda_{as} &= \lambda_{\alpha s} + \lambda_{xs} = L_s i_{\alpha s} + L_m i_{\alpha r} + L_{ls} i_{xs} \\ BackEmf_a &= \frac{d}{dt}\left[L_s i_{\alpha s} + L_m i_{\alpha r} - L_{ls} i_{\alpha s}\right] \end{aligned} \tag{45}$$

As a result, the stator phase voltage matrix (43) must be modified considering the faulty phase back-emf in the phase voltage equilibrium equations and must guarantee sinusoidal flux [4]. Consequently, taking into account the faulty phase voltage and the absence of current in the open-phase, the stator phase voltage matrix can be written as in (46).

$$\begin{bmatrix} v_{bs} \\ v_{cs} \\ v_{ds} \\ v_{es} \end{bmatrix} = \frac{V_{DC}}{4} \cdot \begin{bmatrix} 3 & -1 & -1 & -1 \\ -1 & 3 & -1 & -1 \\ -1 & -1 & 3 & -1 \\ -1 & -1 & -1 & 3 \end{bmatrix} \cdot \begin{bmatrix} S_b \\ S_c \\ S_d \\ S_e \end{bmatrix} - \frac{L_m \cdot \frac{di_{\alpha s}}{dt} + L_m \cdot \frac{di_{\alpha r}}{dt}}{4} \cdot \left[I_4\right] \tag{46}$$

http://dx.doi.org/10.5772/60810

where $[I_4]$ is the identity matrix of order 4 and the second term on the right hand side is the counter electromotive force (45).

4.2. B. Implemented fault-tolerant control methods

In what follows, the two implemented open-phase fault-tolerant controllers are presented. Different control criteria can be implemented depending on the overall electrical drive aim. However, only field oriented control methods have been recently used to manage postfault (open-phase type) operations. The inner current controllers of the field oriented controller have been implemented using linear or predictive control techniques. Both methods require a redefinition of the stator current references in the postfault operation to ensure minimum copper losses, a minimum derating strategy or minimum torque ripples in the drive [4, 31-35]. The maximum achievable α - β currents in the electrical drive vary depending on the selected control criteria. In general, the minimum copper loss criteria is used in applications where efficiency is of special interest and, consequently, Joules losses need to be minimized, while the minimum derating or the minimum torque ripple strategies are preferred when the faulty electrical drive must provide the maximum achievable torque or ensure smooth, vibration-free operation, respectively. From the postfault operation control performance and controller perspective, all the techniques behave in a similar way in the multiphase drive. In our case, the minimum copper loss criteria will be used for comparison purposes for the sake of simplicity.

5. Minimum copper loss criteria

The minimum copper loss (MCL) criterion focuses on reducing the drive losses. The α - β stator current references are then calculated in order to ensure proper torque/flux control while imposing a rotating circle-shaped MMF and maintaining the amplitudes of the phase currents bellow the rated values of the drive (these maximum values are established by the power semiconductors of converter and the stator windings of the electrical machine). As a result, the drive needs to be derated in such a way that the remaining healthy phases do not exceed their nominal current value (I_n) and the maximum reference currents in the α - β subspace are [36]:

$$i_{\alpha s}^{max} = 0.6813 \cdot I_n i_{\beta s}^{max} = -0.6813 \cdot I_n \tag{47}$$

The non-torque contributing y -current reference is set to zero $\left(i_y^* = 0\right)$ in order to minimize Joules losses, whereas the x -current component is not anymore an independent variable for the controller (it is inherently fixed to the α -current component after the fault occurrence). Notice that the procedure to manage the postfault operation effectively minimizes the electrical drive losses, at the expense of reducing the maximum obtainable postfault torque and generating unequal peaks of the phase currents [36, 37].

6. Predictive Current Control (PCC)

The first fault-tolerant control scheme is the PCC method, based on Finite-Control Set (FCS) Model-Based Predictive Techniques [38]. An accurate discrete system model is required in order to predict the machines' operation for every VSI state. The implemented controller is based on an outer PI-based closed-loop speed control and an inner fault-tolerant PCC method, as shown in Figure 2. During every sampling period (k), the speed and the stator currents of the machine are measured. Then, stator currents are mapped into the stationary α - β - x - y subspaces by means of the modified Clarke transformation (T_{PCC}) for postfault operation. The postfault available voltage vectors ($2^4=16$) are used afterwards to predict the stator currents evolution for the next sampling period ($k+1$). These current references are finally evaluated in a cost function (J) to determine which voltage vector produces the minimum values of J. This voltage vector is referred as the optimum switching state ($S_i^{optimum}(k+1)$) to be applied in the power converter of the electrical drive to minimize the cost function (equivalent to the control law). Notice that different cost functions can be defined in order to include different control criteria. This can be easily done by setting weighting factors in the definition of the cost function, as shown in (48) where the A, B, C and D terms multiply errors between the reference (i_{si}^*) and the predicted ($\hat{i}_{si}$) stator currents in the α - β - x - y reference frame (49)-(50).

$$J = A\left|\overline{\overline{i}}_{s\alpha}\right| + B\left|\overline{\overline{i}}_{s\beta}\right| + C\left|\overline{\overline{i}}_{s\beta}\right| + D\left|\overline{\overline{i}}_{sy}\right| \quad (48)$$

$$\overline{\overline{i}}_{s\alpha} = i_{s\alpha}^*(k+1) - \hat{i}_{s\alpha}(k+1),\ \overline{\overline{i}}_{s\beta} = i_{s\beta}^*(k+1) - \hat{i}_{s\beta}(k+1) \quad (49)$$

$$\overline{\overline{i}}_{sx} = i_{sx}^*(k+1) - \hat{i}_{sx}(k+1),\ \overline{\overline{i}}_{sy} = i_{sy}^*(k+1) - \hat{i}_{sy}(k+1) \quad (50)$$

The main control criterion in healthy operation is to maintain a desired electrical torque, while ensuring sinusoidal stator current references in phase coordinates (*a-b-c-d-e*). This objective is met under normal drive operation by setting a constant circular stator current reference vector in the α - β plane and a zero reference stator current vector in the x - y plane. In postfault operation, the x -axis stator current is inherently fixed to the α -axis stator current. Then, the α - β stator current references can be set following a circular trajectory but with a derating factor in its maximum value, while the y-axis stator current is now controlled to be null.

Successively, the α-β current components are mapped in the rotating d-q reference frame by means of the Park rotating transformation (51) and the field-oriented control position estimator (52).

$$\begin{bmatrix} i_{s\alpha} \\ i_{s\beta} \end{bmatrix} = \begin{bmatrix} \cos(\theta) & -\sin(\theta) \\ \sin(\theta) & \cos(\theta) \end{bmatrix} \cdot \begin{bmatrix} i_{sd} \\ i_{sq} \end{bmatrix} \quad (51)$$

$$\theta = \int \left(\omega_r + \frac{i_{sq}^*}{\tau_r \cdot i_{sd}^*} \right) \cdot dt \tag{52}$$

The implementation of PCC techniques for multiphase fault-tolerant drives requires the same control scheme for pre- and postfault operation, as long as the following considerations are addressed after the fault occurrence detection:

- The weight of the x - y currents has to be changed in the cost function. The x current weight will be set to zero and the y current will be the same as for the α-β currents, i.e. $C=0$ and $A=B=D$ in (48).
- The y current reference has to be changed, depending on the selected postfault control criteria.
- The limitation of the α - β currents need to be changed to (47), changing the settings of the saturated anti wind-up PI speed controllers.
- The transformation matrix that relates the switching functions with the phase voltages need to be modified as in (46) and the Clarke transformation of (42).

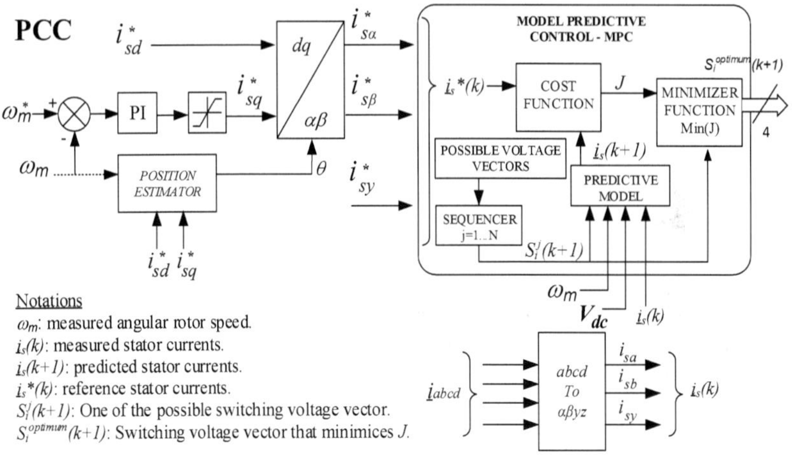

Figure 2. Postfault controller based on the PCC technique.

7. Proportional Resonant Control (PR)

The second open-phase fault-tolerant implemented control scheme is based on stator current x-y Proportional Resonant (PR) regulators, as it is explained in [37]. The control technique is detailed in Figure 3. It is based on a rotor flux controller, where the speed and flux control are implemented in a rotor-flux-oriented reference frame (d-q coordinates) using PI regulators. For simplicity, the d-current reference is set to a constant value while the q-current reference is obtained from the speed error and a PI controller. The phase currents of the machine can then be mapped in the stationary α-β-x-y planes using the classic Clarke transformation and a position estimator. In order to improve the controllers' performance, two feedforward terms e_d and e_q, which depend on the machine model (it is used a rotor-flux estimator based on the speed measurement and the d-current component [37]), are included in the control loop:

$$e_d = \sigma \cdot L_s \cdot i_{sq}^* \cdot \omega_e \tag{53}$$

$$e_q = L_s \cdot \frac{\lambda_r^*}{L_m} \cdot \omega_e \tag{54}$$

$$\frac{d}{dt}\lambda_r + \left(\frac{1}{\tau_r} - j\omega_m\right)\lambda_r = \frac{L_m}{\tau_r} i_{sd} \tag{55}$$

Traditional PI regulators are capable of following the constant x-y current references under normal operation. However, PR regulators are required under postfault operation to appropriately track the oscillating x-y reference current components [39], where the x-current component is forced to track the stator current in the α-current component and the y-current reference is set depending on the postfault control method.

The PR controller is implemented using two PI regulators in two different reference frames to track positive and negative stator current sequences [37], one rotating in the direction of the field-oriented reference frame (ω_1) and the other in the opposite direction ($-\omega_1$). These PI regulators are capable of appropriately following the current references with nonoscillating terms. When their outputs are summed, and the action of the PR control is generated, the controller is capable of effectively driving to zero the total tracking error.

The main advantage of implementing linear controllers in open-phase fault-tolerant drives is that the asymmetry in the impedance terms in the α-β plane does not affect the controller performance. Then, there is no need to consider the back-emf of the faulty phase in the voltage equilibrium equations, and the same electrical drive model can be used for control purposes in normal or abnormal operation. Nonetheless, it must be considered that due to the low bandwidth that PI regulators possess, the parameters of the utilized PI must be tuned for different operating points in pre- or postfault operating conditions, increasing the complexity of the implemented controller.

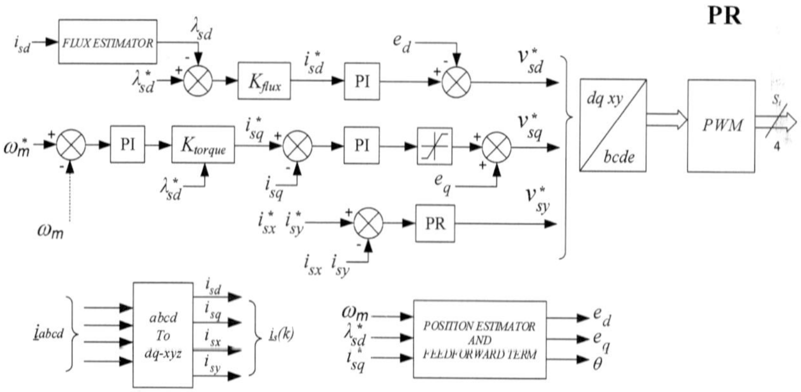

Figure 3. Postfault controller based on PR technique.

8. Experimental and simulation results

In this section, simulation and experimental results will be presented to show the behavior in healthy and faulty states of the five-phase induction machine with symmetrical and distributed windings. Simulation results were obtained using the mathematical model of the machine and a Matlab & Simulink based simulation tool described in [40], whereas the experimentation was done using an electrical drive test-bench designed and implemented in a lab. To start with, the test bench will be described.

8.1. A. Simulation Environment and Test-bench

The developed Matlab & Simulink simulation environments are shown in Figure 4. Each simulation model is composed of three main parts: the controller algorithm (PR and PCC based, respectively), the voltage source converter and the five-phase induction machine model. Depending on the selected postfault control criteria, appropriate current references must be provided to the controller. The minimum copper loss is used during postfault operation and, consequently, the y-current reference is set to zero.

The experimental test-bench is shown in Figure 5. The five-phase machine was built based on a conventional three-phase induction machine (IM) that has been rewound to obtain a symmetrical five-phase induction motor with distributed windings. This five-phase machine is driven by two conventional SEMIKRON (SKS22F) three-phase two-level voltage source inverters (VSI's), connected to an independent external DC power supply as the DC-Link. The IM is mechanically connected to a DC motor, which can provide a programmable mechanical load torque to the five-phase drive. The rotational speed is measured by means of an incremental encoder from the manufacturer Hohner with reference 10-11657-2500, coupled to the

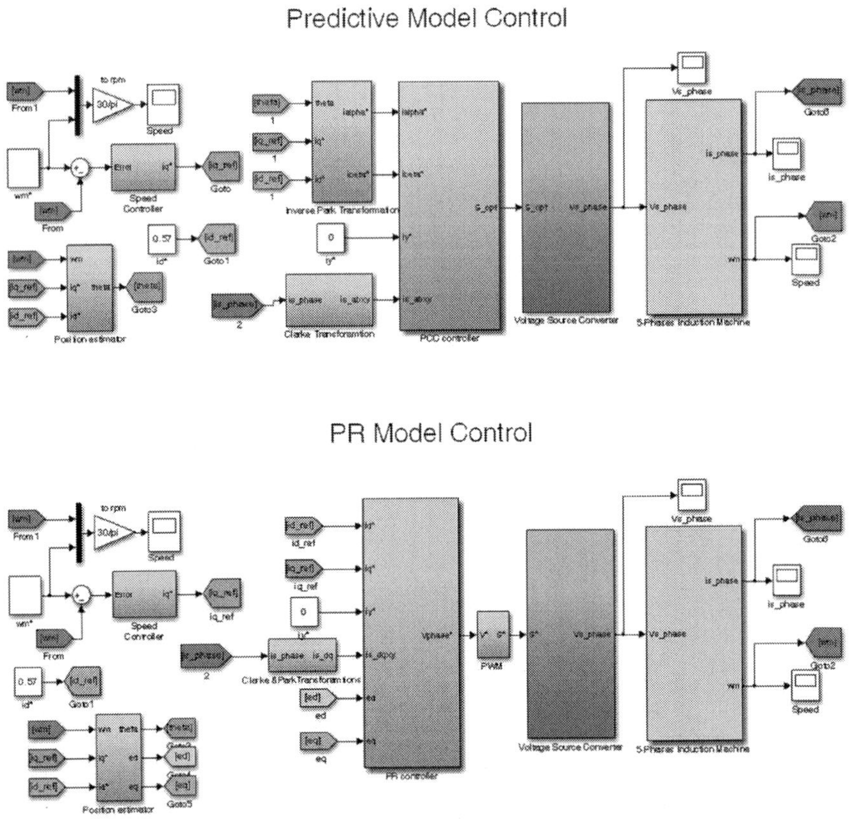

Figure 4. Developed MATLAB/Simulink model, including both PCC (upper figure) and PR controllers (lower figure).

shaft. For control purposes, four phase hall-effect current sensors are used to measure the stator phase currents. The control actions are performed using a DSP-based Electronic Control Unit (ECU) connected to a personal computer (this PC acts as a Human Interface Unit which manages the entire test bench) using a standard RS232 cable. The user of the system can program the control algorithm using the Texas Instruments proprietary software called Code Composer Studio. This software runs in the DSP and configures the ECU's internal peripherals, the communication protocol and the data acquisition system.

The PCC and PR control strategies are implemented in the DSP to analyze and compare the behavior of the real system. Regardless of the control strategy, the experimental tests that follow are performed setting a constant d-axis stator current reference of 0.57 A for constant-flux operation, while the q-axis stator current reference is obtained from the PI-based speed controller (Figure 2 and Figure 3). The VSI's DC-link voltage was set to 300 V. The fixed switching and sampling frequency for PR is set to 2.5 kHz, whereas the sampling period for PCC is set to 0.1 ms, providing around 2.5 kHz of average switching frequency. The postfault operation of the multiphase drive considers always an open-phase fault in leg '*a*'.

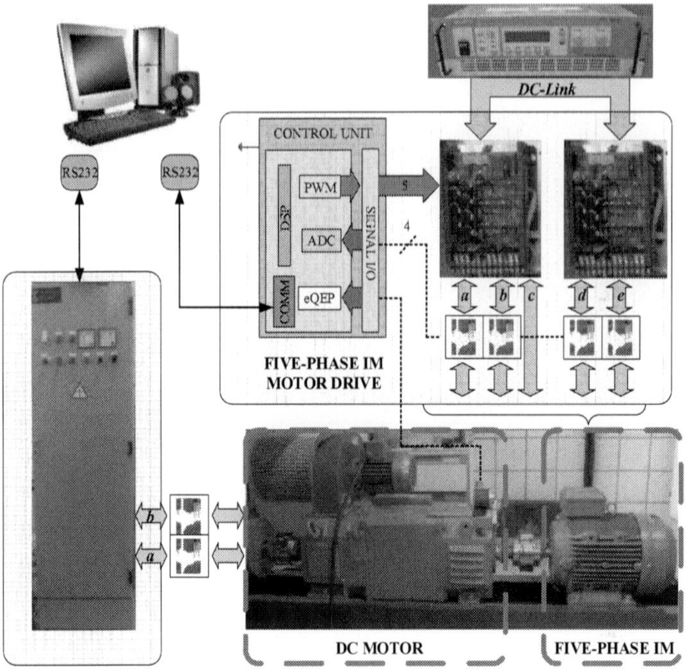

Figure 5. Experimental test bench.

8.2. B. Steady-state performance in postfault operation

The steady-state performance can be easily studied using the simulation environment. First, the postfault model of the system, and the PCC and PR controllers are implemented using aforementioned Matlab & Simulink environments. The behavior of the system is evaluated driving the motor at a reference speed of 500 rpm and applying a load torque of 56% of the nominal one (T_n). The obtained stator phase currents are shown in steady-state in Figures 6 and 7 for PCC and PR controllers, respectively. The minimum copper loss operation is applied. Then, phase currents possess unequal peak values, with phases *b-e* equal in magnitude and higher than those of phases *c-d*. It is observed that the fault-tolerant PCC produces higher current ripple than the PR control method, even though the sampling frequency of the predictive controller is set at four times the value of the PR method. This is due to the intrinsic property of FCS predictive controllers, where the switching frequency is not fixed and depends on the electrical drive operating point. The behavior of the entire system offers faster response and lower switching frequency using PCC than PR-based controllers. The α - β current vector describes a circular trajectory and the x-y terms present the same behavior using both postfault controllers, being the x-current term fixed to $-\alpha$ and the y-current term null.

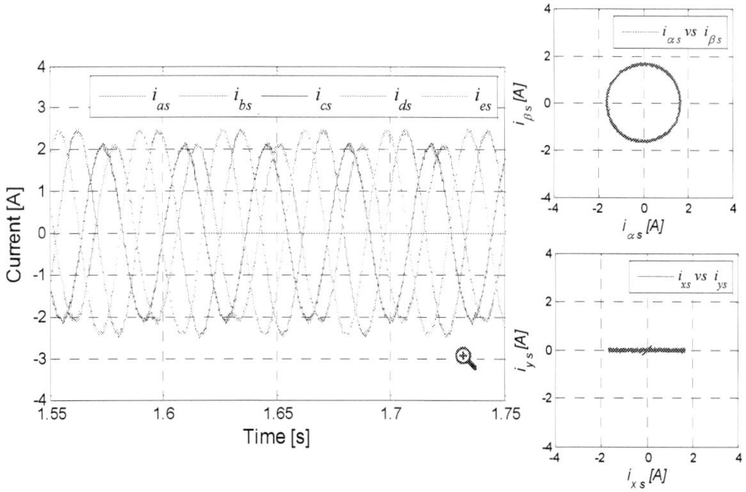

Figure 6. Phase current evolution in different subspaces using the PCC controller and the minimum copper loss criterion.

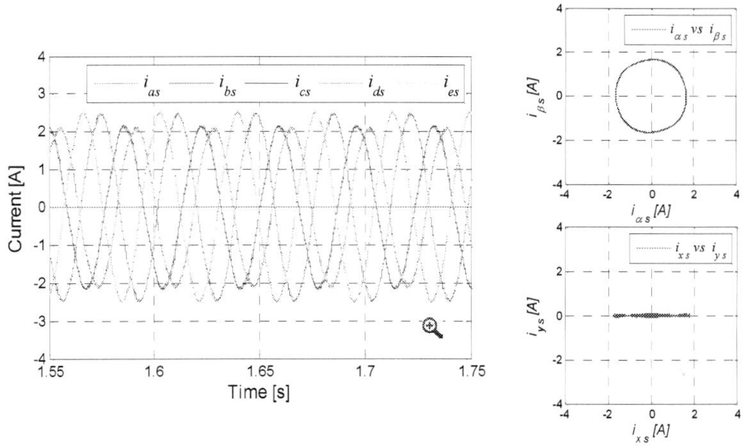

Figure 7. Phase current evolution in different subspaces using the PR controller and the minimum copper loss criterion.

http://dx.doi.org/10.5772/60810

The same steady-state test is performed experimentally using the real test bench. The experimentally obtained results in postfault situation are presented in Figure 8. Notice that simulation and experimental results agree, and the fault-tolerant system using the PCC controller provides higher current ripple than using the PR control method. Nonetheless, both controllers appropriately track the current references in all subspaces, producing a circular trajectory similar to the one obtained in healthy operation.

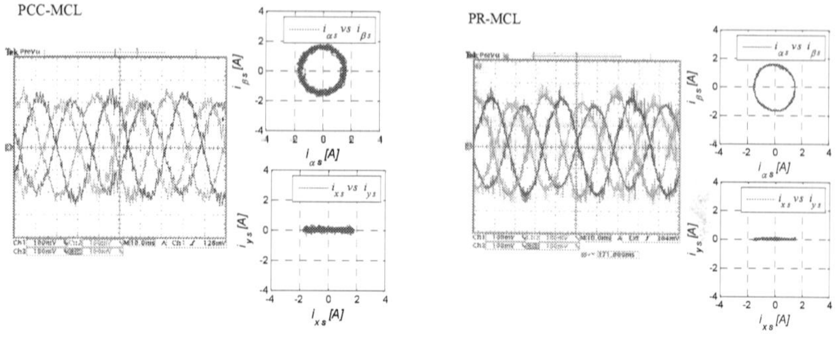

Figure 8. Stator phase current evolution in different subspaces using the PCC (left side) and the PR (right side) controller and the minimum copper loss (MCL) criterion.

8.3. C. Dynamic operation: From pre- to postfault operation

The pre- and postfault operations are now analyzed and compared. In order to provide a more realistic insight, tests have been conducted considering a fault detection delay. Consequently, a delay between the fault occurrence and the control action is observed.

The results provided in Figure 9 show the pre- to postfault transition with a fault detection delay of 40 ms between the fault occurrence in phase 'a' at $t=0.2$ s and the control software reconfiguration. The results obtained when the PCC is implemented are presented in the left column whereas results obtained with PR are shown in the right side. The speed reference is set to 500 rpm, as in previous tests, while a constant load torque of $(0.56T_n)$ is demanded.

In the case of PCC, the q -current waveform clearly indicates that the control is completely lost during the fault detection delay (Figure 9, second row), and as a result a speed drop is observed (Figure 9, first row). Notice that the β-current component is not affected during the fault detection delay because the faulty phase 'a' does not contribute to the β component (Figure 9, third row). Conversely, the α and x stator current components are both driven to zero (Figure 9, third and fourth rows), causing torque oscillations. This abnormal operation is observed during the fault detection delay due to the absence of an accurate system model for the PCC to provide an adequate control. After the fault detection delay, the control scheme is reconfigured and a more accurate system model is considered. As a result, the α-current reference is immediately tracked (Figure 9, third row), the x-current becomes sinusoidal $(i_x=-i_{\alpha s})$ and

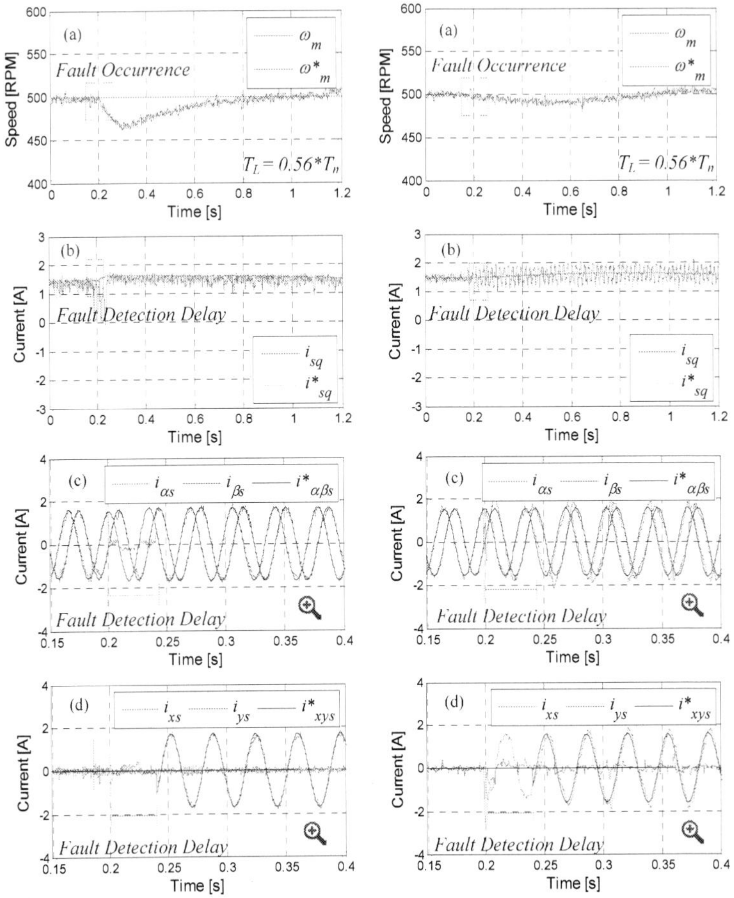

Figure 9. Transition from pre- to postfault operation considering fault detection delay. The motor is driven at 500 rpm with a constant load torque $0.56T_n$. The minimum copper loss strategy is used in postfault operation, and PCC (left plots) and PR-based (right figures) controllers are applied. The fault occurs at t = 0.2 s but it is detected 40 ms after its occurrence. The speed response and q-current component in pre- and postfault situations are shown in rows (a) and (b), while the zoomed-in postfault α-β and x-y currents are presented in rows (c) and (d), respectively.

the y-current is null according to the minimum copper loss criterion (Figure 9, fourth row). On the other hand, the q-current waveform when PR control is implemented shows a slight drop in the moment when the phase is open (Figure 9, second row), but the control action is maintained during the fault detection delay and the motor speed is only slightly affected (Figure 9, first row). Thus, the effect of the delay and the control reconfiguration is noticeably less severe in the case of PR compared to PCC. This can be explained by the fact that the prefault PR control scheme is essentially similar to the postfault scheme except for the transition from

PI to PR in the x - y current controllers. Once the postfault current references have been properly tracked, PR control can effectively provide the reference torque and regulate the speed, but an important current oscillation appears at double the fundamental frequency due to some negative sequence current that cannot be regulated by the d - q controllers. The speed is, however, not affected, so the system is regulated with minimum modifications in postfault situation.

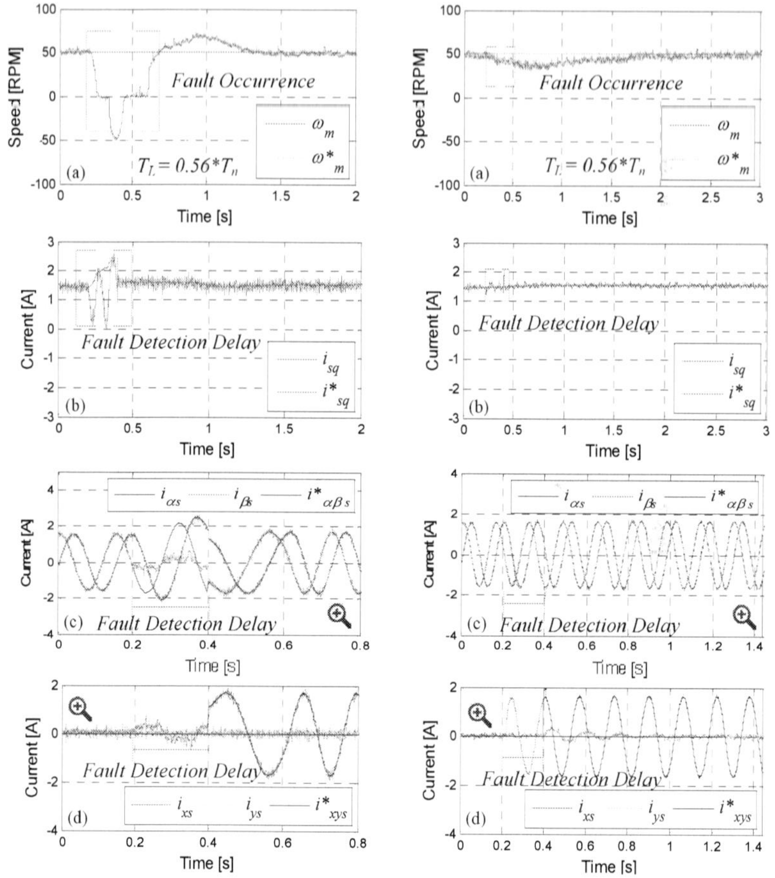

Figure 10. Transition from pre- to postfault operation considering fault detection delay. The motor is driven at 50 rpm with a constant load torque $0.56T_n$. The minimum copper loss strategy is used in postfault operation, and PCC (left plots) and PR-based (right figures) controllers are applied. The fault occurs at t = 0.2 s but it is detected 200 ms after its occurrence. The speed response and q-current component in pre- and postfault situations are shown in rows (a) and (b), while the zoomed-in postfault α-β and x-y currents are presented in rows (c) and (d), respectively.

The transition from pre- to postfault is also tested under low speed operation (Figure 10). The fault detection delay is considered equal to 200 ms and an instantaneous control reconfiguration of the system after the fault occurrence is not considered. As in previous tests, the fault occurs at $t = 0.2$s and constant load and speed references are maintained from pre- to postfault operation.

The machine is driven at 50 rpm, and a 56% of the nominal torque is applied during the test. This value of torque matches with the maximum quantity that the minimum copper loss criteria can manage in a postfault situation. As it is observed, the speed reference tracking is slightly affected after the fault occurrence with PR controller (Figure 10, first row); however, this effect is much more noticeable when the PCC controller is implemented (Figure 10, first row). Despite this considerable drop of speed, the system reaches the reference speed using the PCC controller sooner than when using the PR technique. Then, PCC controllers present again faster responses compared with the PR controllers.

9. Conclusions

This chapter focuses on the management of open-phase faults in multiphase electrical drives. First of all, the different types of faults that appear in conventional and multiphase drives are presented. The ability to continue operating in the event of a fault, which is one of the main advantages of multiphase drives compared to standard three-phase ones, is discussed next. The open-phase fault being the most common type of fault, it is next analyzed in a generic multiphase drive with an odd number of phases. The analysis is particularized for one of the most common multiphase drives, the five-phase induction machine with symmetrical and distributed windings. The considered open-circuit is located in phase 'a', but the result is general due to the spatial symmetry of stator windings. Two recently proposed controllers based on the field oriented control technique, the PR and PCC-based methods, are described as alternatives to manage the pre- and postfault operation with a minimum cost in the redesign and performance of the controllers. Both methods must share the strategy to operate in postfault operation, which must change the limits of the impressed stator currents to guarantee the safety operation of the entire system. This is the case of the minimum copper loss criterion, described in the document and applied with PCC and PR techniques to study the performance of a five-phase IM using simulation and experimental results. These results not only show the behavior of the system in steady and transient states, but also compare the ability of predictive and linear controllers to manage the fault appearance. Provided results show that speed control in postfault operation is viable using either PCC or PR control methods, with nearly similar performance. Speed response of the predictive technique is faster than using a PR controller at the expense of a higher steady-state current ripple. Additionally, PCC proves to be more affected in the transition from pre- to postfault modes of operation because the high dependence on the model accuracy provides less robustness during the unavoidable fault detection delay. Both control methods, however, ensure safe operation within the postfault current ratings, and proper postfault current reference tracking.

http://dx.doi.org/10.5772/60810

Author details

Hugo Guzman[1*], Ignacio Gonzalez[2], Federico Barrero[2] and Mario Durán[1]

*Address all correspondence to: hugguzjim@uma.es

1 Universidad de Málaga, Spain

2 Universidad de Sevilla, Spain

References

[1] A. Consoli, "Special Section on Robust Operation of Electrical Drives,"*IEEE Transactions on Power Electronics*, vol. 27, no. 2, pp. 472-478, 2012.

[2] F. Barrero, M. J. Duran, "Recent Advances in the Design, Modeling and Control of Multiphase Machines,"*IEEE Transactions on Industrial Electronics*, accepted for publication, 2015.

[3] L. Lillo, L. Empringham, P.W. Wheeler, S. Khwan-On, C. Gerada, M.N. Othman, X. Huang, "Multiphase Power Converter Drive for Fault-Tolerant Machine Development in Aerospace Applications," *IEEE Transactions on Industrial Electronics*, vol. 57, no. 2, pp. 575-583, 2010.

[4] L. Parsa, H. A. Toliyat, "Fault-Tolerant Interior-Permanent-Magnet Machines for Hybrid Electric Vehicle Applications,"*IEEE Transactions on Vehicular Technology*, vol. 56, no. 4, pp. 1546-1552, 2007.

[5] X. Huang, A. Goodman, C. Gerada, Y. Fang, Q. Lu, "Design of a Five-Phase Brushless DC Motor for a Safety Critical Aerospace Application,"*IEEE Transactions on Industrial Electronics*, vol. 59, no. 9, pp. 3532-3541, 2012.

[6] H.S. Che, E. Levi, M. Jones, M.J. Duran, W.P. Hew, N.A. Rahim,"Operation of a Six-Phase Induction Machine Using Series-Connected Machine-Side Converters,"*IEEE Transactions on Industrial Electronics*, vol. 61, no. 1, pp. 164-176, 2014.

[7] A. Stefani, "Induction Motor Diagnosis in Variable Speed Drives," PhD Thesis, Department of Electrical Engineering, University of Bologna, 2010.

[8] Pinjia Zhang, Yi Du, T.G. Habetler, Bin Lu, "A Survey of Condition Monitoring and Protection Methods for Medium-Voltage Induction Motors,"*IEEE Transactions on Industry Applications*, vol. 47, no. 1, pp. 34-46, 2011.

[9] A. H. Bonnett, C. Yung, "Increased Efficiency Versus Increased Reliability,"*IEEE Industry Applications Magazine*, vol. 14, no. 1, pp. 29-36, 2008.

[10] S. Nandi, H. A. Toliyat, and X. Li, "Condition Monitoring and Fault Diagnosis of Electrical Motors - A Review,"*IEEE Transactions on Energy Conversion*, vol. 20, pp. 719-729, 2005.

[11] A. M. da Silva, "Induction Motor Fault Diagnostic and Monitoring Methods," MSc Thesis, Marquette University, 2006.

[12] L. Zarri, M. Mengoni, Y. Gritli, A. Tani, F. Filippetti, G. Serra, D. Casadei, "Detection and Localization of Stator Resistance Dissymmetry Based on Multiple Reference Frame Controllers in Multiphase Induction Motor Drives,"*IEEE Transactions on Industrial Electronics*, vol. 60, no. 8, pp. 3506-3518, 2013.

[13] M. Barcaro, N. Bianchi, F. Magnussen, "Faulty Operations of a PM Fractional-Slot Machine with a Dual Three-Phase Winding,"*IEEE Transactions on Industrial Electronics*, vol. 58, no. 9, pp. 3825-3832, 2011.

[14] A.S. Abdel-Khalik, M.I. Masoud, S. Ahmed, A. Massoud, "Calculation of Derating Factors Based On Steady-State Unbalanced Multiphase Induction Machine Model Under Open Phase(s) and Optimal Winding Currents,"*Elsevier Electric Power System Research*, vol. 106, pp. 214-225, 2014.

[15] S. Dwari, L. Parsa, "An Optimal Control Technique for Multiphase PM Machines Under Open-Circuit Faults,"*IEEE Transactions on Industrial Electronics*, vol. 55, no. 5, pp. 1988-1995, 2008.

[16] M. E. H. Benbouzid, "A Review of Induction Motors Signature Analysis as a Medium for Faults Detection,"*IEEE Transactions on Industrial Electronics*, vol. 47, no. 5, pp. 984-993, 2000.

[17] G.F.H. Beng, X. Zhang, D.M. Vilathgamuwa, "Sensor Fault-Resilient Control of Interior Permanent-Magnet Synchronous Motor Drives,"*IEEE/ASME Transactions on Mechatronics*, vol. 20, no. 2, pp. 855-864, 2015.

[18] S.M. Bennett, R.J. Patton, S. Daley, "Rapid Prototyping of a Sensor Fault Tolerant Traction Control System,"*IEE Colloquium on Fault Diagnosis in Process Systems* (Digest No: 1997/174), 1997.

[19] H. Wang, S. Pekarek, B. Fahimi, "Multilayer Control of an Induction Motor Drive: A Strategic Step for Automotive Applications,"*IEEE Transactions on Power Electronics*, vol. 21, no. 3, pp. 676-686, 2006.

[20] D. Chakraborty, V. Verma, "Speed and Current Sensor Fault Detection and Isolation Technique for Induction Motor Drive Using Axes Transformation,"*IEEE Transactions on Industrial Electronics*, vol. 62, no. 3, pp. 1943-1954, 2015.

[21] L. Parsa, H.A. Toliyat, "Sensorless Direct Torque Control of Five-Phase Interior Permanent-Magnet Motor Drives,"*IEEE Transactions on Industry Applications*, vol. 43, no. 4, pp. 952-959, 2007.

[22] C. Hung-Chi, H. Chih-Hao, C. Da-Kai, "Position Sensorless Control for Five-Phase Permanent-Magnet Synchronous Motors,"*IEEE/ASME International Conference on Advanced Intelligent Mechatronics (AIM2014)*, 2014.

[23] A.S. Morsy, A.S. Abdel-khalik, S. Ahmed, A.M. Massoud, "Sensorless Speed Control of a Five-Phase Induction Machine Under Open-Phase Condition", *The Journal of Engineering, IET Open Access*, 2014.

[24] F. Zidani, M. E. H. Benbouzid, D. Diallo, A. Benchaib, "Active Fault-Tolerant Control of Induction Motor Drives in EV and HEV Against Sensor Failures Using a Fuzzy Decision System,"*IEEE International Conference on Electric Machines and Drives, IEMDC'03.*, vol. 2, pp. 677-683, 2003.

[25] F. Meinguet, N. Ngac-Ky, P. Sandulescu, X. Kestelyn, E. Semail, "Fault-Tolerant Operation of an Open-end Winding Five-Phase PMSM Drive with Inverter Faults," *39th Annual Conference of the IEEE Industrial Electronics Society (IECON 2013)*, 2013.

[26] A. Mohammadpour, L. Parsa, "Global Fault-Tolerant Control Technique for Multi-Phase Permanent-Magnet Machines,"*IEEE Transactions on Industry Applications*, vol. 51, no. 1, pp. 178-186, 2015.

[27] N. Bianchi, S. Bolognani, M.D. Pré, "Strategies for the Fault-Tolerant Current Control of a Five-Phase Permanent-Magnet Motor,"*IEEE Transactions on Industry Applications*, vol. 43, no. 4, pp. 960-970, 2007.

[28] L. Alberti, N. Bianchi, "Experimental Tests of Dual Three-Phase Induction Motor Under Faulty Operating Condition,"*IEEE Transactions on Industrial Electronics*, vol. 59, no. 5, pp. 2041-2048, 2012.

[29] M.O.E. Aboelhassan, T. Raminosoa, A. Goodman, L. De Lillo, C. Gerada, "Performance Evaluation of a Vector-Control Fault-Tolerant Flux-Switching Motor Drive,"*IEEE Transactions on Industrial Electronics*, vol. 60, no. 8, pp. 2997-3006, 2013.

[30] A. Mohammadpour, L. Parsa, "A Unified Fault-Tolerant Current Control Approach for Five-Phase PM Motors with Trapezoidal Back EMF under Different Stator Winding Connections,"*IEEE Transactions on Power Electronics*, vol. 28, no. 7, pp. 3517-3527, 2013.

[31] H. Guzman, M.J. Duran, F. Barrero, "A Comprehensive Fault Analysis in a Five-Phase Induction Motor Drive with an Open Phase,"*15th International Power Electronics and Motion Control Conference (EPE-PEMC 2012)*.

[32] M. E. H. Benbouzid, D. Diallo, M. Zeraoulia, "Advanced Fault-Tolerant Control of Induction-Motor Drive for EV/HEV Traction Applications: From Conventional to Modern and Intelligent Control Techniques,"*IEEE Trans. On Vehicular Tech.*, vol. 56, no. 2, 2007.

[33] A. Abdel-Khalik, A. Morsy, S. Ahmed, A. Massoud, "Effect of Stator Winding Connection on Performance of Five-phase Induction Machines,"*IEEE Trans. on Industrial Electronics*, vol. 61, no. 1, pp. 3-19, 2014.

[34] H. Guzman, M. Duran, F. Barrero, B. Bogado, S. Toral, "Speed Control of Five-Phase Induction Motors with Integrated Open-Phase Fault Operation using Model-Based Predictive Current Control Techniques,"*IEEE Transactions on Industrial Electronics*, accepted for publication, DOI: 10.1109/TIE.2013.2289882.

[35] E. Levi, "Multiphase Electric Machines for Variable-Speed Applications,"*IEEE Transactions on Industrial Electronics*, vol. 55, no. 5, pp. 1893-1909, 2008.

[36] J.R. Fu, T.A. Lipo, "Disturbance Free Operation of a Multiphase Current Regulated Motor Drive with an Opened Phase,"*IEEE Trans. on Industry Applications*, vol. 30, no. 5, pp. 1267-1274, 1994.

[37] A. Tani, M. Mengoni, L. Zarri, G. Serra, D. Casadei, "Control of Multi-Phase Induction Motors with an Odd Number of Phases Under Open-Circuit Phase Faults,"*IEEE Trans. on Power Electronics*, vol. 27, no. 2, pp. 565-577, 2012.

[38] F. Locment, E. Semail, X. Kestelyn, "Vectorial Approach-Based Control of a Seven-Phase Axial Flux Machine Designed for Fault Operation,"*IEEE Trans. on Industrial Electronics*, vol. 55, no. 10, pp. 3682-3691, 2008.

[39] R. Kianinezhad, B. Nahid-Mobarakeh, L. Baghli, F. Betin, GA. Capolino, "Modeling and Control of Six-Phase Symmetrical Induction Machine Under Fault Condition Due to Open Phase,"*IEEE Trans. on Industrial Electronics*, vol. 55, no. 5, pp. 1966-1977, 2008.

[40] H. Guzman, J.A. Riveros, M.J. Duran, F. Barrero, "Modeling of a Five-Phase Induction Motor Drive with a Faulty Phase,"*15th Internat*ional Power Electronics and Motion Control Conference (EPE-PEMC 2012).

Chapter 13

Reduced-order Observer Analysis in MBPC Techniques Applied to the Six-phase Induction Motor Drives

Raúl Gregor, Jorge Rodas, Derlis Gregor and Federico Barrero

Additional information is available at the end of the chapter

Abstract

Model-based predictive control techniques have been recently applied with success in power electronics, particularly in the fields of current control applied to AC multiphase electrical drives. In AC electrical drives control, most of state variables (i.e., rotor currents, rotor fluxes, etc.) cannot be measured, so they must be estimated. As a result of this issue, this chapter proposes a comparative study of reduced-order observers used to estimate the rotor currents in an model-based predictive current control applied to the six-phase induction motor. The proposed control techniques are evaluated using the Luenberger observer and the optimal estimator based on Kalman filter. Different operation modes are analyzed and are further compared in terms of statistical parameters of performance (i.e., covariance, standard deviation, mean square error, etc.). The effectiveness of proposed methods is verified by a set of comparative experiments obtained by using a six-phase induction motor system experimental setup.

Keywords: Model-based predictive control (MBPC), Kalman filter (KF), Luenberger observer (LO), Six-phase induction motor (SpIM)

1. Introduction

While the first variable speed drives back to the late 1960s, multiphase drives have only gained the special attention of the research community during the past few years in comparison with the traditional three-phase scheme for various applications - especially in those where high reliability and fault tolerance are needed, as cases of ship propulsion, locomotive traction, electric and hybrid electric vehicles, more-electric aircraft, and high-power industrial applications - and recently in wind energy applications [1]-[3]. Different types of multiphase machines have been recently developed mainly for high-power applications where the increase of the number of phases enables reduction of power per phase, which leads to a reduction of the power per inverter leg. Often the multiphase machines can be classified according to the phase numbers in 5-phase [4]-[8], 6-phase [9]-[11], 9-phase [13]-[14], 12-phase [15]-[17], and 18-phase [18] and by the spatial distribution of the phases within the stator winding symmetrically or asymmetrically. The six-phase induction motor (SpIM) fed by two sets of voltage source inverters was investigated since 1993. Because of the configuration of induction motor having two sets of balanced windings, with phase shift of 30 electrical degrees, six harmonic torque pulsations produced by two sets of windings, respectively, are antiphase and therefore can be completely eliminated. Nowadays, numerous control strategies such as direct torque control (DTC), model-based predictive control (MBPC), and vector control have been developed for SpIM. The DTC technique has the advantages of low machine parameter dependence and fast dynamic torque response. Moreover, the main advantage of the MBPC technique is it focuses on flexibility to define different control criteria, changing only a cost function, a reason why this control technique has been recently applied to the SpIM [19]. MBPC is a control theory developed at the end of the 1970s but has been recently introduced as a viable alternative in power converters and drives. Various control schemes based on MBPC, including current, flux and torque, speed, and sensorless speed control, have been recently reported. Developed schemes have demonstrated good performance in the current and torque control of conventional drives, at the expense of a high computational burden. It is a more flexible control scheme than DTC, and it also provides faster torque response than the field-oriented control (FOC). The interest in predictive control approach and multiphase drives has grown during the last few years, when the development of modern microelectronics devices has removed the computational barriers in their implementation. However, predictive control techniques have been only proved as a viable alternative to conventional controllers in the current regulation of the multiphase power converter. Predictive torque control (PTC), as a variation of the predictive current control methods, has been recently analyzed as an alternative to classic DTC at a theoretical level [20].

In this work, the predictive model of the SpIM is obtained from the vector space decomposition (VSD) approach using the state-space representation method where the two state variables are the stator and rotor currents. As the rotor currents are not measurable parameters, these must be estimated. This chapter hence focuses in the efficiency analysis of the MBPC techniques using the Luenberger Observer (LO) and the optimal estimator based on Kalman Filter (KF). The chapter provides a background material about model-based predictive current control applied to SpIM and includes experimental results by using an experimental setup based on a digital signal controller (DSC). Finally, the main results are discussed in the conclusion section.

2. The SpIM mathematical model

The asymmetrical SpIM with two sets of three-phase stator windings spatially shifted by 30 electrical degrees and isolated neutral points as seen on Figure 1 (a) is one of the most widely discussed topologies. The asymmetrical SpIM is a continuous system which can be described by a set of differential equations. The model can be simplified by using the VSD theory introduced introduced in [21], [22], [26] which enables to transform the

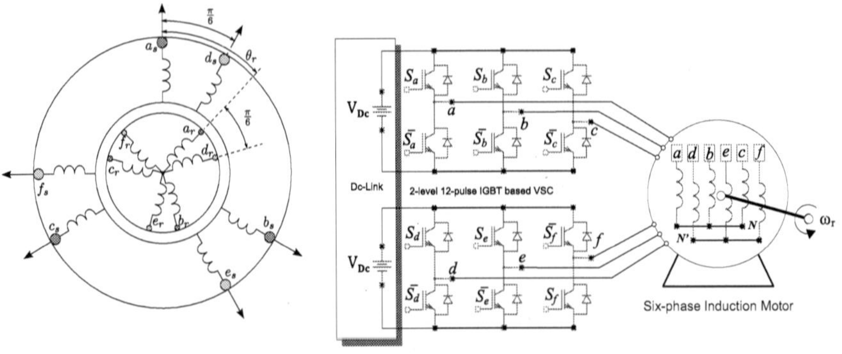

(a) Cross section of an SpIM (b) General scheme of an asymmetrical SpIM drive

Figure 1. Asymmetrical SpIM feed topology and winding configuration

original six-dimensional space of the motor model into three two-dimensional orthogonal subspaces in stationary reference frames $(\alpha - \beta)$, $(x - y)$, and $(z_1 - z_2)$ by means of a 6×6 transformation matrix using an amplitude-invariant criterion. This matrix, namely, $\mathbf{T}$, is defined as:

$$\mathbf{T} = \frac{1}{3}\begin{bmatrix} 1 & \frac{\sqrt{3}}{2} & -\frac{1}{2} & -\frac{\sqrt{3}}{2} & -\frac{1}{2} & 0 \\ 0 & \frac{1}{2} & \frac{\sqrt{3}}{2} & \frac{1}{2} & -\frac{\sqrt{3}}{2} & -1 \\ 1 & -\frac{\sqrt{3}}{2} & -\frac{1}{2} & \frac{\sqrt{3}}{2} & -\frac{1}{2} & 0 \\ 0 & \frac{1}{2} & -\frac{\sqrt{3}}{2} & \frac{1}{2} & \frac{\sqrt{3}}{2} & -1 \\ 1 & 0 & 1 & 0 & 1 & 0 \\ 0 & 1 & 0 & 1 & 0 & 1 \end{bmatrix}. \tag{1}$$

It is worth remarking that, according to the VSD approach, the electromechanical energy conversion variables are mapped in the $(\alpha - \beta)$ subspace, meanwhile the current components in the $(x - y)$ subspace represent supply harmonics of order $6n \pm 1$ $(n = 1, 3, 5, ...)$ and only produce losses. The voltage vectors in the $(z_1 - z_2)$ subspace are zero due to the isolated neutral points configuration [23]. Moreover, the SpIM is supplied by a 2-level 12-pulse IGBT based VSC and a Dc-Link ($\mathbf{V}_{\mathbf{Dc}}$), as shown in Figure 1 (b).

The VSC has a discrete nature with a total number of $2^6 = 64$ different switching state vectors defined by six switching functions corresponding to the six inverter legs $(S_a, S_d, S_b, S_e, S_c, S_f)$, where $S_{a-f} \in \{0, 1\}$. The different switching state vectors and the $\mathbf{V}_{\mathbf{Dc}}$ voltage define the phase voltages which can in turn be mapped to the $(\alpha - \beta) - (x - y)$ space according to the VSD approach [24]. To represent the stationary reference frame $(\alpha - \beta)$ in the dynamic reference frame $(d - q)$, a rotation transformation can be used. This transformation matrix, namely, $\mathbf{T}_{dq}$ is represented as:

$$\mathbf{T}_{dq} = \begin{bmatrix} cos\,(\theta_r) & sin\,(\theta_r) \\ -sin\,(\theta_r) & cos\,(\theta_r) \end{bmatrix}, \tag{2}$$

where θ_r is the rotor angular position referred to the stator as shown in Figure 1 (a).

From the VSD approach, the following conclusions should be emphasized:

1. The electromechanical energy conversion variables are mapped to the $(\alpha - \beta)$ subspace. Therefore, the fundamental supply component and the supply harmonics of order $12n \pm 1$ $(n = 1, 2, 3, ...)$ are represented in this subspace.
2. The current components in the $(x - y)$ subspace do not contribute to the air-gap flux and are limited only by the stator resistance and stator leakage inductance. These components represent the supply harmonics of the order $6n \pm 1$ $(n = 1, 3, 5, ...)$ and only produce losses, so consequently they should be controlled to be as small as possible.
3. The voltage vectors in the $(z_1 - z_2)$ are zero due to the separated neutral configuration of the machine.

The VSI with isolated neutrals is depicted in Figure 1 (b), being the gating signal represented by $[S_a, ..., S_f]$ and their complementary values by $[\overline{S}_a, ..., \overline{S}_f]$, where $\mathbf{S}_i \in \{0, 1\}$. The discrete nature of the VSI defines the phase voltages which can be mapped in the $(\alpha - \beta) - (x - y)$ according to the VSD approach. Figure 2 shows the active vectors in the $(\alpha - \beta)$ and $(x - y)$ subspaces, where each switching vector state is identified using the switching function by two octal numbers corresponding to the binary numbers $[S_a S_b S_c]$ and $[S_d S_e S_f]$, respectively. Stator voltages are related to the input control signals through the VSI model. An ideal inverter converts gating signals into stator voltages that can be projected to $(\alpha - \beta)$ and $(x - y)$ subspaces and gathered in a row vector $\mathbf{U}_{\alpha\beta xys}$ computed as

$$\mathbf{U}_{\alpha\beta xys} = \begin{bmatrix} u_{\alpha s}, & u_{\beta s}, & u_{xs}, & u_{ys}, & 0, & 0 \end{bmatrix}^T = Vdc\,\mathbf{T}\,\mathbf{M}, \tag{3}$$

where $(^T)$ indicates the transposed matrix and $\mathbf{M}$ represents the model of the VSI that can be expressed as function to the switching vectors as follows:

$$\mathbf{M} = \frac{1}{3} \begin{bmatrix} 2 & 0 & -1 & 0 & -1 & 0 \\ 0 & 2 & 0 & -1 & 0 & -1 \\ -1 & 0 & 2 & 0 & -1 & 0 \\ 0 & -1 & 0 & 2 & 0 & -1 \\ -1 & 0 & -1 & 0 & 2 & 0 \\ 0 & -1 & 0 & -1 & 0 & 2 \end{bmatrix} \mathbf{S}^T. \tag{4}$$

As shown in Figure 2, the 64 possible voltage vectors lead to only 49 different vectors in the $(\alpha - \beta)$ and $(x - y)$ subspaces. Applying the transformation matrix, the mathematical model

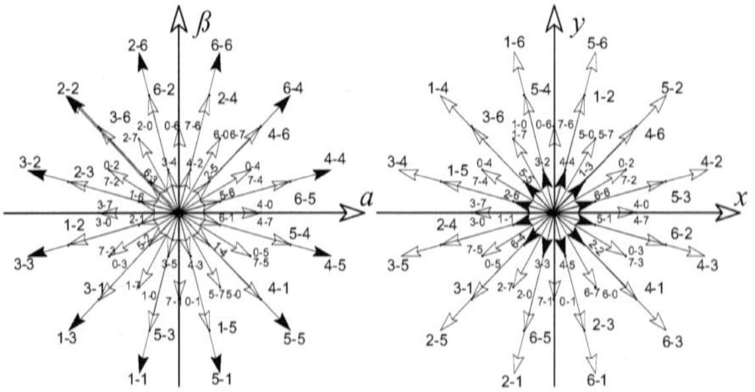

Figure 2. Voltage space vectors and switching states in the $(\alpha - \beta)$ and $(x - y)$ subspaces for a six-phase asymmetrical VSI

of the SpIM can be written using the state-space (SS) representation as follows:

$$[\mathbf{u}]_{\alpha\beta} = [\mathbf{G}] \frac{d}{dt} [\mathbf{x}]_{\alpha\beta} + [\mathbf{F}] [\mathbf{x}]_{\alpha\beta}, \tag{5}$$

where $[\mathbf{u}]_{\alpha\beta} = \left[u_{\alpha s} \; u_{\beta s} \; 0 \; 0 \right]^T$ represents the input vector, $[\mathbf{x}]_{\alpha\beta} = \left[i_{\alpha s} \; i_{\beta s} \; i_{\alpha r} \; i_{\beta r} \right]^T$ denotes the state vector, and $[\mathbf{F}]$ and $[\mathbf{G}]$ are matrices that define the dynamics of the drive that for the particular case of the SpIM are represented as follows:

$$[\mathbf{F}] = \begin{bmatrix} R_s & 0 & 0 & 0 \\ 0 & R_s & 0 & 0 \\ 0 & \omega_r L_m & R_r & \omega_r L_r \\ -\omega_r L_m & 0 & -\omega_r L_r & R_r \end{bmatrix}, \tag{6}$$

$$[\mathbf{G}] = \begin{bmatrix} L_s & 0 & L_m & 0 \\ 0 & L_s & 0 & L_m \\ L_m & 0 & L_r & 0 \\ 0 & L_m & 0 & L_r \end{bmatrix}, \tag{7}$$

where R_s and R_r are the stator and rotor resistance, ω_r is the rotor angular speed, and $L_s = L_{ls} + 3\,L_m$, $L_r = L_{lr} + 3\,L_m$, and L_m are the stator, rotor, and magnetizing inductances, respectively. For a machine with P pairs of poles, the mechanical part of the drive is given by the following equations:

$$T_e = 3\frac{P}{2}\left(\psi_{\beta r} i_{\alpha r} - \psi_{\alpha r} i_{\beta r}\right), \tag{8}$$

$$J_i \frac{d}{dt}\omega_r + B_i \omega_r = \frac{P}{2}\left(T_e - T_L\right), \tag{9}$$

where T_L denotes the load torque, J_i the inertia, $\psi_{\alpha\beta r}$ the rotor flux, and B_i the friction coefficient.

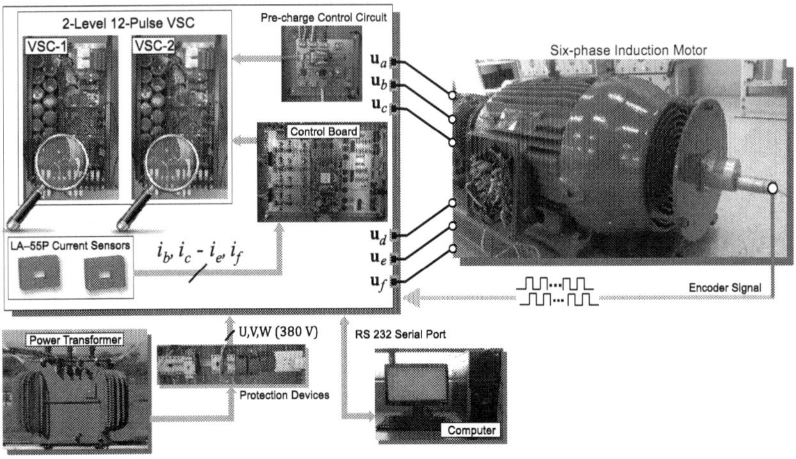

Figure 3. Scheme of the experimental setup

The equations in $(x-y)$ subspace do not link to the rotor side and consequently do not contribute to the air-gap flux; however, they are an important source of Joule losses. Using the SS representation, these equations can be written as:

$$[\mathbf{u}]_{xy} = \begin{bmatrix} L_{ls} & 0 \\ 0 & L_{ls} \end{bmatrix} \frac{d}{dt} [\mathbf{i}]_{xy} + \begin{bmatrix} R_s & 0 \\ 0 & R_s \end{bmatrix} [\mathbf{i}]_{xy}, \tag{10}$$

where L_{ls} represents the stator leakage inductance.

3. SpIM parameter identification

A commercial three-phase induction machine with three pairs of poles, 72 slots, and 15 kW of rated power has been rewound to obtain an asymmetrical six-phase winding (configured with two isolated neutral points) with the same pairs of poles and power with the original three-phase machine. Conventional test (blocked rotor and no-load tests) procedures have been applied to determine experimentally the electrical and mechanical parameters of the SpIM. The obtained values are shown in Table 1.

Two three-phase VSC modules manufactured by Semikron SKS 35F B6U+E1CIF+B6CI21V series are used to generate the six-phase stator voltages and to obtain the experimental results. A hardware timer based on the LM555 device operating in monostable mode is implemented to control the internal pre-charge circuit of both the SKS 35F modules. The Dc-Link voltage is V_{Dc} = 585 V. The implementation of the control system is based on the DSC TMS320LF28335 manufactured by Texas Instruments and the MSK28335 board from Technosoft which has 12 pulse-width modulation (PWM) outputs. The PWM is configured with a 10 kHz of switching frequency. Stator currents are measured by using Hall effect sensors (LA-55P from LEM). The analog-to-digital (A/D) converter peripherals of the MSK28335 board with 16 parallel channels are used to capture all the measured signals. On

Parameter	Six-phase induction motor		
	Symbol	Value	Unit
Stator resistance	R_s	0.62	Ω
Rotor resistance	R_r	0.63	Ω
Stator inductance	L_s	0.2062	H
Rotor inductance	L_r	0.2033	H
Mutual inductance	L_m	0.0666	H
System inertia	J_i	0.27	kg·m^2
Viscous friction coefficient	B_i	0.012	kg·m^2/s
Nominal frequency	f_a	50	Hz
Load torque	T_L	0	N·m
Pair of poles	P	3	–

Table 1. Electrical and mechanical parameters

the other hand, the mechanical speed is measured by employing a Hengstler RI 58-O digital incremental encoder with a resolution of 10,000 pulses per revolution and the enhanced quadrature encoder pulse (eQEP) peripheral of the DSC. To preserve the system integrity, input, and output, digital outputs of the control board are galvanically isolated by means of a Texas Instruments ISO7230CDW isolator. Figure 3 shows a picture of the different parts of the experimental test bench. In order to validate the electrical and mechanical parameters, a PLL software implementation is used to calculate the stator current angle (θ). Finally, the angle is used to calculate the stator current in dynamic reference frame $(i_{ds} - i_{qs})$ using the transformation matrix shown in Eq. (2). Statistical parameters of performance (taking as reference the experimental evolution of stator currents in dynamic reference frame) are quantifiable for two different implementations: the SpIM model based on MatLab/Simulink simulation environment and a real SpIM using the experimental setup.

3.1. Digital PLL implementation

Figure 4 (a) shows that the dynamic performance of the proposed PLL is highly influenced by the compensator G(z). Considering that the reference signal is the stator current in d axis and since the loop gain includes an integral term, θ must track the constant component of the reference signal with zero steady-state error. However, to ensure zero steady-state error, the loop gain must include at least two integrators. Therefore, G(z) must include at least one integral term, that is, one pole at $z = 1$. The other poles and zeros of G(z) are determined mainly by the closed-loop bandwidth requirements of the PLL and stability indices such as phase margin and gain margin, according with the procedure described in [25]. Due to the fact that G(z) is controllable, the transfer function can be expressed into controllable canonical form as follows:

$$\mathbf{x}_{(k+1|k)} = [\mathbf{F}]_{5\times5} \left[\mathbf{x}_{(k|k)}\right]_{5\times1} + [\mathbf{D}]_{5\times1} \left[e_{(k|k)}\right]_{5\times1}', \tag{11}$$

$$\omega^*_{(k|k)} = [\mathbf{C}]_{1\times5} \left[\mathbf{x}_{(k|k)}\right]_{5\times1}', \tag{12}$$

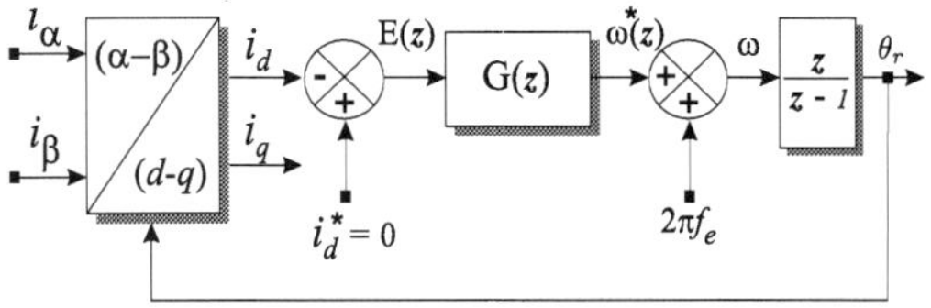

(a) Block diagram of a PLL with special design of the compensator

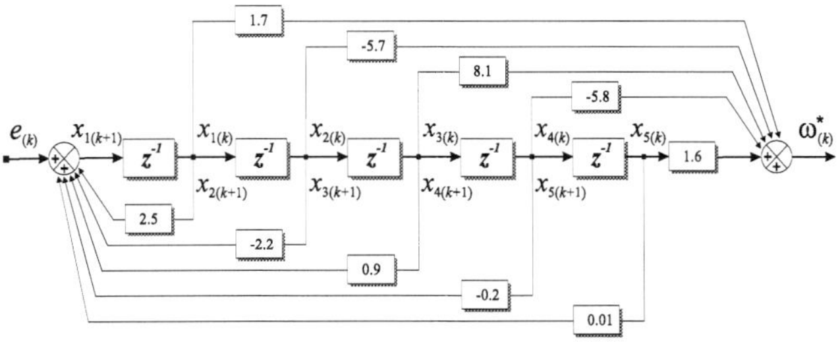

(b) Representation of G(z) transfer function on controllable canonical form

Figure 4. PLL software implementation block diagram

where the matrix $[\mathbf{F}]_{5\times5}$ and the vectors $[\mathbf{D}]_{5\times1}$ and $[\mathbf{C}]_{1\times5}$ define the dynamics of the PLL compensator [G(z)], which for the set of state variables shown in Figure 4 (b) are as follows:

$$[\mathbf{F}]_{5\times5} = \begin{bmatrix} 2.5 & -2.2 & 0.9 & -0.2 & 0.01 \\ 1 & 0 & 0 & 0 & 0 \\ 0 & 1 & 0 & 0 & 0 \\ 0 & 0 & 1 & 0 & 0 \\ 0 & 0 & 0 & 1 & 0 \end{bmatrix}, \tag{13}$$

$$[\mathbf{D}]_{5\times1} = \begin{bmatrix} 1\,0\,0\,0\,0 \end{bmatrix}^T, \tag{14}$$

$$[\mathbf{C}]_{1\times5} = \begin{bmatrix} 1.7 & -5.7 & 8.1 & -5.8 & 1.6 \end{bmatrix}. \tag{15}$$

This state-space realization is called controllable canonical form because the resulting model is guaranteed to be controllable. Since the control enters a chain of integrators, it has the ability to move every state as shown in Figure 4 (b).

The proposed PLL architecture has been implemented by using the TMS320LF28335 DSC, considering floating-point arithmetic and 10 kHz sampling frequency. The PLL algorithm is executed as an interrupt service routine (ISR), which is triggered by one of the general-purpose timer circuits available on chip. The same timer also triggers the acquisition of input signals, simultaneously with the sampling interrupt. As the on-chip A/D converters have a fast conversion rate (approximately 106-ns conversion time), input data are made available at the beginning of the ISR with negligible time delay. The current components in

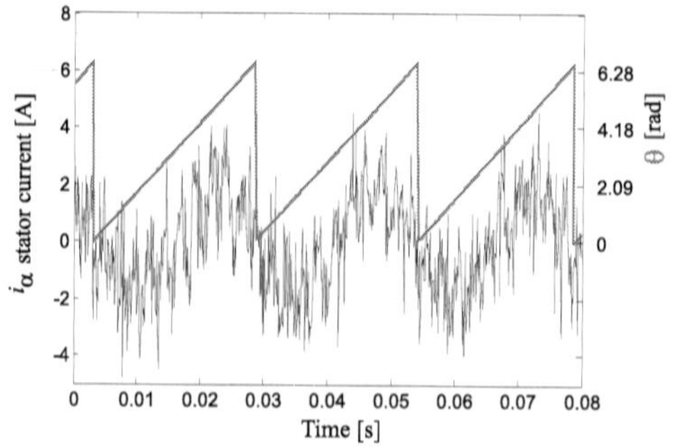

Figure 5. Stator current angle evolution obtained experimentally by using the proposed PLL with special design of the compensator

stationary references frame (α-β) are calculated at each sampling time from the measured phase stator currents ($i_{bs}, i_{cs}, i_{ds}, i_{fs}$) by using Eq. (1), immediately after performing A/D conversion.

Figure 5 shows the stator current angle evolution obtained experimentally by using the proposed PLL architecture, when the SpIM is fed with electrical frequency voltages (f_e) of 40 Hz. It can be seen that the angle evolves from 0 to 2π during a single period of the stator current wave. It is also possible to observe that the result is satisfactory even when the stator currents in stationary reference frame are distorted due to electrical noise.

3.2. SpIM parameter validation

SpIM electrical and mechanical parameters have been analyzed and validated using the experimental setup as well as a SpIM MatLab/Simulink model where a fourth-order Runge-Kutta numerical integration method has been applied to compute the evolution of the state variables step by step in the time domain. Table 1 shows the electrical and mechanical parameters of the asymmetrical SpIM which have been considered during the simulation. The validation of the measured parameters has been evaluated under no-load conditions.

Figure 6 shows the stator current start-up characteristics when a VSC supplied with 585 V of Dc-Link is considered and when a sinusoidal modulation index of 0.275 and 40 Hz of frequency is applied. Figure 6 (a) shows the i_β current evolution of the SpIM provided by the MatLab/Simulink model. In this case, the VSC, the PWM scheme, and the AC motor are simulated within the MatLab/Simulink model. The stator current evolution is compared with the i_d current obtained using the experimental setup in order to verify the analogy between the MatLab/Simulink model simulation results and the experimental results especially with respect to the time constants associated with the SpIM (start-up current, speed, steady-state current, etc.). It can be seen that the time constant converges to the value obtained experimentally both in transient and steady-state conditions where it is possible to quantify a steady-state current of approximately 2 A. Moreover, Figure 6 (b) shows the results obtained experimentally. These results have been compared with the i_d current obtained experimentally. It can be seen that the start-up current evolution converges to a common value for the MatLab/Simulink-based simulations as well as for the experimental

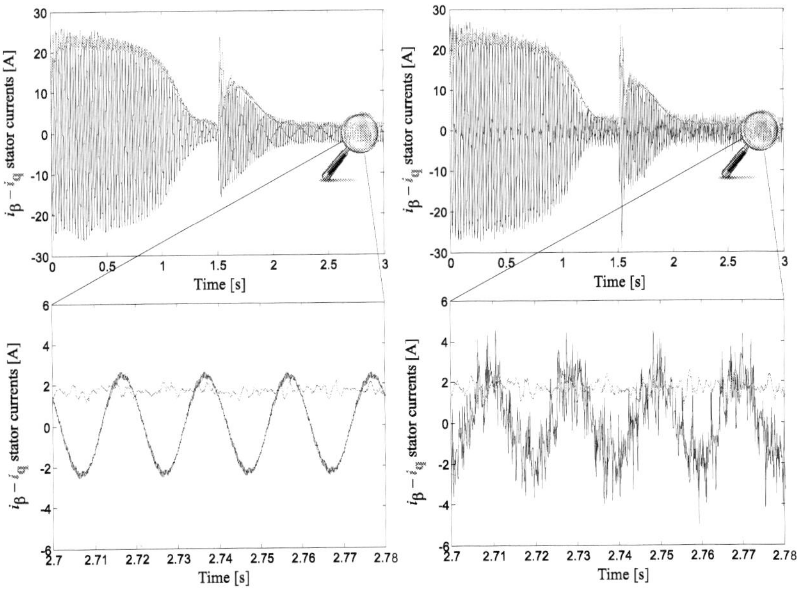

(a) Stator current (i_β) obtained by using the MatLab/Simulink simulation environment

(b) Stator current (i_β) current obtained by experimental tests

Figure 6. Stator current start-up characteristics

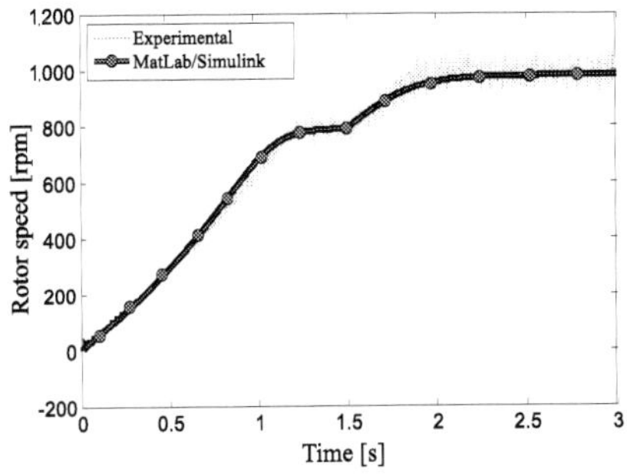

Figure 7. Transient rotor speed evolution

setup, with a start-up transient of approximately 1.15 s. After 1.5 s, the reference frequency is changed from 40 to 50 Hz, while the modulation index is kept constant at 0.275.

Statistical performance parameters such as the covariance, the standard deviation (SD), and the mean square error (MSE) are used in order to evaluate the accuracy of the parameters, taking as reference the results obtained through simulations, as well as those obtained by means of experimental tests. The envelope of the fundamental frequency component of the

PARAMETER	Statistical parameters of performance COVARIANCE	SD	MSE
MatLab/Simulink	1.60×10^{-26}	1.025	1.064
Experimental	5.23×10^{-26}	1.004	1.030

Table 2. Performance analysis.

stator currents in stationary reference frame can be calculated using the Hilbert transform (HT) method. This envelope detection method involves creating the analytic signal of the stator current using the HT. An analytic signal is a complex signal, where the real part ($i_{\alpha s}$) is considered the original signal and the imaginary part ($ji_{\beta s}$) is the HT of the original signal. A discrete time analytic signal ($\hbar(k)$) can be defined as follows:

$$\hbar(k) = i_{\alpha s}(k) + ji_{\beta s}(k), \tag{16}$$

while the envelope of the signal can be determined by computing the modulus of the analytic signal from the following equation:

$$|\hbar(k)| = \sqrt{\left[\sum_{i=0}^{n} i_{\alpha s}(k)\right]^2 + \left[\sum_{i=0}^{n} i_{\beta s}(k)\right]^2}. \tag{17}$$

Using the above equation, it is possible to determine the envelope evolution of the stator current, which is used to evaluate those aforementioned statistical performance parameters. This analysis enables to determine the degree of dispersion of the envelope (of the stator current) with respect to the value obtained experimentally through the PLL software implementation (which is shown in red color in Figure 6). The statistic relationship between the curves (i_{qs} and stator current envelope) and the MSE has been analyzed under steady-state conditions. Table 2 details the obtained results for the two different SpIM implementation methods considered in Figure 6. Notice that the obtained performance results are similar for both cases (MatLab/Simulink model and experimental). Moreover, Figure 7 shows the rotor speed evolution for the two cases analyzed before. It can be seen that the results provided by the MatLab/Simulink model in steady state converge to the values obtained experimentally using a motor having three pairs of poles and 50 Hz of nominal frequency (close to 1,000 rpm).

Further analysis has been done to validate the parameters under different test conditions. For example, a change in the modulation index from 0.275 to 0.481 was considered at t = 1.5 s, while a constant voltage frequency of 40 Hz was considered. Figure 8 (a) shows the trajectory of the $i_{\alpha s}$ *vs.* $i_{\beta s}$ as well as ($i_{ds} - i_{qs}$) current evolution considering at least four current periods in steady-state operation, where it is also possible to observe the effect of the change of the modulation index in the reference voltages. Figure 8 (a) shows the results obtained using the MatLab/Simulink model, and Figure 8 (b) shows the experimental results. As in the previous case, it can be seen that the simulated current converges to values equivalent to those obtained experimentally and exhibiting similar dynamic behavior. Finally, Figure 8 (c)

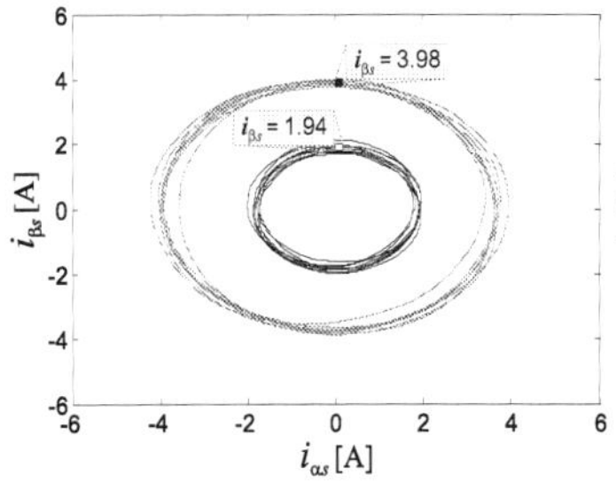

(a) $i_{\alpha s}$ *vs.* $i_{\beta s}$ current obtained by using the MatLab/Simulink simulation environment

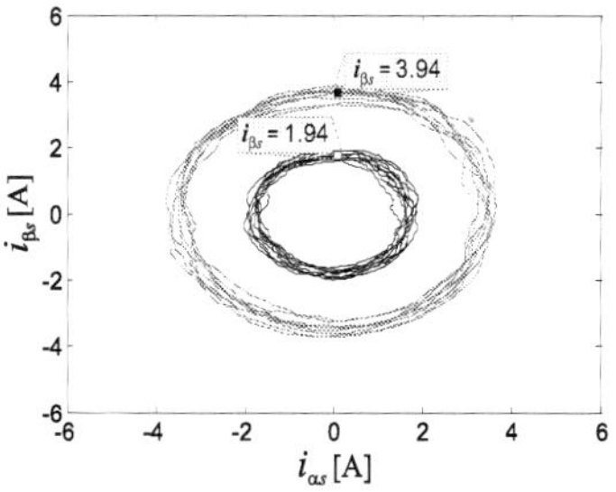

(b) $i_{\alpha s}$ *vs.* $i_{\beta s}$ current obtained by experimental tests

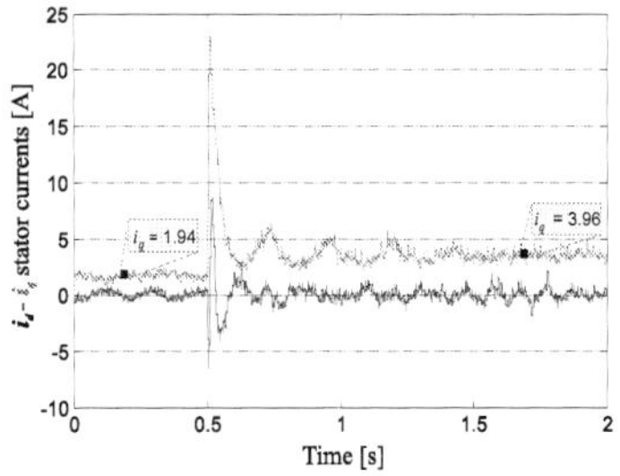

(c) Stator current evolution in dynamic reference frame

Figure 8. Dynamic reference frame characteristics

shows the stator current evolution in the dynamic reference frame $(d-q)$ obtained by means of Eq. (2) using the angle values calculated by the PLL software implementation. It can be seen that the steady-state current values converge to those values shown in **Figure 8 (a)** and Figure 8 (b) before and after applying the change in the modulation index from 0.275 to 0.481, being these values close to 2 and 4 A, respectively. These results validate the SpIM electrical and mechanical parameters shown in Table 1.

4. Predictive model

Assuming the mathematical model expressed by Eq. (5) and using the state variables defined by the vector $[\mathbf{x}]_{\alpha\beta}$, the derivative of states can be defined as follows:

$$\begin{aligned}
\dot{x_1} &= c_3\left(R_r x_3 + \omega_r x_4 L_r + \omega_r x_2 L_m\right) + c_2\left(u_{\alpha s} - R_s x_1\right), \\
\dot{x_2} &= c_3\left(R_r x_4 - \omega_r x_3 L_r - \omega_r x_1 L_m\right) + c_2\left(u_{\beta s} - R_s x_2\right), \\
\dot{x_3} &= c_4\left(-R_r x_3 - \omega_r x_4 L_r - \omega_r x_2 L_m\right) + c_3\left(-u_{\alpha s} + R_s x_1\right), \\
\dot{x_4} &= c_4\left(-R_r x_4 + \omega_r x_3 L_r + \omega_r x_1 L_m\right) + c_3\left(-u_{\beta s} + R_s x_2\right),
\end{aligned} \tag{18}$$

where c_i $(i = 1, 2, 3, 4)$ are constants defined as:

$$c_1 = L_s L_r - L_m^2,\ c_2 = \frac{L_r}{c_1},\ c_3 = \frac{L_m}{c_1},\ c_4 = \frac{L_s}{c_1}. \tag{19}$$

This set of differential equations can be represented in the state-space form as follows:

$$\begin{aligned} \dot{\mathbf{X}}(t) &= f\left(\mathbf{X}(t), \mathbf{U}(t)\right), \\ \mathbf{Y}(t) &= \mathbf{C}\mathbf{X}(t), \end{aligned} \tag{20}$$

with state vector $\mathbf{X}(t) = [x_1, x_2, x_3, x_4]^T$, input vector $\mathbf{U}(t) = \left[u_{\alpha s}, u_{\beta s}\right]$, and output vector $\mathbf{Y}(t) = [x_1, x_2]^T$. The components of the vectorial function f and matrix $\mathbf{C}$ are obtained in a straightforward manner from Eq. (18) and the definitions of state and output vector.

The continuous time model represented by Eq. (20) can be discretized in order to be used for the predictive controller using the forward difference approximation method of the first derived, also known as the forward Euler method. Thus, a prediction of the future next-sample state $\hat{\mathbf{X}}(k+1|k)$ is expressed as:

$$\hat{\mathbf{X}}(k+1|k) = \mathbf{X}(k) + T_m f\left(\mathbf{X}(k), \mathbf{U}(k)\right), \tag{21}$$

where (k) is the current sample and T_m the sampling time. In Eq. (21), currents and voltages of the stator and the mechanical speed are measurable variables; however, the rotor currents cannot be measured directly. This difficulty can be overcome by means of estimating the rotor current using the reduced-order estimator concept. Figure 9 shows the proposed predictive current control technique for the asymmetrical SpIM.

4.1. The Estimator Based on the State Variables (SV)

The state variables evolution in discrete time can be represented using the following equations:

$$\begin{aligned} \begin{bmatrix} \hat{\mathbf{X}}_a(k+1) \\ \hat{\mathbf{X}}_b(k+1) \end{bmatrix} &= \begin{bmatrix} \overline{\mathbf{A}}_{11} & \overline{\mathbf{A}}_{12} \\ \overline{\mathbf{A}}_{21} & \overline{\mathbf{A}}_{22} \end{bmatrix} \begin{bmatrix} \mathbf{X}_a(k) \\ \mathbf{X}_b(k) \end{bmatrix} + \begin{bmatrix} \overline{\mathbf{B}}_1 \\ \overline{\mathbf{B}}_2 \end{bmatrix} \mathbf{U}_{\alpha\beta s}, \\ \mathbf{Y}(k) &= \begin{bmatrix} \overline{\mathbf{I}} & \overline{\mathbf{0}} \end{bmatrix} \begin{bmatrix} \mathbf{X}_a(k) \\ \mathbf{X}_b(k) \end{bmatrix}, \end{aligned} \tag{22}$$

where $\mathbf{X}_a = \left[i_{\alpha s}(k)\ i_{\beta s}(k)\right]^T$ is the vector directly measured which is $\mathbf{Y}$, $\mathbf{X}_b = \left[i_{\alpha r}(k)\ i_{\beta r}(k)\right]^T$ is the remaining portion to be estimated, $\overline{\mathbf{I}}$ represents the identity matrix, and $\overline{\mathbf{A}}$ and $\overline{\mathbf{B}}$ are matrices whose components are obtained in the following equations:

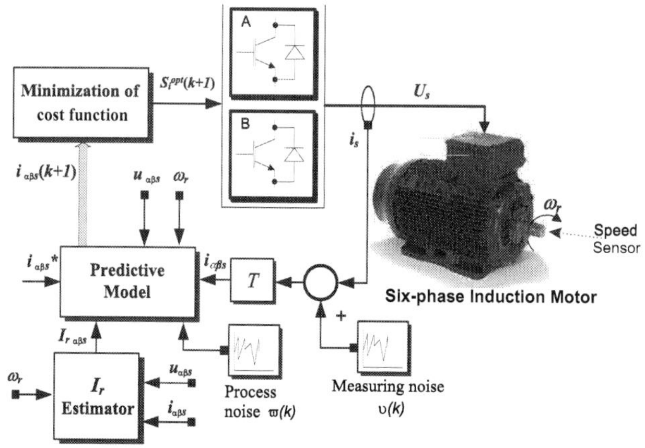

Figure 9. Proposed predictive current control technique for the asymmetrical SpIM

$$
\overline{\mathbf{A}} = \begin{bmatrix} (1 - T_m c_2 R_s) & T_m c_3 L_m \omega_r & \vdots & T_m c_3 R_r & T_m c_3 L_r \omega_r \\ -T_m c_3 L_m \omega_r & (1 - T_m c_2 R_s) & \vdots & -T_m c_3 L_r \omega_r & T_m c_3 R_r \\ \cdots & \cdots & \cdots & \cdots & \cdots \\ T_m c_3 R_s & -T_m c_4 L_m \omega_r & \vdots & (1 - T_m c_4 R_r) & -T_m c_4 L_r \omega_r \\ T_m c_4 L_m \omega_r & T_m c_3 R_s & \vdots & T_m c_4 L_r \omega_r & (1 - T_m c_4 R_r) \end{bmatrix},
$$

$$
\overline{\mathbf{B}} = \begin{bmatrix} T_m c_2 & 0 \\ 0 & T_m c_2 \\ \cdots & \cdots \\ -T_m c_3 & 0 \\ 0 & -T_m c_3 \end{bmatrix}. \tag{23}
$$

The prediction of the stator currents can be calculated as follows:

$$
\hat{i}_{\alpha s}(k+1|k) = (1 - T_m c_2 R_s) i_{\alpha s}(k) + T_m c_3 L_m \omega_r(k) i_{\beta s}(k) + T_m c_2 u_{\alpha s}(k) + T_m c_3 \xi_{\alpha s}(k), \tag{24}
$$

where $\xi_{\alpha s}(k) = \left(R_r i_{\alpha r}(k) + L_r \omega_r(k) i_{\beta r}(k)\right)$.

On the other hand, the quadrature current can be written as follows:

$$
\hat{i}_{\beta s}(k+1|k) = -T_m c_3 L_m \omega_r(k) i_{\alpha s}(k) + (1 - T_m c_2 R_s) i_{\beta s}(k) + T_m c_2 u_{\beta s}(k) + T_m c_3 \xi_{\beta s}(k), \tag{25}
$$

where $\xi_{\beta s}(k) = \left(R_r i_{\beta r}(k) + L_r \omega_r i_{\alpha r}(k)\right)$.

It can be seen from the above equations that the prediction of the stator currents has a measurable $(m(k) = [m_\alpha(k), m_\beta(k)])$ and unmeasured $(e(k) = [e_\alpha(k), e_\beta(k)])$ parts. Assuming this, the prediction equations can be rewritten as follows:

$$\hat{i}_{\alpha s}(k+1|k) = m_\alpha(k) + e_\alpha(k), \tag{26}$$

$$\hat{i}_{\beta s}(k+1|k) = m_\beta(k) + e_\beta(k), \tag{27}$$

where

$$m_\alpha(k) = (1 - T_m c_2 R_s) i_{\alpha s}(k) + T_m c_3 L_m \omega_r(k) i_{\beta s}(k) + T_m c_2 u_{\alpha s}(k), \tag{28}$$

$$m_\beta(k) = -T_m c_3 L_m \omega_r(k) i_{\alpha s}(k) + (1 - T_m c_2 R_s) i_{\beta s}(k) + T_m c_2 u_{\beta s}(k), \tag{29}$$

$$e_\alpha(k) = T_m c_3 \xi_{\alpha s}(k), \tag{30}$$

$$e_\beta(k) = T_m c_3 \xi_{\beta s}(k). \tag{31}$$

Analyzing Eqs. (26) and (27), which establish a prediction of the stator currents in the $(\alpha - \beta)$ subspace for a $(k+1)$ sampling time using the measurements of the (k) sampling time, it can be noted that the term $m(k)$ contains measurable variables, such as stator currents, rotor speed, and the stator voltages, while the term $e(k)$ contains unmeasurable variables of the asymmetrical SpIM, for this particular case are the rotor currents in the $(\alpha - \beta)$ subspace. Consequently, to solve the equations, it is necessary to obtain an estimate of the value of $\hat{e}(k|k)$, since the rotor currents are not measurable states of the system. This can be solved using the following equations:

$$\hat{e}_\alpha(k|k) = \hat{e}_\alpha(k-1) = i_{\alpha s}(k) - m_\alpha(k-1), \tag{32}$$

$$\hat{e}_\beta(k|k) = \hat{e}_\beta(k-1) = i_{\beta s}(k) - m_\beta(k-1). \tag{33}$$

Considering null initial conditions $\hat{e}_\alpha(0) = 0$ and $\hat{e}_\beta(0) = 0$, the estimated portion that represented the rotor currents can be calculated from a recursive formula given by:

$$\hat{e}_\alpha(k|k) = \hat{e}_\alpha(k-1) + (i_{\alpha s}(k) - \hat{i}_{\alpha s}(k-1)), \tag{34}$$

$$\hat{e}_\beta(k|k) = \hat{e}_\beta(k-1) + (i_{\beta s}(k) - \hat{i}_{\beta s}(k-1)). \tag{35}$$

4.2. The estimator based on a Luenberger Observer

The dynamics of the unmeasured part of the state vector defined by Eq. (22) is described as:

$$\mathbf{X}_b(k+1) = \overline{\mathbf{A}}_{22}\mathbf{X}_b(k) + \overline{\mathbf{A}}_{21}\mathbf{X}_a(k) + \overline{\mathbf{B}}_2\mathbf{U}_{\alpha\beta s}, \tag{36}$$

where the last two terms are known and can be considered as an input for the $\mathbf{X}_b$ dynamics. The $\mathbf{X}_a$ part may be expressed as:

$$\mathbf{X}_a(k+1) - \overline{\mathbf{A}}_{11}\mathbf{X}_a(k) - \overline{\mathbf{B}}_1\mathbf{U}_{\alpha\beta s} = \overline{\mathbf{A}}_{12}\mathbf{X}_b(k). \tag{37}$$

Note that Eq. (37) represents a relationship between a measured quantity on the left and the unknown state vector on the right. Assuming this, Eq. (36) can be rewritten as follows:

$$\begin{aligned}\hat{\mathbf{X}}_b(k+1) = (\overline{\mathbf{A}}_{22} - \mathbf{K}_l\overline{\mathbf{A}}_{12})\hat{\mathbf{X}}_b(k) + \mathbf{K}_l\mathbf{Y}(k+1) + \\ (\overline{\mathbf{A}}_{21} - \mathbf{K}_l\overline{\mathbf{A}}_{11})\mathbf{Y}(k) + (\overline{\mathbf{B}}_2 - \mathbf{K}_l\overline{\mathbf{B}}_1)\mathbf{U}_{\alpha\beta s}(k),\end{aligned} \tag{38}$$

where $\mathbf{K}_l$ is the Luenberger gain matrix. Therefore, Eqs. (37) and (38) describe the dynamics of the reduced-order estimators for Luenberger observer [27].

4.3. The Estimator Based on a Kalman Filter

Considering uncorrelated process and measurement of Gaussian noises, Eq. (22) can be also written as follows:

$$\begin{aligned}\hat{\mathbf{X}}(k+1|k) = \overline{\mathbf{A}}\mathbf{X}(k) + \overline{\mathbf{B}}\mathbf{U}(k) + \mathbf{H}\varpi(k), \\ \mathbf{Y}(k) = \mathbf{C}\mathbf{X}(k) + \nu(k),\end{aligned} \tag{39}$$

where $\mathbf{H}$ is the noise weight matrix, $\varpi(k)$ is the noise matrix of the system model (process noise), and $\nu(k)$ is the matrix noise of measurement. The covariance matrices R_ϖ and R_ν of these noises are defined as:

$$\begin{aligned}R_\varpi = cov(\varpi) = E\left\{\varpi \cdot \varpi^T\right\}, \\ R_\nu = cov(\nu) = E\left\{\nu \cdot \nu^T\right\},\end{aligned} \tag{40}$$

where $E\{.\}$ denotes the expected value. Thus, the dynamics of the reduced-order estimator equations are:

$$\begin{aligned}\hat{\mathbf{X}}_b(k+1|k) = (\overline{\mathbf{A}}_{22} - \mathbf{K}_k\overline{\mathbf{A}}_{12})\hat{\mathbf{X}}_b(k) + \mathbf{K}_k\mathbf{Y}(k+1) + \\ (\overline{\mathbf{A}}_{21} - \mathbf{K}_k\overline{\mathbf{A}}_{11})\mathbf{Y}(k) + (\overline{\mathbf{B}}_2 - \mathbf{K}_k\overline{\mathbf{B}}_1)\mathbf{U}_{\alpha\beta s}(k),\end{aligned} \tag{41}$$

where $\mathbf{K}_k$ represents the KF gain matrix that is calculated at each sampling time in a recursive manner from the covariance of the noises as:

$$\mathbf{K}_k(k) = \mathbf{\Gamma}(k) \cdot \mathbf{C}^T R_v^{-1}, \tag{42}$$

where $\mathbf{\Gamma}$ is the covariance of the new estimation, as a function of the old covariance estimation (φ) as follows:

$$\mathbf{\Gamma}(k) = \varphi(k) - \varphi(k) \cdot \mathbf{C}^T(\mathbf{C} \cdot \varphi(k) \cdot \mathbf{C}^T + R_v)^{-1} \cdot \mathbf{C} \cdot \varphi(k). \tag{43}$$

From the state equation, which includes the process noise, it is possible to obtain a correction of the covariance of the estimated state as:

$$\varphi(k+1) = \mathbf{A}\mathbf{\Gamma}(k) \cdot \mathbf{A}^T + \mathbf{H}R_{\varpi} \cdot \mathbf{H}^T; \tag{44}$$

this completes the required relations for the optimal state estimation. Thus, $\mathbf{K}_k$ provides the minimum estimation errors, given a knowledge of the process noise magnitude (R_{ϖ}), the measurement noise magnitude (R_v), and the covariance initial condition ($\varphi(0)$) [28].

4.4. Cost function

The cost function should include all aspects to be optimized. In the current predictive control applied to the asymmetrical six-phase induction motor, the most important features to be optimized are the tracking errors of the stator currents in the $(\alpha - \beta)$ subspace for a next sampling time, since these variables are related to the electromechanical conversion. To minimize the prediction errors at each sampling time k, it is enough to utilize a simple term as:

$$J = \| \hat{e}_{i\alpha s}(k+1|k) \|^2 + \| \hat{e}_{i\beta s}(k+1|k) \|^2 \leftrightarrow \begin{cases} \hat{e}_{i\alpha s}(k+1|k) = i^*_{\alpha s}(k+1) - \hat{i}_{\alpha s}(k+1|k), \\ \hat{e}_{i\beta s}(k+1|k) = i^*_{\beta s}(k+1) - \hat{i}_{\beta s}(k+1|k), \end{cases} \tag{45}$$

where $\| \,.\, \|$ denotes the vector modulus, i^*_s is a vector containing the reference for the stator currents, and $\hat{i}_s(k+1|k)$ is the prediction of the stator currents calculated from measured and estimated states and the voltage vector $U_{\alpha\beta s}(k)$. Figure 10 (a) shows all projections of the stator current predictions calculated from the prediction model. The current control selects the control vector that minimizes the cost function at each sampling time. Figure 10 (b) shows the selection of the optimal vector based on a minimization of prediction errors.

More complicated cost functions can be devised, for instance, to minimize harmonic content, VSI switching losses, torque and flux, and/or active and reactive power. Also, in multiphase drives, stator current can be decomposed in subspaces in different ways. An appropriate decomposition allows to put more emphasis on harmonic reduction as will be shown in the case study for a six-phase motor drive [29,30]. The most relevant cost functions are shown in Table 3. The superscript $(^*)$ denotes the reference value, and the terms involved in each cost function are detailed in Table 4.

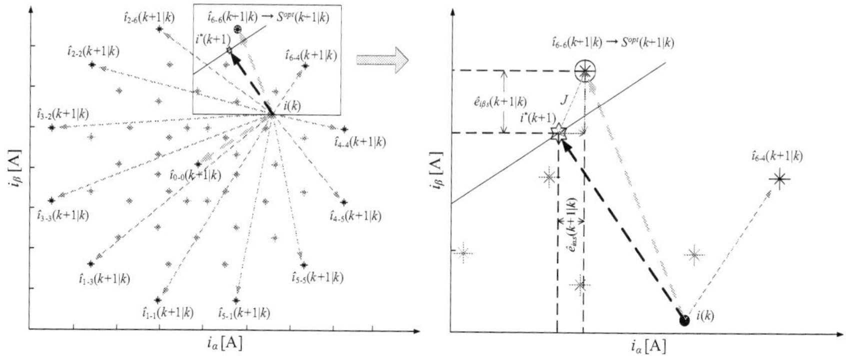

(a) Projection of the stator current prediction in stationary reference frame $(\alpha - \beta)$

(b) Evaluation of the cost function (J) and selection of the optimal vector (S^{opt})

Figure 10. Minimization of tracking error in stator currents in stationary reference frame $(\alpha - \beta)$

Controlled variables	Cost functions (J)
Currents (α-β) and harmonic $(x-y)$	$\|\|i_\alpha^* - i_\alpha\| + \|i_\beta^* - i_\beta\|\|^2 + \lambda\|\|i_x^* - i_x\| + \|i_y^* - i_y\|\|$
Active and reactive power	$\|Q_{in}\| + \|P_{in}^* - P_{in}\|$
Torque and flux	$\|T_e^* - T_e\| + \lambda\|\|\psi_s^*\| - \|\psi_s\|\|$
Currents (α-β) and voltage balance	$\|\|i_\alpha^* - i_\alpha\| + \|i_\beta^* - i_\beta\| + \lambda\|V_{c1} - V_{c2}\|$
Currents (α-β) and VSI switching losses	$\|\|i_\alpha^* - i_\alpha\| + \|i_\beta^* - i_\beta\|\| + \lambda N_s$

Table 3. Possible cost functions in function to the controlled variables

Variable	description
i_α	Measured α current
i_β	Measured β current
i_x	Measured x current
i_y	Measured y current
Q_{in}	Reactive power
P_{in}	Active power
T_e	Torque
ψ_s	Flux of the stator
λ	Weighting factor
V_{c1}, V_{c2}	Voltages on each capacitor (VSI balanced)
N_s	Number of switches

Table 4. Description of the terms involved in each cost function of Table 3

4.5. Optimizer

The optimization is done by exhaustive search over all possible realizations of the control actions. However, for electrical machines, some combinations of gating signals produce the same stator voltages, as shown in Figure 2. This means that, for prediction purposes, they are equivalent. This reduces the effective number of gating combinations to $\varepsilon = 2^{\Phi} - r$, r

being the number of redundant configurations and ϕ the phase numbers of the machine. For the particular case of the SpIM, assuming the previous consideration, the search space for the optimal solution are 49 different vectors (48 active and 1 null). For a generic multiphase machine, the optimization algorithm produces the optimum gating signal combination (S^{opt}) using the estimator based on the state variables as follows:

Algorithm 1 Optimization algorithm for state variable method

$J_o := \infty, i := 1.$
while $i \leq \varepsilon$ **do**
  $\mathbf{S}_i \leftarrow \mathbf{S}_{i,j} \, \forall j = 1, ..., \phi.$
  comment: Compute stator voltages. Eq. (3).
  comment: Compute the prediction of the states. Eq. (22).
  comment: Compute the cost function. Eq. (45).
  if $J < J_o$ **then**
    $J_o \leftarrow J, \mathbf{S}^{opt} \leftarrow \mathbf{S}_i.$
  end if
  $i := i + 1.$
end while

Algorithms 2 and 3 show the pseudocode for the particular case of the proposed estimation methods, the Luenberger observer and Kalman filter, respectively.

Algorithm 2 Proposed algorithm for Luenberger observer method

comment: Optimization algorithm.
$J_o := \infty, i := 1$
while $i \leq \varepsilon$ **do**
  $\mathbf{S}_i \leftarrow \mathbf{S}_{i,j} \, \forall j = 1, ..., \phi$
  Compute stator voltages. Eq. (3).
  Compute the prediction of the measurement states. Eqs. (36)-(37) assuming null initial conditions $\mathbf{X}_b(0) = 0$.
  Compute the cost function. Eq. (45).
  if $J < J_o$ **then**
    $J_o \leftarrow J, \mathbf{S}^{opt} \leftarrow \mathbf{S}_i$
  end if
  $i := i + 1$
end while
Compute the prediction for $\hat{X}_b(k+1)$ by using Eq. (38).

5. Simulation results and discussion

A MatLab/Simulink simulation environment has been designed to analyze the efficiency of the proposed reduced-order observer applied to the model-based predictive current control of the SpIM considering the electrical and mechanical parameters that are shown in Table 1. Numerical integration using fourth-order Runge-Kutta algorithm has been applied to compute the evolution of the state variables step by step in the time domain. A detailed block diagram of the proposed predictive current control technique is provided in Figure 9.

Algorithm 3 Proposed algorithm for Kalman Filter method

Compute the covariance matrix. Eq. (43).
Compute the Kalman Filter gain matrix. Eq. (42).
comment: Optimization algorithm.
$J_o := \infty, i := 1$
while $i \leq \varepsilon$ **do**
  $\mathbf{S}_i \leftarrow \mathbf{S}_{i,j} \, \forall j = 1, ..., f$
  Compute stator voltages. Eq. (3).
  Compute the prediction of the measurement state. Eq. (39).
  Compute the cost function. Eq. (45).
  if $J < J_o$ **then**
    $J_o \leftarrow J, \mathbf{S}^{opt} \leftarrow \mathbf{S}_i$
  end if
  $i := i + 1$
end while
Compute the correction for the covariance matrix. Eq. (44).

The reduced-order observer efficiency has been analyzed by performing parametric simulations considering a 10 kHz of sampling frequency and non-ideal conditions assuming that the control system has measurement (R_v) and process (R_w) noises. Figure 11 (a) (top) shows the obtained parametric simulation results for the particular case of the estimator based on the SV technique when are considered a constant frequency reference of 50 Hz with 15 A of reference current in stationary reference frame ($i_{\alpha s}$) and different levels of measurement and process noises (from 0 to 0.16) under varying load torque conditions (from 0 to 20 N·m). It can be seen in this figure the evolution of the MSE (measured between the reference and simulated currents) when the load torque and the measurement and process noises simultaneously increase. This behavior is associated with uncertainties in the estimation of the stator current due to the method based on the state variables. It can be observed that the MSE increases in direct proportion under varying load torque conditions (from 0.25 to 0.65 A). Figure 11 (a) (middle) shows the stator current tracking characteristic, where the following parameters are considered: $R_v = R_w = 0.08$ and $T_L = 10$ N·m. The references and simulated and prediction currents are represented in red, black, and green colors, respectively. According to Table 5, under these operating conditions, the performance of the MBPC based on the SV method in terms of MSE$i_{\alpha s}$ and THD$i_{\alpha s}$ can be quantified in 0.69 A and 6 %, respectively. On the other hand, Figure 11 (a) (bottom) shows the rotor current evolution, calculated according to the SV methods under the same conditions described above.

Moreover, Figure 11 (b) (top) and Figure 11 (c) (top) show the parametric simulation of the MBPC technique for cases based on the LO and KF estimators, respectively. It can be noted in these graphs that the proposed MBPC methods based on the LO and KF estimators introduce improvements quantified with respect to the MSE, mainly when these control algorithms are compared with the MBPC method based on state variables. It can be concluded from these graphs that the MBPC based on the LO and KF estimators exhibits low sensitivity (in terms of MSE) to change of the load torque, and the performance is related with the measurement and process noise levels. Under the same test conditions considered above, the MSE measured between the reference and simulated currents in alpha axis are 0.47 A and 0.48 A for

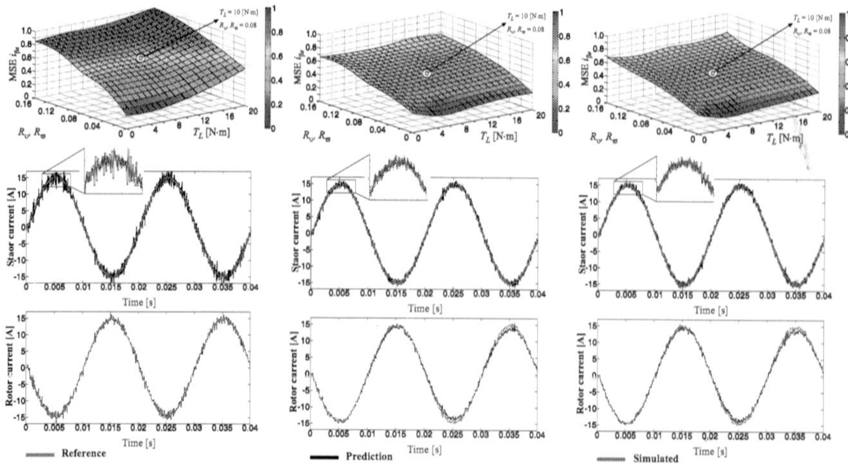

(a) Estimator based on the SV (b) Estimator based on the LO (c) Estimator based on the KF

Figure 11. Performance analysis considering a 15 A and 50 Hz of reference current

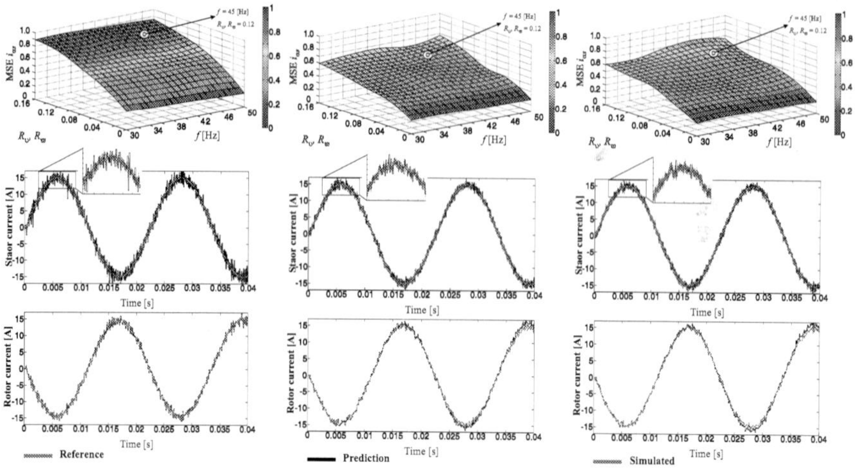

(a) Estimator based on the SV (b) Estimator based on the LO (c) Estimator based on the KF

Figure 12. Performance analysis considering a 15 A of reference current and $T_L = 0$

Figure 11 (b) (middle) and Figure 11 (c) (middle), respectively. Figure 11 (b) (bottom) and Figure 11 (c) (bottom) show the rotor current estimated, for the cases based on the LO and KF estimators, respectively.

The performance of the MBPC based on reduced-order estimators has been evaluated considering a 15 A of reference current with no-load condition and different levels of measurement and process noises (from 0 to 0.16) under varying reference frequencies (from 30 to 50 Hz). Figure 12 (top) shows the performance analysis in terms of MSE, where it is possible to observe from the parametric simulation that the three control methods evaluated has low sensitivity to the frequency variation when are considered no-load conditions. It can be seen that the efficiency strongly depends on the measurement and process noise levels, as in the previous case. Figure 12 (middle) shows the stator current tracking characteristic, where the following parameters are considered: $R_v = R_w = 0.12$ and 45 Hz of reference frequency. Finally, Figure 12 (bottom) shows the rotor current estimated, for the case of study.

	Figure 11 analysis			
	MSE$i_{\alpha s}$	MSE$i_{\beta s}$	THD$i_{\alpha s}$	THD$i_{\beta s}$
State variables	0.6965	0.6571	6.00%	6.06%
Luenberger observer	0.4799	0.4971	4.39%	4.62%
Kalman filter	0.4802	0.5084	4.29%	4.37%
	Figure 12 analysis			
	MSE$i_{\alpha s}$	MSE$i_{\beta s}$	THD$i_{\alpha s}$	THD$i_{\beta s}$
State variables	0.7798	0.7702	7.44%	7.17%
Luenberger observer	0.5622	0.5951	5.13%	5.18%
Kalman filter	0.5897	0.5208	5.10%	5.20%

Table 5. Performance analysis

A similar analysis was performed for the case of beta current component, obtaining similar results as shown in Table 5. These simulation results substantiate the expected performance of the proposed algorithms based on reduced-order observers.

6. Conclusion

In this chapter, an efficiency analysis of two reduced-order observers for rotor current estimator applied to the model-based predictive current control of the SpIM has been presented. The electrical and mechanical parameters of the SpIM have been measured and validated experimentally using an experimental setup. Real 15 kW SpIM parameters have been used to perform simulations using a MatLab/Simulink simulation environment. The simulation results obtained by different operation points under no-load and full-load conditions as well as different measurement and process noises have shown an increase in the efficiency of the proposed current control methods (based on the Kalman filter and Luenberger observer) measured with respect to the mean squared error of the stator currents in stationary reference frame, especially when they are compared with the control method based on state variables. Furthermore, the optimal estimator based on the Kalman filter achieves better performance than the Luenberger observer in terms of THD, mainly because it takes into account the effects of the noises in the control structure, recalculating the state feedback matrix at each sampling time recursively given the covariance of the new estimation as a function of the old covariance estimation. These results show that the experimental implementation of these control techniques are feasible and can be applied to the SpIM to increase the efficiency of the MBPC technique.

Acknowledgment

The authors would like to thank the Paraguayan Government for the economical support they provided by means of a CONACYT grant project 14-INV-101 – Desarrollo y análisis de eficiencia de nuevos algoritmos de control enfocados al generador hexafásico en aplicaciones de energía eólica. In addition, they wish to express their gratitude to the reviewers for their helpful comments and suggestions.

Author details

Raúl Gregor[1*], Jorge Rodas[1], Derlis Gregor[2] and Federico Barrero[3]

*Address all correspondence to: gregor.raul@gmail.com

1 Facultad de Ingeniería, Universidad Nacional de Asuncion, Department of Power and Control Systems, Asuncion, Paraguay

2 Facultad de Ingeniería, Universidad Nacional de Asuncion, Department of Computer Science, Asuncion, Paraguay

3 Escuela Superior de Ingenieros, Universidad de Sevilla, Department of Electronic Engineering, Sevilla, España

References

[1] E. Levi, "Multiphase Electric Machines for Variable-Speed Applications", Industrial Electronics, IEEE Transactions on, vol. 55, no. 5, pp. 1893–1909, 2008.

[2] H. Guzman, M.J. Duran, F. Barrero, B. Bogado, and S. Toral, "Speed Control of Five-Phase Induction Motors With Integrated Open-Phase Fault Operation Using Model-Based Predictive Current Control Techniques", Industrial Electronics, IEEE Transactions on, vol. 61, no. 9, pp. 4474–4484, 2014.

[3] H.S. Che, M.J. Duran, E. Levi, M. Jones, W.P. Hew, and N.A. Rahim, "Postfault Operation of an Asymmetrical Six-Phase Induction Machine With Single and Two Isolated Neutral Points", Power Electronics, IEEE Transactions on, vol. 29, no. 10, pp. 5406–5416, 2014.

[4] M. Salehifar, R. Salehi Arashloo, M. Moreno-Eguilaz, V. Sala, and L. Romeral, "Observer-Based Open Transistor Fault Diagnosis and Fault-Tolerant Control of Five-Phase Permanent Magnet Motor Drive for Application in Electric Vehicles", Power Electronics, IET, vol. 8, no. 1, pp. 76–87, 2015.

[5] Y. Zhao, H. Wang, and L. Xiao, "Investigation of Fault-Tolerant Capability of Five-Phase Doubly Salient Electromagnetic Generator", Electric Power Applications, IET, vol. 9, no. 1, pp. 80–93, 2015.

[6] A. Mohammadpour, and L. Parsa, "Global Fault-Tolerant Control Technique for Multiphase Permanent-Magnet Machines", Industry Applications, IEEE Transactions on, vol. 51, no. 1, pp. 178–186, Jan.–Feb. 2015.

[7] A.S. Abdel-Khalik, S. Ahmed, A.A. Elserougi, and A.M. Massoud, "Effect of Stator Winding Connection of Five-Phase Induction Machines on Torque Ripples Under Open Line Condition", Mechatronics, IEEE/ASME Transactions on, vol. 20, no. 2, pp. 580–593, Apr. 2015.

[8] H. Guzman, F. Barrero, and M.J. Duran, "IGBT-Gating Failure Effect on a Fault-Tolerant Predictive Current-Controlled Five-Phase Induction Motor Drive", Industrial Electronics, IEEE Transactions on, vol. 62, no. 1, pp. 15–20, Jan. 2015.

[9] S.M.J. Rastegar Fatemi, N.R. Abjadi, J. Soltani, and S. Abazari, "Speed sensorless control of a six-phase induction motor drive using backstepping control", Power Electronics, IET, vol. 7, no. 1, pp. 114–123, Jan. 2014.

[10] H. Yashan, Z. Zi-Qiang, and L. Kan, "Current Control for Dual Three-Phase Permanent Magnet Synchronous Motors Accounting for Current Unbalance and Harmonics", Emerging and Selected Topics in Power Electronics, IEEE Journal of, vol. 2, no. 2, pp. 272–284, Jun. 2014.

[11] H.S. Che, E. Levi, M. Jones, H. Wooi-Ping, and N.A. Rahim, "Current Control Methods for an Asymmetrical Six-Phase Induction Motor Drive", Power Electronics, IEEE Transactions on, vol. 29, no. 1, pp. 407–417, Jan. 2014.

[12] L. Feng, H. Wei, C. Ming, and Z. Gan, "Analysis of Fault Tolerant Control for a Nine-Phase Flux-Switching Permanent Magnet Machine", Magnetics, IEEE Transactions on, vol. 50, no. 11, pp. 1–4, Nov. 2014.

[13] E. Jung, Y. Hyunjae, S. Seung-Ki, C. Hong-Soon, and C. Yun-Young, "A Nine-Phase Permanent-Magnet Motor Drive System for an Ultrahigh-Speed Elevator", Industry Applications, IEEE Transactions on, vol. 48, no. 3, pp. 987–995, May.–Jun. 2012.

[14] M. Ruba, and D. Fodorean, "Analysis of Fault-Tolerant Multiphase Power Converter for a Nine-Phase Permanent Magnet Synchronous Machine", Industry Applications, IEEE Transactions on, vol. 48, no. 6, pp. 2092–2101, Nov.–Dec. 2012.

[15] O. Bottauscio, G. Serra, M. Zucca, and M. Chiampi, "Role of Magnetic Materials in a Novel Electrical Motogenerator for the More Electric Aircraft", Magnetics, IEEE Transactions on, vol. 50, no. 4, pp. 1–4, Apr. 2014.

[16] J.M. Apsley, and S. Williamson, "Analysis of Multiphase Induction Machines with Winding Faults", Industry Applications, IEEE Transactions on, vol. 42, no. 2, pp. 465–472, Mar.–Apr. 2006.

[17] J.M. Apsley, S. Williamson, A.C. Smith, and M. Barnes, "Induction Motor Performance as a Function of Phase Number", Electric Power Applications, IEE Proceedings, vol. 153, no. 6, pp. 898–904, Nov. 2006.

[18] O. Grigore-Muler, and M. Barbelian, "The Simulation of a Multi-Phase Induction Motor Drive", Optimization of Electrical and Electronic Equipment (OPTIM), 2010 12th International Conference on, pp. 297–306, 20–22 May. 2010.

[19] Y. Hu, Z.Q. Zhu, and K. Liu, "Current Control for Dual Three-Phase Permanent Magnet Synchronous Motors Accounting for Current Unbalance and Harmonics", Emerging and Selected Topics in Power Electronics, IEEE Journal of, vol. 2, no. 2, pp. 272–284, 2014.

[20] J.A. Riveros, F. Barrero, E. Levi, M.J. Duran, S. Toral, and M. Jones, "Variable-Speed Five-Phase Induction Motor Drive Based on Predictive Torque Control", Industrial Electronics, IEEE Transactions on, vol. 60, no. 8, pp. 2957–2968, Aug. 2013.

[21] H.S. Che, E. Levi, M. Jones, M. Duran, W.P. Hew, and N. Abd Rahim, "Operation of a Six-Phase Induction Machine Using Series-Connected Machine-Side Converters", Industrial Electronics, IEEE Transactions on, vol. 61, no. 1, pp. 164–176, Jan. 2014.

[22] Y. Zhao and T. Lipo, "Space Vector PWM Control of Dual Three-Phase Induction Machine Using Vector Space Decomposition", Industry Applications, IEEE Transactions on, vol. 31, no. 5, pp. 1100–1109, Sep. 1995.

[23] H.S. Che, M. Duran, E. Levi, M. Jones, W.P. Hew, and N. Abd Rahim, "Postfault Operation of an Asymmetrical Six-Phase Induction Machine with Single and Two Isolated Neutral Points", Power Electronics, IEEE Transactions on, vol. 29, no. 10, pp. 5406–5416, Oct. 2014.

[24] F. Barrero, J. Prieto, E. Levi, R. Gregor, S. Toral, M. Duran, and M. Jones, "An Enhanced Predictive Current Control Method for Asymmetrical Six-Phase Motor Drives", Industrial Electronics, IEEE Transactions on, vol. 58, no. 8, pp. 3242–3252, Aug. 2011.

[25] A. Yazdani and R. Iravani, "Voltage-Sourced Converters in Power Systems: Modeling, Control, and Applications", New Jersey, USA: Wiley, 2010, pp. 213–216.

[26] F. Barrero, M.R. Arahal, R. Gregor, S. Toral, and M.J. Duran, "A Proof of Concept Study of Predictive Current Control for VSI-Driven Asymmetrical Dual Three-Phase AC Machines", Industrial Electronics, IEEE Transactions on, vol. 56, no. 6, pp. 1937–1954, Jun. 2009.

[27] M. Arahal, F. Barrero, S. Toral, M.J. Duran, and R. Gregor, "Multi-phase Current Control Using Finite-State Model-Predictive Control", Control Engineering Practice, vol. 17, no. 5, pp. 579–587, Oct. 2008.

[28] R. Vargas, P. Cortes, U. Ammann, J. Rodriguez, and J. Pontt, "Predictive Control of a Three-Phase Neutral-Point-Clamped Inverter", Industrial Electronics, Transactions on, vol. 54, no. 5, pp. 2697–2705, Oct. 2007.